T0329787

Fog, Edge, and Pervasive Computing in Intelligent IoT Driven Applications

Fog, Edge, and Pervasive Computing in Intelligent IoT Driven Applications

Edited by

Deepak Gupta
Aditya Khamparia

IEEE PRESS

WILEY

Published by John Wiley & Sons, Inc., Hoboken, New Jersey.
Published simultaneously in Canada.

For general information on our other products and services or for technical support, please contact our Customer Care Department within the United States at (800) 762-2974, outside the United States at (317) 572-3993 or fax (317) 572-4002.

Wiley also publishes its books in a variety of electronic formats. Some content that appears in print may not be available in electronic formats. For more information about Wiley products, visit our web site at www.wiley.com.

Library of Congress Cataloging-in-Publication Data

Names: Gupta, Deepak, editor. | Khamparia, Aditya, 1988- editor.
Title: Fog, edge, and pervasive computing in intelligent IoT driven
    applications / edited by Deepak Gupta, Aditya Khamparia.
Description: Hoboken, New Jersey : Wiley-IEEE Press, 2020. | Includes
    bibliographical references and index. | Description based on print version
    record and CIP data provided by publisher; resource not viewed.
Identifiers: LCCN 2020025436 (print) | LCCN 2020025437 (ebook) | ISBN
    9781119670100 (epub) | ISBN 9781119670094 (adobe pdf) | ISBN 9781119670070
    (hardback) | ISBN 9781119670070q(hardback) | ISBN 9781119670094q(adobe
    pdf) | ISBN 9781119670100q(epub)
Subjects: LCSH: Internet of things. | Cloud computing. | Distributed
    databases. | UbIquitous computing.
Classification: LCC TK5105.8857 (ebook) | LCC TK5105.8857 .F64 2020 (print) |
    DDC 004.67/8–dc23
LC record available at https://lccn.loc.gov/2020025436 LC record available at https://lccn.loc
.gov/2020025437

Cover Design: Wiley
Cover Image: © Andriy Onufriyenko/Getty Images

Set in 9.5/12.5pt STIXTwoText by SPi Global, Chennai, India
10 9 8 7 6 5 4 3 2 1

Contents

About the Editors

Deepak Gupta is an eminent academician; he has various roles and responsibilities juggling his time between lectures, research, publications, consultancy, community service, PhD and post-doctorate supervision etc. With 12 years of rich expertise in teaching and two years in industry; he focuses on rational and practical learning. He has contributed substantial literature in the fields of Human–Computer Interaction, Intelligent Data Analysis, Nature-Inspired Computing, Machine Learning and Soft Computing. He is working as an Assistant Professor at Maharaja Agrasen Institute of Technology (GGSIPU), Delhi, India. He has served as Editor-in-Chief, Guest Editor, Associate Editor in SCI and various other reputed journals (Elsevier, Springer, Wiley & MDPI). He has actively been involved in the organization of various well reputed International conferences. He is not only backed with a strong reputation but his innovative ideas, end-results of his research and implementation of technology in the medical field is significantly contributing to society. He is currently a Post-Doc researcher at University of Valladolid, Spain. He has completed his Post-Doc from Inatel, Brazil, and Ph.D. from Dr. APJ Abdul Kalam Technical University. He has authored/edited 33 books published internationally (Elsevier, Springer, Wiley, Katson). He has published 103 scientific research works in reputed International Journals and Conferences including 50 SCI Indexed Journals of the IEEE, Elsevier, Springer, Wiley and many more. He has also published one patent. He is Editor-in-Chief of the OA journal *Computers and Quantum Computing and Applications* (QCAA), Associate Editor of *Expert Systems* (Wiley), Intelligent Decision Technologies (IOS Press), *Journal of Computational and Theoretical Nenoscience*, Honorary Editor of ICSES *Transactions on Image Processing and Pattern Recognition*. He is also a series editor of Intelligent *Biomedical Data*

Analysis De Gruyter (Germany), series editor of *Smart Sensor Technologies for Biomedical Engineering* with Elsevier. He is also associated with various professional bodies such as ISTE, IAENG, IACSIT, SCIEI, ICSES, UACEE, Internet Society, SMEI, IAOP, and IAOIP. Invited as a Faculty Resource Person/Session Chair/Reviewer/TPC member in different FDP, conferences and journals. He is the convener of the 'ICICC' and 'ICDAM' springer conference series.

Aditya Khamparia is an eminent academician; he has various roles and responsibilities including lectures, research, publications, consultancy, community service and PhD supervision etc. With seven years of rich expertise in teaching and two years in industry; he focuses on individual centric and practical learning. Currently, he is working as Associate Professor of Computer Science and Engineering at Lovely Professional University, Punjab, India. His research areas are Machine Learning, Soft Computing, Educational Technologies, IoT, Semantic Web and Ontologies. He has published more than 50 scientific research publications in reputed International/National Journals and Conferences, which are indexed in various international databases. Invited as a Faculty Resource Person/Session Chair/Reviewer/TPC member in different FDP, conferences and journals. Dr. Aditya received research excellence award in 2016, 2017, 2018 and 2019 at Lovely Professional University for his research contribution during the academic year. He is member of CSI, IET, ISTE, IAENG, ACM and IACSIT. He is also acting as reviewer and member of various renowned national and international conferences/journals. Invited as a Faculty Resource Person/Session Chair/Reviewer/TPC member in different FDP, conferences and journals.

List of Contributors

Iqrar Ahmad
Department of information system
Community College
King Khalid University Muhayel
Kingdom of Saudi Arabia

Nazir Ahmad
Department of information system
Community College
King Khalid University Muhayel
Kingdom of Saudi Arabia

Afroj Alam
Department of Computer Science
Integral University
Lucknow
India

Pallavi Asthana
Amity University
Uttar Pradesh
India

S. Aswath
Department of Computer Science
PES University
Bangalore, 560008
India

Subrato Bharati
Department of EEE
Ranada Prasad Shaha University
Narayanganj-1400
Bangladesh

Naveen Kumar Bhati
Sunder Deep College of Engineering
and Technology
Uttar Pradesh Technical University
India

Hatim M. Elhassan Ibrahim Dafallaa
Department of information system
Community College
King Khalid University Muhayel
Kingdom of Saudi Arabia

Charu Gandhi
Department of Computer Science
JIIT
Noida
India

Aarti Goel
Department of Information
Technology
Netaji Subhas University of
Technology
New Delhi
India

Sonia Goel
Punjabi University Patiala

Anjali Goyal
Department of Computer Applications
GNIMT
Ludhiana
Punjab
India

Sonia Goyal
Department of Electronics and
Communication Engineering
Punjabi University Patiala

Adnan Hasanović
University of Novi Pazar
36300 Novi Pazar
Dimitrija Tucovica bb
Serbia

Bramah Hazela
Amity University
Uttar Pradesh
India

Naiyar Iqbal
Department of Computer Science
and IT
Maulana Azad National Urdu
University
Hyderabad

Isha
Associate Professor
Lovely Professional University
Punjab
India

Maria Jamal
Department of Mathematics
Birla Institute of Technology
India

Balwinder Kaur
School of Computer Science and
Engineering
Lovely Professional University
Phagwara
India

Inderpreet Kaur
Reseach Scholar
CU Gharaun
Dept. of Computer Applications
CGC landran

Prabjot Kaur
Department of Mathematics
Birla Institute of Technology
India

Ranjit Kaur
Department of Electronics and
Communication Engineering
Punjabi University Patiala

Shweta Kaushik
Department of Computer Science
Engineering
ABES Institute of Technology
New Delhi
India

Jaiteg Singh Khaira
Dept. of Computer Applications
Chitkara University
Punjab Campus

Aditya Khamparia
School of Computer Science and
Engineering
Lovely Professional University
Phagwara
Punjab

and

School of Computer Science and
Engineering
Lovely Professional University
Punjab
India

Rizwan khan
Al-Barkaat College of Graduate
Studies
Aligarh
India

Anil Kumar
Amity University
Uttar Pradesh
India

Priyanka Rajan Kumar
Punjabi University Patiala

Vijay Laxmi
Professor
Guru Kashi University
Punjab
India

Zoran Lončarević
ITS - Studies for Information
Technologies
11000 Belgrade
Savski nasip 7
Serbia

Arun Malik
Associate Professor
Lovely Professional University
Punjab
India

Pragun Mangla
Department of Electronics and
Communication
Netaji Subhas Institute of Technology
New Delhi
India

Ashish Mishra
Gyan Ganga Institute of Technology
Jabalpur
Madhya Pradesh
India

Bhabani Shankar Prasad Mishra
School of Computer Engineering
KIIT University
Bhubaneswar
Odisha

Subhashree Mishra
School of Electronics Engineering
KIIT University
Bhubaneswar
Odisha

Sumita Mishra
Amity University
Uttar Pradesh
India

Satinder Singh Mohar
Department of Electronics and
Communication Engineering
Punjabi University Patiala

M. Rubaiyat Hossain Mondal
Institute of ICT
Bangladesh University of Engineering
and Technology
Dhaka
Bangladesh

Rajit Nair
Jagran Lakecity University

Nikita
School of Engineering & Technology
CT University
Ludhiana
India

Priyanka Pattnaik
School of Computer Engineering
KIIT University
Bhubaneswar
Odisha

Murali Mallikarjuna Rao Perumalla
School of Computer Science and
Engineering
Lovely Professional University
Phagwara
Punjab

Prajoy Podder
Institute of ICT
Bangladesh University of Engineering
and Technology
Dhaka
Bangladesh

Sahar Qazi
Department of Computer Science
Jamia Millia Islamia
New Delhi

Mamoon Rashid
Assistant Professor
School of Computer Science and
Engineering
Lovely Professional University
Jalandhar
India

Khalid Raza
Department of Computer Science
Jamia Millia Islamia
New Delhi

Mohammed Burhanur Rehman
Department of information system
Community College
King Khalid University Muhayel
Kingdom of Saudi Arabia

Md Robiul Alam Robel
Department of CSE
Cumilla University
Cumilla
Bangladesh

Harsh Sadawarti
School of Engineering & Technology
CT University
Ludhiana
India

Kamaljit Singh Saini
Dept. of Computer Applications
Chandigarh University
Gharuan

Muzafer Saračević
University of Novi Pazar
36300 Novi Pazar
Dimitrija Tucovica bb
Serbia

Deepak Kumar Sharma
Department of Information
Technology
Netaji Subhas University of
Technology
New Delhi
India

Preeti Sharma
Bansal College of Engineering

Dileep Kumar Singh
Jagran Lakecity University

Harjit Singh
Research Scholar
Guru Kashi University
Punjab
India

Jaiteg Singh
Chitkara University Institute of
Engineering and Technology
Chitkara University
140401 Punjab
Rajpura
India

Sanjay Kumar Singh
School of Computer Science and
Engineering
Lovely Professional University
Phagwara
Punjab

Saravjeet Singh
Chitkara University Institute of
Engineering and Technology
Chitkara University
140401 Punjab
Rajpura
India

Jimmy Singla
School of Computer Science and
Engineering
Lovely Professional University
Phagwara
India

Sonia Singla
University of Leicester
U.K.

Sofia
Research Scholar
Lovely Professional University
Punjab
India

Natasha Tiwari
University of Oxford
UK

Umer Iqbal Wani
Assistant Professor
School of Computer Science and
Engineering
Lovely Professional University
Jalandhar
India

Preface

This book focuses on recent advances, roles and benefits of fog, edge, and pervasive computing for intelligent and smart Internet of Things (IoT) enabled applications, aimed at narrowing the increasing gap. This book aims to describe the different techniques of intelligent systems from a practical point of view: solving common life problems. But this book also brings a valuable point of view to engineers and businessmen, trying to solve practical, economical, or technical problems in the field of their company activities or expertise. The purely practical approach helps to transmit the idea and the aim of the author is to communicate the way to approach and to cope with problems that would be intractable in any other way. This book solicits contributions which include theory, applications, and design methods of intelligent systems, Ubiquitous techniques, trends of fog, edge, and cloud applications as embedded in the fields of engineering, computer science, mathematics, and life sciences, as well as the methodologies behind them.

Book Objectives

With the rapid growth and emerging development in artificial technology, novel hybrid and intelligent IoT, edge, fog driven, and pervasive computing techniques are an important part of our daily lives. These technologies are utilized in various engineering, industrial, smart farming, video security surveillance, VANETs and vision augmented driven applications. These applications required real time processing of associated data and work on the principle of computational resource oriented meta heuristic and machine learning algorithms. Due to physical size limitations, small computing IoT and mobile devices are having resource limited constraints with low computing power and are unable to manage good quality of service and related parameters for distinguished applications. To overcome the limitations of such mobile devices edge/fog and pervasive computing have been proposed as a promising research area to carry out high end infrastructure usage and provide computation, storage and task execution effectively for end device users. As edge/fog computing is implemented at network edges, it promises low latency as well as agile computation augmenting services for device users. To successfully support intelligent IoT applications, therefore, there is a significant need for (1) exploring the efficient deployment of edge/fog/pervasive computing services at the network nodes level, (2) identifying the novel algorithm related to fog/edge/pervasive computing for resource allocation with low constraint and power usage, and (3) designing collaborative and distributed architectures specialized for edge/fog/pervasive computing.

Target Audience

The target audience of book is professionals and practitioners in the field of intelligent system, edge computing and cloud enabled applications and ubiquitous computing science paradigm may benefit directly from others' experiences. Graduate and master students of final projects and particular courses in intelligent system, edge and fog based real-life applications or medical domain can take advantage, making the book interesting for engineering and medical university teaching purposes. The research community of intelligent systems, sensor applications and intelligent sensor-based applications, consisting of many conferences, workshops, journals and other books, will take this as a reference book.

Organization

Chapter 1 describes the Internet of Things, fog, edge and pervasive computing are emerging technologies, having several promising applications including healthcare. These technologies are witnessing a paradigm shift in the healthcare sector moving out from traditional ways of visiting hospitals. It connects the doctors, patients, and nurses through smart intelligent sensor devices at low cost with high bandwidth network. In this chapter, authors discussed new computing paradigms precisely and present their applications in ubiquitous healthcare. This chapter also covers various problems and challenges that have been faced by the practitioners in the last few years in the field of cloud computing and IoT that has been solved by fog, edge and pervasive computing.

Chapter 2 discusses difficulties and future headings to investigate the role of fog, edge and pervasive computing. Studies have revealed that fog/edge computing (FEC) based organizations can expect an essential activity in expanding the cloud by means of finishing go-between organizations at the edge of the framework. Dimness/edge computing-based IoT's (FECIoT) appropriated configuration overhauls organization provisioning along the cloud-to-things continuum, thus making it sensible for key applications. Edge and fog registering are firmly related – both allude to the capacity to process information closer to the requester/buyer to lessen idleness cost and increment client experience. Both can channel information before it ehitse a major information lake for further utilization, lessening the measure of information that should be handled.

Chapter 3 addresses the technique selection issue encountered during the requirements elicitation stage, through a proposed machine learning model to transfer the experts' knowledge of elicitation technique selection to the less experienced. Based on the system analysts, stakeholders automate various techniques to provide the best optimization technique nomination.

Chapter 4 covers the advantages and disadvantages of using machine learning in edge/fog/pervasive computing. The various studies carried out by researchers is also covered. Every field has numerous applications, and in this chapter we discuss a few possible applications in this fog era using machine learning techniques. By the end of the chapter you should know about ML frameworks and the various machine learning algorithms used for fog/edge computing.

Chapter 5 provides a description of the software which has three modules: student, librarian and admin. These modules have unique features for searching for library books with the title, author's name, subject, ISBN/ISSN, etc. Within the chapter the interfaces of the software are shown as images which is an abstraction that may be developed on available mobile operating system like iOS, Android, etc. The interfaces are designed bearing in mind that it will be used on cross platform environments fulfilling minimum requirements using the IoT available in the market. Furthermore, overall information is preserved with the help of cloud storage while keeping parallel options for physical storage on the destination master computer. The cloud-based system has given library management a new dimension while giving a new feature referred as emanagement on the goɛ as a web or abstract GUI.

Chapter 6 describes a systematic review that was conducted to determine work done by various publishers on kidney cancer and to spot the research gaps between the studies so far. The outcome of this study permitted the effective diagnose of kidney cancer or renal cancer carried out using an adaptive neuro fuzzy method with 94% accuracy. Although, many data mining techniques were applied by researchers, the accuracy of these methods was less than the adaptive neuro fuzzy method. This method is worthwhile to identify the diagnosis of renal cancer better and more rigorously.

Chapter 7 explains a proposed approach to use edge computing in a transportation and route-finding process in order to handle performance issues. Huge demand for centralized cloud computing poses severe challenges such as degraded spectral efficiency, high latency, poor connection, and security issues. To handle these issues, fog computing and edge computing has come into existence.

One application of cloud computing is location based services (LBS). Intelligent transport systems being the important application of LBS rely on GPS, sensors, and spatial databases for convenient transport facilities. These location-based applications are highly dependent on external systems like GPS devices and map API's (cloud support) for the spatial data and location information. These applications acquire spatial data using API's from different proprietary service providers. The dependency on the API's and GPS devices, create challenges for effective fleet management and routing process in dead zones. Dead zones are areas where no cellular coverage exists.

Chapter 8 describes the simulation and design of an optimized low-cost comb drive based acoustic MEMS sensor. These sensors would be useful for condition monitoring of automobiles on the basis of changes in sound waves emerging from malfunctioning or defective parts of automobiles. These sensors can be developed from silicon substrates. Simulation is done using COMSOL Multiphysics simulation software based on finite element analysis. This optimized sensor is sensitive for the frequency range of 30–300 Hz. This frequency range was obtained after the FFT analysis of various signals received from engines using MATLAB software.

Chapter 9 offers an outline of developing the Internet of Things (IoT) technology in the area of healthcare as a flourishing research and experimental trend at the present time. The main advantages and benefits are considered in this chapter. In recent times, several studies in the healthcare information system proposed that the disintegration of health information is one of the most significant challenges in the arrangement of patient medical records. As a result, in this chapter, we provide an detailed design and overview of IoT healthcare systems along with its architecture.

Chapter 10 presents the combination of VANET and fog computing offering a range of options for cloud computing applications and facilities. Fog computing deals with high-virtualized VANET software and communication systems, where dynamic-speed vehicles travel. Mobile Adhoc Netoworks may also require low-latency fog computing in VANET and local connections within short distances. The modern state of the work and upcoming viewpoints of VANET fog computing are explored in this chapter. In addition, this chapter outlines the features of fog computing and fog-based services for VANETs. In addition to this, fog and cloud computing-based technology applications in VANET are discussed. Some possibilities for challenges and issues associated connected with fog computing are also discussed in this chapter.

Chapter 11 outlines an idea to design an efficient data collection method in the IoT network. IoT technology deals with smart devices to collect data as well as to provide useful and accurate data to users. Many data collection methods already exist, but they still have some drawbacks and need more enhancements.

This chapter outlines detailed information about the design of a novel data collection method using fog computing in the IoT network. The main reason for using fog computing over the cloud computing is to provide security to data which is completely lacking in cloud computing. Now-a-days, security of data is one of the most important requirements.

Chapter 12 provides an overview of using a fog computing platform for analyzing data generated by IoT devices. A fog computing platform will be compared with state-of-the-art to differentiate its impact in terms of analytics. Lots of data is being generated in IoT based devices used in smart homes, traffic sensors, smart cities, and various connected appliances. Fog computing is one area which is quite popular in processing this huge amount of IoT data. However, there are challenges in these models for performing real time analytics in such data for quick analytics and insights. A fog analytics pipeline is one such area which could be a possible solution to address these challenges.

Chapter 13 proposes a method of diagnosis based on the relationship between patients and symptoms, and between symptoms and diagnosis using linguistic variables by intuitionistic fuzzy sets. It then describes the state of some patients after knowing the results of their medical tests by degree of membership and degree of non-membership based on the relationship between patients and symptoms, and symptoms and diagnosis. Later a max-min-max composition and formula is applied to calculate the Hamming distance to identify the disease with the least Hamming distance for various patients. A revised max-min average composition is applied to identify the disease with the maximum score. Finally, it shows how urethral stricture in various patients is mathematically diagnosed.

Chapter 14 discusses the types of attacks involved in the IoT network with their counter measures, also covering the different layers, protocols, and the security challenges related to the IoT. It is the capability of the device that makes the IoT brilliant and this has been achieved by placing the intelligence into the devices. The intelligence in the sensors is developed by adding sensors and actuators which can collect information and pass it to the cloud through Wi-Fi, Bluetooth, ZigBee, and so on. But these IoT network are vulnerable to different types of attack: physical, network, software, and encryption, and these attacks actually stop the IoT devices performing their normal operations. So, it required that we must overcome these attacks.

Chapter 15 discusses the domain of IoT integrated telehealth or telemedical services with various segments where there is a need to work on advanced technologies to achieve a higher degree of accuracy and performance. As shown by the research reports and analytics from Allied Market Research, the global value and market of the IoT in telehealth and medical services will exceed 13 billion dollars. From the extracts of Statista, the Statistical Research Portal the huge usage of IoT based deployments is quite prominent and is increasing very frequently because of the usage patterns in various domains. In another research report, the usage patterns of tablets in various locations from 2014 to 2019 shows that the figures are growing. Because of these data, it is necessary to enforce the security mechanisms for the IoT and wireless based environment.

Chapter 16 explores optimization in the IoT which is used to improve the performance of network by enhancing the efficiency of the network, reducing the overheads and energy consumption, and increasing the rate of deployment of various devices in the IoT. The applications of IoT are smart cities, augmented maps, IoT in health care etc. and various issues in IoT such as security, addressing schemes etc. are discussed. Various optimization techniques such as heuristic and bio-inspired algorithms, evolutionary algorithms, and their applicability in IoT are described. Further the fruit fly optimization algorithm (FOA) and flow chart of FOA is explored in detail. Finally the applications of FOA in IoT and node deployment using FOA are explained. On the basis of observation FOA can be used to increase the coverage rate of sensor nodes.

Chapter 17 chapter presents an overall and in-depth study of different optimization algorithms inspired from nature's behaviour. Today optimization is a powerful tool for the engineer in virtually every discipline. It provides a rigorous, systematic method for rapidly zeroing in on the most innovative, cost-effective solutions to some of today's most challenging engineering design problems. The IoT is the concept of connecting everyday devices to the internet allowing the devices to send and receive data. With the IoT, devices can constantly report their status to a receiving computer that uses information to optimize decision making. IoT network optimization many benefits for improving traffic management, operating efficiency, energy conservation, reduction in latency, higher throughput, and faster rates in scaling up or deploying IoT services and devices in the network.

Chapter 18 outlines the optimization techniques for intelligent IoT applications in transport processes. The travelling salesman problem (TSP) has an important role in operational research and in this case, it was implemented in the design of the IoT application. The chapter describes some specific methods of solving, analysing and implementing a possible solution with an emphasis on a technique based on genetic algorithms. In this chapter we connect the TSP optimization problem in transport and traffic with IoT-enabled applications for a smart city. In the experimental part of the chapter we present specific development and implementation of the application for TSP with testing and experimental results.

Chapter 19 describes the impact of the Internet of everything solutions which are connecting every object. This has generated a large amount of data. This amount of data cannot be processed by a centralized cloud environment. There are applications where data needs real time response and low latency. The data being sent to the cloud for processing and then coming back to the application generating data can seriously impact the performance. This delay can cause delay in decision making and this is not acceptable in real time applications. To handle such scenarios, fog computing has emerged as a solution. Fog computing extends the cloud near to the edge of the network to decrease latency as well as bandwidth requirements. It acts as an intermediate layer between the cloud and devices generating data.

Chapter 20 is concerned with security and privacy handling issues occurred in pervasive and edge boundary system for recognizing voice, sound using intelligent IoT mining techniques. Fog computing is a promising registering worldview that extends distributed computing to the edge of systems. Like distributed computing yet with unique qualities, fog computing faces new security challenges other than those acquired from distributed computing. This chapter studies existing writing on fog figuring applications to recognize basic security holes. Comparable innovations like edge figuring, cloudlets, and micro-server farms have additionally been incorporated to give an all-encompassing survey process.

Chapter 21 focuses on fog computing and the second section deals with edge computing. In this the author first introduces the basics of fog and edge computing, its architecture, working, advantages and use cases, and then primarily focuses on their security and privacy issues separately. In the end solutions and research opportunities in both fields are discussed.

Closing Remarks

In conclusion, we would like to sum up here with few lines. This book is a small step towards the enhancement of academic research through motivating the research community and research organizations to think about the impact of fog, edge and IoT computing frameworks, networking principles and its applications for augmenting the academic research. This book is giving insight on the various aspects of academic computing research and the need for knowledge sharing and prediction of relationships through several links and their usages. This includes research studies, experiments, and literature reviews about pervasive, fog computational activities and to disseminate cutting-edge research results, highlight research challenges and open issues, and promote further research interest and activities in identifying missing links in cloud computing. We hope that research scholars, educationalists and students alike will find significance in this book and continue to use it to expand their perspectives in the field of edge, fog and pervasive computing and its future challenges.

Deepak Gupta
Maharaja Agrasen Institute of Technology, India

Aditya Khamparia
Lovely Professional University, India

Acknowledgments

We would like to thank the many people; those who contributed, supported and guided us through this book by different means. This book would not have been possible without their guidance and help.

First and foremost, we want to express heartfelt gratitude to our Guru for spiritual empathy and incessant blessings, to all teachers and friends for their continued guidance and inspiration throughout the period of our studies and career.

We would like to thank Wiley-IEEE Press publisher who gave us an opportunity to publish with them. We would like to express our appreciation to all contributors including the accepted chapters' authors, and many other contributors who submitted their chapters that cannot be included in the book. Special thanks to Mary Hatcher, Victoria Bradshaw, Teresa Netzler and Louis Vasanth Manoharan from Wiley-IEEE Press for their kind support and great efforts in bringing the book to completion. The encouragement of the Editorial Advisory Board (EAB) cannot be overstated. These are renowned experts who took time from their busy schedules to review chapters, provide constructive feedback, and improve the overall quality of the chapters.

We would like to thank our dear friends and colleagues for their continuous support and countless efforts throughout the process of publication of this book.

We express our personal and special thanks to our family members for supporting us throughout our careers, for love, the tremendous support and inspiration which they gave throughout the years.

Last but not least: we request forgiveness of all those who have been with us over the course of the years and whose names we have failed to mention.

Dr. Deepak Gupta
Maharaja Agrasen Institute of Technology, India

Dr. Aditya Khamparia
Lovely Professional University, India

1

Fog, Edge and Pervasive Computing in Intelligent Internet of Things Driven Applications in Healthcare: Challenges, Limitations and Future Use

Afroj Alam[1], Sahar Qazi[2], Naiyar Iqbal[3], and Khalid Raza[2*]

[1]*Department of Computer Science, Integral University, Lucknow, India*
[2]*Department of Computer Science, Jamia Millia Islamia, New Delhi*
[3]*Department of Computer Science and IT, Maulana Azad National Urdu University, Hyderabad*

Abstract

The Internet of Things (IoT), fog, edge and pervasive computing are all emerging technologies, which have several promising applications including healthcare. This technology is witnessing a paradigm shift in the healthcare sector, moving out from traditional ways of visiting hospitals. It connects the doctors, patients, and nurses through smart intelligent sensor devices at low cost with high bandwidth networks. In this chapter, we discuss these new computing paradigms precisely and present their applications in ubiquitous healthcare. This chapter also covers various problems and challenges that have been faced by practitioners in the last few years in the field of cloud computing and the IoT that have been solved by fog, edge and pervasive computing.

Keywords *Fog Computing; Edge Computing; Pervasive Computing; IoT; Healthcare*

1.1 Introduction

Today, the Internet of Things (IoT), fog, edge and pervasive computing are buzzwords, which have pivotal applications in different fields of studies including healthcare, engineering, and other intelligent applications. Cloud computing and the IoT have emerged as a new paradigm in the field of information and communication technology (ICT) as a revolution of the 21st century. It was a long-awaited dream of to use computing as a utility. Traditional computing extends the model to

*Corresponding Author: kraza@jmi.ac.in

Fog, Edge, and Pervasive Computing in Intelligent IoT Driven Applications, First Edition.
Edited by Deepak Gupta and Aditya Khamparia.

a cloud computing paradigm which has the capability to renovate a huge portion of the information technology industry, making the software even more interesting as a service that customers can access on-demand. The IoT acts as an interconnection between various gadgets and the Internet, including mobile phones, vehicles, farms, factories, home automation systems, and wearable devices from the viewpoint of the enhancement of the competence of real-life computing usage. This new technology, especially in the healthcare sector, is a change from the conventional approach of visiting clinics or hospitals. It links doctors, patients, and nurses by means of intelligent, affordable sensor gadgets with the support of cloud computing (Qi *et al.*, 2019). Unfortunately, a number of IoT based intelligent sensor gadgets are developing at a rapid rate. On the basis of evaluation, if the pace of extension proceeds constantly from 2020, the number of wearable gadgets on the planet will reach to around 26 billion (Imran and Qadeer, 2019). The volume of data generated using these IoT gadgets is very large. The capability of the present cloud model is not adequate to deal with the requirements of the IoT, *i.e.*, the current cloud has issues regarding volume, latency, and bandwidth. The current cloud cannot fulfill every one of the prerequisites of QoS (Quality of Service) in the IoT, therefore the goal is that another framework, fog computing, is introduced that will solve the issues of volume, latency and bandwidth (Shi *et al.*, 2015).

Fog computing has appeared with a new computation model which is placed between the cloud and intelligence sensor-based IoT devices through which an assortment of heterogeneous gadgets are pervasively associated as the terminal of a network which provides communication facilities to ease the execution of relevant IoT services (Chang *et al.*, 2019). Fog computing covers the cloud computing approach in the direction of the edge of the network, which has many advantages over cloud computing. Fog computing is appropriate for the applications by which real-time, high response time, and less latency are important issues, specifically in healthcare utilization (Mutlag *et al.*, 2019). It is enabling new or mutated applications and facilities with a productive transaction between cloud and fog, especially with the issues of volume, latency, and bandwidth regarding data management (JoSEP *et al.*, 2010).

In this chapter, we propose to explain new trends of computing models to understand the evolving IoT applications, exclusively fog and edge computing, their background, features, model architecture and current challenges. This chapter also covers various problems and challenges that have been faced by the practitioners in previous years in the field of cloud computing associated with the IoT that has been solved by fog, edge and pervasive computing (De Donno *et al.*, 2019). Further, because the Cybercrime Report 2016 suggests that cybercrime damages will be around $6 trillion every year by 2021, up from $3 trillion in 2015 it will cover how to secure the privacy of IoT based sensor devices and private data in the cloud using machine learning. Further, we will demonstrate in this chapter that fog computing definitely reduces latency as opposed to cloud computing. The low latency is significant for the medical IoT framework because of real-time requirements. Although the Cloud-based IoT (CIoT) structure is a typical way to deal with executing IoT frameworks, it is, however, confronting developing difficulties in the IoT. Specifically, CIoT deals with current challenges such as data transmission rate, latency rate, interruption, limitation of resource and secure system. The developing difficulties of CIoT have brought up an issue – what is needed to conquer the barrier of current cloud-driven architecture? Fog computing architecture is a visionary model that includes all probabilities to encompass the cloud to the edge network of CIoT, from the distant central cloud datacenter, the interim system hubs to the far edge where the front-end IoT gadgets are situated.

1.2 Why Fog, Edge, and Pervasive Computing?

Fog Computing
Fog computing is a distributed paradigm that provides computation, storage and network facilities between client gadgets and cloud datacenters mostly but not specifically situated on edge networks (Inbaraj, 2020). In such a way a cloud-based facility can be enlarged nearer to the IoT gadgets/centers. In this scenario, fog acts as a middle layer between IoT based sensor machines and cloud datacenters (Bangui *et al.*, 2018). The idea of fog computing was first created by Cisco in 2012 to report the difficulties of the IoT applications in traditional cloud computing. The challenges of fog computing are the facilitation and enhancement of mobility, real-time interaction, privacy, security, low latency, low energy consumption and network bandwidth for real-life applications where we need a quick response from the cloud, especially in the healthcare sector.

One of the benefits of fog computing is that, in place of transferring the entire data of IoT devices to the cloud, the fog will filter the data and then send a summary of the data. Another benefit is that fog computing processes the data before transferring to the cloud and will lead to reducing the communication period rate along with reducing the requirement of storage of massive data at the cloud. The key role of fog computing is data gathering from IoT sensors gadgets, data processing, data filtering and then finally sending a summary of the data to the cloud (Mehdipour *et al.*, 2019).

The Need for Fog Computing
Over the past decade, we have seen that the trend of storage, computing, controlling, and network management function over the data has been shifted from traditional computing to the cloud computing paradigm. On the basis of evaluation, if the pace of extension proceeds constantly from 2020, the number of wearable

gadgets on the planet will reach around 26 billion (Imran and Qadeer, 2019). The volume of data generated using these IoT gadgets is very large. The capability of the present cloud model is insufficient to deal with the necessities of IoT, *i.e.*, the current cloud has issues of volume, latency, and bandwidth that will have a detrimental effect on the way the IoT works. This problem will need a real-time answer, especially in the healthcare system (Mutlag *et al.*, 2019; Kelly, 2016). So, another platform is expected to meet these needs; a stage that we call fog computing or just the fog because fog is a cloud near the ground (JoSEP *et al.*, 2010).

Edge Computing

Edge computing is a developing model that allows computation to be executed at the edge of network devices in support of a cloud downstream and upstream for the benefit of IoT. An edge network is essentially comprised of client gadgets (such as cell phones, smart devices, *etc.*), edge gadgets (such as routers, set-top boxes, bridges, workstations, wireless access points *etc.*), edge servers, *etc.* and these elements can be furnished with the essential capabilities for supporting edge computation. If we deploy computing resources at the edge of the network, that will improve the quality of service in a different application (*e.g.* in intelligent and real-time healthcare system) in terms of the low latency of time, low bandwidth of the network, optimizing cost, increasing privacy and providing high energy efficiency. For example, fitness trackers act as an edge between bodies and clouds.

Research has proved that there are many benefits of edge computing, *e.g.* user face recognition and response time is reduced by 900 ms to 169 ms when the computation is deployed at the edge rather than the cloud. Also, energy consumption reduced by 30%–40% by edge computing (Chun *et al.*, 2011).

Edge computing brings two major improvements to the existing cloud computing. The first one is that edge nodes can preprocess large amounts of input data and then filter the data to reduce the size before transferring it to the central servers in the cloud. The other improvement is that the cloud resources are optimized by enabling edge nodes with computing ability.

The Need for Edge Computing

Edge computing offers several advantages over traditional architectures, including cost-saving, storing, and processing data faster. Some of the reasons for the need are as follows.

Push from cloud services: In the last decade the IoT and Cloud of Things (CoT)) have started contributing to healthcare domains (Uddin, 2019). But, due to the exponential increase of IoT based sensor devices in the large system, we have a massive amount of data in smart healthcare (Tolentino *et al.*, 2010). It is forecast by the Cisco Internet Business Solutions Group that there will be 50 billion things to be connected to the Internet by 2020 (Bonomi *et al.*, 2012). If all the data needs to be sent to the cloud for processing, the response time will be too long and so we will need huge network bandwidth. Some IoT applications, especially in healthcare, might require a very short response time. In this case, the data needs to be processed at the edge for shorter response time, more efficient processing and smaller network pressure.

Change from Data Consumer to Producer: Nowadays most electrical devices are IoT enabled which not only consume data but also produce data. For example: in cloud computing, devices such as mobile phones at the network edge traditionally only consume data, *e.g.*, enabling a user to watch a video. Now, though, users are also producing data with their mobile devices, such as by uploading posts and photos to social networking sites. This change requires more functionality at the network edge. Another example would be wearable health devices. Since the physical data collected by the things at the edge of the network is usually private, processing the data at the edge could protect user privacy better than uploading raw data to the cloud (Mutlag *et al.*, 2019).

Pull from IoT: as we know in the future most electronic devices will be IoT enabled. These IoT based sensor devices generate a huge amount of data. For example, fitness trackers, ECG, microwave-ovens, LED bars. These data we have to send and request for processing to the cloud will be a huge load on the cloud. Finally, we will face many problems like time latency, high bandwidth, *etc.* These problems can be offloaded on the cloud by edge computing devices like smartphones which act as an edge between humans and the cloud. By using edge machines, we can reduce the burden of data, time latency and bandwidth on the cloud.

Pervasive Computing

Pervasive computing, also called ubiquitous computing, was conceived by Mark Weiser in 1988 at Xerox PARC. It is an embedded computational technology in the form of a microprocessor in every object, these objects can communicate with each other effectively and perform useful tasks. So, we can say that ubiquitous computing makes our lives convenient by creating digital atmospheres that are conscious, robust, and receptive to human requirements. Pervasive computing can occur with any device, any time, any place and in any data format across the network (Aazam and Huh, 2015).

Pervasive computing has evolved not only in laptops and smart mobiles but also in IoT based wearable devices, intelligent sensors, lighting systems and so on. Nowadays pervasive computing has a big role in real-time healthcare systems (Gia *et al.*, 2015). Example: according to the current survey in China approximately 1 500 000 people die due to heart disease (Raza, 2019). What is happening is that generally, it has been seen that heart disease patients stay at home and do not ask for a doctor until they feel sick which occurs at a very late stage of the disease. So, most patients die before getting any treatment.

Therefore, there should be a new paradigm for healthcare performance that decreases the death rate which brings changes from the passive healthcare to pervasive healthcare model which will be a real-time monitoring system. Nowadays we have healthcare services which are available only in hospitals. So, it is very difficult for elderly people or disabled people to fulfill their healthcare demands under emergency conditions. In these cases, they have proposed pervasive healthcare to deliver health services to everyone, everywhere, all of the time.

1.3 Technologies Related to Fog and Edge Computing

Mobile Edge Computing

According to the new Gartner prediction, that there will be 26 billion gadgets that will become internet-enabled by 2020. The data generated through these devices will be digitally representing the state of the physical world. Sharing this much data in the network will have serious issues of security, privacy and misusing of data by unauthorized users (Elhayatmy *et al.*, 2018).

Also, to transmit this much huge data to the cloud for analysis and computation will be a heavy burden on the network, especially at peak times and it is possible that network delays will increase exponentially for transmitting the data to the cloud. Hence the service quality for various IoT applications will decrease.

A novel approach to mobile edge computing shifts the computing analysis and storage capacity from the remote data center to the Mobile Edge (ME) so that we can reduce the latency time between end devices and computing devices. This novel approach can free their workloads from computing devices to the edge (Tabas and Glass, 2013). By integrating IoT with Mobile Edge Computing (MEC), we can also enhance the Quality of Service (QoS) for IoT applications. As we know IoT applications mostly obtain data from a variety of IoT gadgets and produce business intelligence information by evaluating the obtained data, which would be installed at the mobile edge (Ansari *et al.*, 2018). Therefore, IoT devices data will be transferred to the IoT applications without travel over the mobile core web system. It can potentially speed up IoT applications in processing big IoT data streams (Yousefpour *et al.*, 2018).

Edge Cognitive Computing-based Smart Healthcare System

The industrial development and global environmental changes in a knowledge-driven economy have increased the percentage of chronic disease, which is a threat to human health (Tabas and Glass, 2013). In a traditional healthcare system, the IoT devices transfer the clinical data to the cloud for diagnosis of disease, that increases the time latency due to the heavy burden on the network because of huge data on the cloud, and failure to produce real-time remedial analysis and services in critical situation (Zhou *et al.*, 2016a).

To overcome these problems, cognitive computing is exploited by medical professionals. In an intelligent IoT application and a real-time healthcare system, it is important to install the cognitive computing competency to the network edge and follow up a cognitive study of the patient's physical health and network devices. This will reduce the latency time and allow many corresponding resources for clients in a critical situation (Chen *et al.*, 2018).

Similarities of Edge Computing and Fog Computing

Support for mobility: fog and edge computing applications both have features of communication directly with mobile devices *i.e.* mobility support technique.

Real-time interaction with IoT devices: both applications have the capability of real-time interaction *i.e.* quick response time which is very important in healthcare.

Energy consumption: in fog as well as at edge devices the consumption of energy for computation is less compared to the cloud.

Heterogeneity support: both fog and edge computing nodes are available in diverse forms and that will be established in a different variation of situations.

Low latency: fog and edge computing both reduce latency in comparison to cloud computing.

Hardware Connectivity: both fog and edge computing support LAN, WAN, WLAN, WiFi technology for communication with the cloud.

The Main Difference Between Edge Computing and Fog Computing

Edge computing mostly occurs upon edge devices (smartphone, smart object) directly, through which sensors are attached or we can say that edge computing acts as a gateway that is placed near to the sensors. Whereas fog computing shifts the edge computing process towards the sensors that are connected to the LAN or into the LAN hardware itself, so computing may be physically very far from sensors and actuators (Zhou *et al.*, 2016b).

In fog computing the computation of data is done at fog nodes or inside the LAN which is situated at the IoT gateway. On the other hand, computation of data into the edge computing is done upon the devices (smartphone, smart object) or on the sensors (body fitness tracker) itself rather than transferred elsewhere (Vora *et al.*, 2017).

In fog computing the intelligent location and computation power are placed at the local area network whereas edge computing is placed in the device itself (IoT based sensor devices like ECG, smartwatch, smartphone) (Korzun *et al.*, 2019).

Fog computing decreases a load of data on the cloud through filtering, which means fog computing works with clouds, whereas edge computing exists without the presence of a cloud. According to current edge computing, fog computing is a superset of edge computing.

Fog computing has computation layer leverage devices known as Fog Computing Nodes (FCNs), for example many-to-many gateways and wireless routers which are useful for computation and storage of data from end devices locally before sending it to the cloud. On the other hand, mobile edge computing says that the arrangement of intermediary hubs with capacity and preparing abilities in the base stations of wireless devices in this manner offers cloud computing capacities inside the Radio Area Network (RAN) (Bonomi *et al.*, 2012). Fog computing follows de-centralized or hierarchical architecture whereas edge computing follows distributed or localized architecture.

Characteristics of Fog in Comparison to Cloud

Features of fog computing are similar to cloud computing because both provide the services as on-demand for computation, storage, and network gadgets. Nevertheless, compared to cloud computation, fog computation implements near IoT gadgets (Yousefpour *et al.*, 2018). Some characteristics of fog computation over the cloud are explained below.

Geographically cloud computing services are centralized, on the other hand we need widespread deployment or locally deployment for fog computing applications, objectives and services that will be helpful for better administration to clients on the edge of the network (Priyadarshini *et al.*, 2019).

The proximity of fog computing near to the clients is one other feature of fog computing compared to the cloud which is helpful for predicting the user required services and provide real-time interaction for speedy services especially in the healthcare system (Zhang *et al.*, 2018).

There are lots of areas in which fog computing can play a better role than cloud computing. Example: smart-traffic lights for enabling traffic indicators to clear lanes on sensing the flashing light of an ambulance. Because in fog computing we have real-time interactions with edge devices, these interactions are not as fast due to high latency in the case of cloud computing because edge devices are very far from the cloud (Peter, 2015).

1.4 Concept of Intelligent IoT Application in Smart (Fog) Computing Era

According to the new Gartner prediction, that there will be 26 billion internet-enabled gadgets by 2020. These devices are health monitoring sensors, wearable body sensor networks, consumer electronic devices, smart vehicles, and all types of sensors coming under IoT (Mittal *et al.*, 2017). This is a new paradigm-shift from traditional interactions between sensor devices and humans which provides the ubiquitous computing environment and realization of smart hospitals, smart cities, smart healthcare, smart homes, and smart vehicles that improve lifestyles (Dastjerdi *et al.*, 2016a).

Scientists and investigators have estimated that by 2025 the number of economic impacts will be reduced by 11 trillion annually, which would represent around 11 percent of the world economy IoT devices; and which requires the installation of 1 trillion IoT gadgets. It is predicted that after 2025 with the help of the IoT there will be savings of $11 trillion annually that is a good economic impact because that represents approximately 11% of the world economy (Dastjerdi *et al.*, 2016a).

As we know most of the IoT based applications produce an unprecedented amount of data, which needs computing resources, network bandwidth, storage capability, heterogeneity, and others that have triggered a technological revolution and are supplied through cloud computing. Most of the smart healthcare, smart cities, and smart vehicles are connected with cloud computing, and this cloud provides the services as on-demand and scalable storage, along with processing facilities according to IoT application needs. However, cloud computing has disadvantages in view of the large delay that negatively impacts IoT activity which requires a real-time response (Mutlag *et al.*, 2019). For example, for health-controlling, critical response, and other inactivity sensitive applications, the high delay activated for the data to the cloud and back to the application is unsuitable.

According to a current survey of IoT applications related to the healthcare sector, approximately 30 million clients are transferring data up to 25 000 records every second which is not efficient for the cloud in terms of storage, computing and bandwidth for real-time applications (Mukherjee *et al.*, 2017).

Simultaneous with the increase in quantity there are different kinds of dynamic end-user and access gadgets: tablets, smartphones, edge routers, intelligent building controllers, smart meters, and many more smart clients. So, it is very important to understand "what we have to do near to the end-users' devices?" Should your car be your primary data store? What can a crowd of close to intelligent end-points and network edge sensor gadgets together complete over a distributed and self-structured network on the edge (Chiang and Zhang, 2016)?

To address these issues fog computing technology was introduced which integrated edge devices and cloud resources which helped to remove these limitations in terms of storing, computing, and communication to edge gadgets, that provide and boost mobility, privacy, safety, less latency and network bandwidth of fog computing. Real-time observing (*e.g.*, heart attack ailments) is one of the significant attributes in medical applications that involve less latency and greater response time, consequently, fog computing plays the finest resolution for these types of applications (Vora *et al.*, 2017). In a fog computing infrastructure, there must be at least one fog or many computing fogs which are interconnected to each other. This interconnected fog can increase the ability to scale, redundant and flexible, and when additional computing is needed, it is conceivable to include extra fog nodes. These given stated features fulfill the requirement of healthcare applications (Mutlag *et al.*, 2019; Kraemer *et al.*, 2017).

Though, fog nodes (such as, intelligent network gadgets like a router, gateway, server, base station, and so forth) can't make up these requirements until the structure of fog nodes is reshaped to be perfect with medicinal services applications which are explained below.

Challenges in IoT Requires New Architectural Computing

With the evolving of very intelligent IoT devices, they introduce numerous new challenges that can't be easily resolved through the current cloud and host computing architectures. So, we describe some of the basic challenges.

Stringent latency requirements: as we know, most industrial controlling systems, such as smart grids, good packaging systems, traffic control cabinets along the roadside, smart vehicle and oil–gas systems, generally demand very low latencies between the sensor and the control node and stay within a few milliseconds. Rather than this, we have many other IoT applications smartphones, smart home appliances, tablets, edge routers, smart meters and energy controllers in a smart power grid, smart building controllers, virtual reality applications, and gaming applications, may require latencies below a few tens of milliseconds. These requirements force us to switch from mainstream cloud services to fog services. There are some other IoT applications like AR, VR, which require very low latency constraints in the order of tens of milliseconds. By enabling intelligence features such as data analytics at the network edge instead of at the cloud, the end-to-end delay could be decreased significantly. This allows the network to accomplish more, efficiently controlling and responding faster to environmental stimuli (Weiner *et al.*, 2014).

Network bandwidth constraints: with the vast and rapid growth of the ubiquitous computing environment of interconnected IoT devices, sensors like healthcare systems and wireless sensor networks produce data at an exponential rate. For example, IoT based enabled smart cars may produce tens of megabytes of data per second. The included data in the smart car is (i) the route and speed of the car, (ii) all software and hardware have a concept of wear and tear facilities. The operating conditions of the car such as the wear and tear on its components (Kelly, 2016), (iii) during running on the road these smarts cars must have interaction with the surrounding environment such as the condition of weather, traffic on road. The data produced by an intelligent IoT based sensor automated car is even greater; it is predicted that the data is about one gigabyte per second. It is predicted that the smart grid in a smart city in the US will generate 1000 petabytes of data each year. These IoT devices have essentially challenged a lot of characteristics one of which is network bandwidth constraints. This will have a negative effect on the QoS at the edge devices in real-time IoT applications with this amount of huge data at the cloud, which also needs a high quality of network bandwidth at the cloud for sending this much data. So, part of the pre-processing, filtering, computational analysis, *etc.* could be executed on fog nodes and even the edge devices, easing the network of the huge volume of traffic loads.

New security challenges: the incorporation of cloud computing and emerging IoT applications have provided great opportunities for developing a smart city or smart hospital. But challenges still exist in some applications like in healthcare where different types of diseases occur from time to time. Regardless of the existent success, there is vagueness and challenges in an effective healthcare system regarding the security and privacy of cloud client's private data which is stored in the central cloud (Darwish *et al.*, 2019). Cloud clients need guarantees regarding the privacy and security of the personal data because cloud providers process personal data of clients from multiple sources in healthcare which increases the threat of infringing the privacy, security, and confidentiality of patient data.

Fog computing has gained so much attention in various applications in a diverse scenario like the healthcare system, smart grid, smart cities (Mehdipour, *et al.*, 2019). Fog computing is becoming day by day more powerful and popular, it needs a lot of effort to enhance the reliability of this computing model

Liu *et al.*, (2017) proposed that bioinformatics-based authentication would be helpful to secure privacy in fog computing. It emphasized the need to execute algorithms to preserve privacy homomorphic cryptography between fog and cloud to protect privacy. Mukherjee *et al.* (2017) emphasized the requirement for a novel security and data protection mechanism for the fog because existing solutions for the cloud cannot be directly applied to the fog. They have found the six research challenges in fog computing regarding security and privacy, which are, faith, privacy protection, authentication, and key agreement, intrusion detection systems, dynamic join and leave of fog nodes, and cross-issue and fog forensic (Ni *et al.*, 2017).

Cloud Access Security Broker (CASB) can apply in fog computing to improve the privacy and security of the cloud client's data at the strand of the network. A fog framework can be the intermediary for asset compelled gadgets to help oversee and update the security credentials and software on IoT devices at the edge network also perform a wide range of security functions, such as malware scanning and filtering the data before sending to the cloud (Chiang and Zhang, 2016).

Limitation of Space in Current Cloud Computing:
The integrated IoT application and cloud data centers are widely implemented through prominent IT firms like Google, Microsoft, Amazon, Cisco, and others to facilitate computation and storage devices as a service to companies or individuals through Infrastructure-as-a-Service (IaaS), Platform-as-a-Service (PaaS), Software-as-a-Service (SaaS) (Peng *et al.*, 2018). Although the cloud computing paradigm is able to tackle huge volumes of data from IoT clusters, transferring huge amounts of data to and from cloud computing remains a challenge of limited bandwidth (Iqbal and Islam, 2016). Fog computing provides one of the promising solutions to this problem. Computing at fog is one of the new trends in the IT sector that is aimed to compute data near the data origin. It displaces applications, facilities, data, processing power, and decision controlling from central nodes to logical ends of a network (Chen *et al.*, 2018). Fog processing considerably reduces the data size which has to be transferred between the endpoints and the cloud, and it enables data analysis and information reproduction from the data source (de Macedo *et al.*, 2019). Fog computing has the capability to manage different types of gadgets and sensors, as well as provide local processing and storage which is required in healthcare. Hence fog computing is one of the appropriate approaches for the healthcare IoT framework.

1.5 The Hierarchical Architecture of Fog/Edge Computing

Fog computing is a new breed in the area of computation paradigm, which provides the services facilities to the edge of network alternative to the cloud in traditional computing. Also, fog computing has good collaboration with the cloud, mainly when there are data storage management and analytics (Kraemer *et al.*, 2017). Fog and cloud use similar devices for implementation, such as communication, computation, controlling, storing, and services competencies and share a

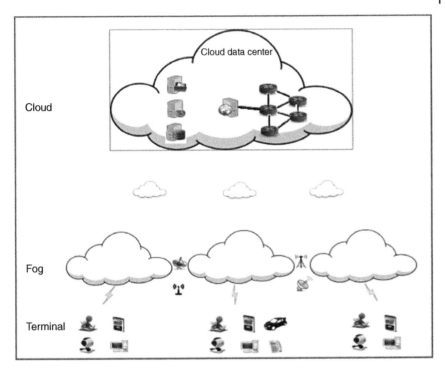

Figure 1.1 Cloud and fog computing architecture.

lot of the same concepts and features like virtualization software multi-tenancy. In this section, we look at the hierarchical architecture of fog computing.

In a couple of years, a lot of architecture has been introduced regarding fog computing which has mainly originated from a three-layer structure. Fog computing has shifted some of the services partially on account of the cloud computing to the edge computing by proposing one layer called the fog layer that is placed between end devices and the cloud (Chen *et al.*, 2018). Figure 1.1 depicts the three-layer hierarchy (*i.e.* terminal layer, fog layer, and cloud layer) of fog computing architecture that is described below.

Terminal Layer: is the layer that is located at the user or client-end and physical atmosphere. It mainly contains a lot of IoT devices such as smart grid, smart reader, smart traffic system, smart sensors, smart vehicles, and mobile phones. Most of the time, however, smartphones and intelligent vehicles have the computation capability which users only use as intelligent sensing gadgets (Hu *et al.*, 2017). These gadgets are broadly available and commonly used. This layer is used to sense data from physical things and transfer the resulting sensed data to the next higher layer for computing and storing. In this layer, the fog user equipment can directly interact with the adjacent fog user equipment in the device to device mode without the assistance of fog-APs, in which the high-power nodes are used to deliver overall control signaling for the D2D paired fog user equipment.

Fog layer: is positioned at the edge network. The fog computation layer contains a large number of fog nodes, that commonly includes network devices like router, gateway, switch, access points, and base stations (Liu *et al.*, 2017). Rather than this the fog layer also includes fog servers whose job is to compute and store the huge data of the edge layer that divest to the local fog server, access control routers that regulate or move the input data stream which is important because sometimes the input data stream can be so big, packets can be transferred to another idle local fog server, virtual machines, adaptive load dispatcher, cloud servers, coordinator server (Zhang *et al.*, 2017).

These fog nodes are generally placed among the user-end gadgets and cloud for smart cities, cafeteria, malls, bus stops, hospitals, roads, gardens, *etc.* It is predicted that some complex facilities such as water-supply, power-stations, and transportation are handled by greedy approaches, organizing them in a fog layer with a ranking technique that reduces the latency despite the absence of connection to the cloud (Varshney and Simmhan, 2017; Zhang *et al.*, 2018).

The storage capability of the fog layer is a lot more than edge resources but lesser than the cloud. Also offering computing resources but less than the cloud. Fog nodes can be static, which will be fixed on a given area, or moveable on a mobile sensor gadget. Fog nodes are linked with the terminal layer as well as the cloud. When linked by means of the cloud data center through the network then fog is accountable for communication and collaboration *via* the cloud which results in excellent proficiencies in terms of storage and computing. The real-time data investigation and delay complexity task can be performed at the fog layer (Saad, 2018).

Cloud Layer: the cloud server layer is the uppermost stage of the fog computing hierarchal architecture. This tier/layer contains a huge number of high-performance servers and storage devices, which have the capability to store entire data transferred from the fog nodes and offers many application facilities like smart cities, smart healthcare, smart transportation, smart workshops, *etc.* (Jia *et al.*, 2019). The cloud layer has strong potential abilities of computing and storage for supporting extensive computational analysis. This cloud layer is different from traditional cloud computing architecture. In the fog computing model entire services for computation and storage will not send to the cloud. Fog will filter specific data and then send it to the cloud-based on-demand and load. The cloud layer/component of fog computing is well managed both with respect to time and space and planned through control approaches to increase the use of the cloud resources (Hu *et al.*, 2017).

This computing architecture improves performance and decreases the latency power consumption of mobile devices in a dynamic atmosphere. The below architecture may facilitate the technical care aimed at IoT, Cyber-Physical System and Mobile Internet to facilitate effective (less time and less space) for client data processing and storing services.

1.6 Applications of Fog, Edge and Pervasive Computing in IoT-based Healthcare

The world has and is still significantly evolving due to the advent of *Intelligent Computing* shaping the nature of world interaction and response. It is reiterated here that the past decade has observed a gigantic leap from traditional computation to cloud computation which is being employed for myriad applications. However, cloud computation is limited to a few applications only. In order to fuel the demerit, the *Internet of things* (IoT) is filling the gap and is known as one of the novel technological trendsetters. Both cloud and fog computing make the backbone of IoT (Happ, 2018; Paul *et al.*, 2018). Cloud computing helps researchers and businesses to employ and manage remote resources on a low budget while IoT employs many embedded devices, sensors, and actuators which produce humongous data (big data) that requires apex computation in order to derive useful knowledge (Al-Fuqaha *et al.*, 2015; Evans *et al.*, 2011).

In medical and healthcare applications, we term IoT as the *Internet of Living Things* (IoLT) which encapsulates *wireless sensor networks* (WSN) that are mainly involved in monitoring patients enduring several diseases. These wireless sensors are *wearable, easy to use, miniature-sized,* with *durable performance* making them a part-and-parcel of every household today. These wireless sensors are attached to the patient so that medical data can be utilized for their healthcare management (Raza and Qazi, 2019). Fog computing has played an enormously good role in IoLT. It is known for its efficient gathering, storing, and organizing of medical data. Figure 1.2 displays a simple approach to data collection, storing, and managing patient data.

The three major pillars of fog computing in IoLT have been mentioned as follows (Kumar and Mahajan, 2019): Figure 1.3 is a schematic description of the same.

(A) Data Filtering: the primary step towards management and compression of medical data. Filtering, as the name suggests, is to make data clean, without any noises or redundancies. To get the patient's medical situation, many biological signals such as ECG, EMG, and photo platysma gram are first filtered for further analysis.

(B) Data Compression: to handle the quantity of medical data, vital information can be compressed without any important information loss. This compressed data is then stored and managed accordingly.

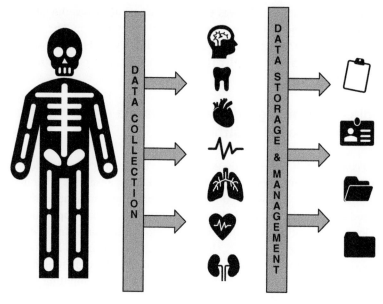

Figure 1.2 Efficient IoLT approach for collection, storage, and management of patient data

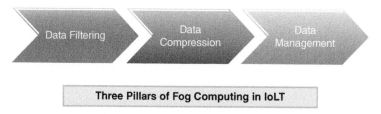

Figure 1.3 The three pillars of fog computing on the Internet of Living Things (IoLT)

(C) Data Management: is one of the crucial components of IoLT. Medical data is noisy and tends to be redundant. In system architecture, fog computation continuously gathers a big volume of important medical facts within an instant time frame *via* the sensor system, thus giving a quick response about the operator and system circumstances. This task is important in healthcare fraternities as it helps to make important medical decisions for the patients.

Many researchers have tried to employ fog computing to achieve better results in IoLT for efficient point of care (PoC) management. Ahmad *et al.* (2016) devised a fog computing-oriented healthcare framework named – *HealthFog*, wherein, each fog layer is a middle layer between end-user and cloud and was primarily developed to improve management of healthcare data concerns (Verma and Sood,

2018). Nandyala and Kim in the same year (Nandyala and Kim, 2016) showcased an IoT based ubiquitous-healthcare (U-healthcare) management system which focusses on using fog computing for developing smart hospitals and homes for efficient healthcare monitoring. Gia *et al.* (2015) discerned a fog-computing system for analyzing biosignals for real-time applications at the fog server end. A smart e-health gateway system was proposed which employs a fog computing layer that mainly connects to such gateways in both hospitals and homes (Negash *et al.*, 2017). Rahmani *et al.* (2018) devised a way to incorporate an intelligent middle layer between smart sensor points and the cloud thus developing an IoLT-fog based gateway system to practically manage medical cases easily (Hassan *et al.*, 2019). Suh *et al.* (2011) presented a wireless sensor system for cardiac patients wherein they prescribed a *3-tier architecture* that encapsulates a database, sensor, and web servers. The system helps in detecting various heart-related problems. Mobile-based healthcare management using IoLT was devised by Jara *et al.* (2013) which utilizes internet potential to monitor the medical healthcare of patients easily in their homes. Banaee *et al.* (2013) represented myriad algorithms for studying medical data using sensors which thus, helps in making better medical decisions for effective treatment.

1.7 Issues, Challenges, and Opportunity

Fog computing is an extension of cloud data centers having a lot of advantages in real-time IoT applications. That is allocated between the cloud and end-users which eliminates the problems of delay, traffic, computational speed, time latency, storage capabilities, data analysis. Hence fog computing has many opportunities in IoT application or fog is the future of the cloud. Also, in real-life healthcare IoT based application needs very frequent updated data. Hence the analysis of data is required every minute to stay updated which is possible very easily at fog servers (Prakash *et al.*, 2017).

For protecting the environment from severe deterioration and making it convenient, the medical sector should emphasize service as a quality. The medical sector classified up to 1.0 to 4.0 era. According to the current survey, the medical industry is still in its emerging phase because its evolution started around 1970. To run healthcare 4.0 in India, the expected budget is estimated to be 6000 million US dollars by 2020. A large number of medical IoT gadgets are generating a huge volume of data at frequent periods, which is why storing and securing these huge data are the most important problems. Logically, keeping the data in the cloud may not be feasible in a lot of situations so, new driven technology, Fog Computing (FC) can handle the situation easily (Kumari *et al.*, 2018).

1.7.1 Security and Privacy Issues

In fog computing the fog nodes (gadgets) are installed in an environment having an internet connection. So, fog requires additional storage resources as a service at the edge to process the requirements which are leading to management and maintenance cost issues. There are extensive types of research issues that are continuing inside the fog component. Here our primary focus is on privacy and security issues. Rather this lot of issues like network management, fog servers deployment, computation latency, power consumption *etc.* (AbdElhalim *et al.*, 2019).

Security is an important issue that must exist in a different layer of fog computing. Important securities are network security, service infrastructure, and authentication issues (Yi *et al.*, 2015). Here briefly explained about network security. Fog computing is dominated by wireless networks networking and issues with this are one of the important issues regarding fog networking because the network acts as a connector among edge devices, fog nodes, and cloud data centers. Also, the security of the network guarantees the security and reliability of the entire system. Network isolation techniques can ensure that malicious networks can't send threats to the connected layers (Abbasi and Shah, 2017).

Privacy and security are important matters and end-users information should not be accessed by illegal users. Most of the time this is risky for end-user's personal safety. So, precaution is necessary for all layers (data layer, fog layer, and cloud layer) when integrating these layers (Abbasi and Shah, 2017). The objective of fog computing is to not reduce the independence of its users. Users should have their right to share their data with whom they want and not to other users. In the same manner, they must have the authority to protect their private data from others such as what services they are using at any specific period. Users should have the authority to keep their site hidden from others as it could reveal information about them (Rahmanand Chuah, 2018). Fog computing has the ability to guarantee all these rights to make this system a reliable and secure one by a standard procedure which is given below.

Fog devices manipulate the IoT based sensor devices data at the fog servers then transfer it to the cloud data center. In this way, we can filter the confidential part of the private data, which does not need to be sent to the cloud. In this way, we can protect the privacy of personal data especially in healthcare medical data (Kumari *et al.*, 2018). To protect the privacy of data, we need some standard procedures and regulation monitoring systems especially in healthcare for medical data. That will help to reduce the interconnection overburden and protect the privacy of patient data. Chakraborty *et al.* (2016) offered a vibrant fog, that is enormously, geospatially dispersed and a potential perceptive high-level programming model for time-conscious usage. The frame was intended for heart pulse rate data. These data are also time-sensitive in medicine and manage life and death conditions (Kumari *et al.*, 2018).

1.7.2 Resource Management

Fog driven IoT application is a new paradigm of traditional cloud computing. So, fog computing has many research challenges. One of these is resource management. Because fog computing uses three different technologies (IoT, fog, cloud) all three are implemented together in one system. The big issues and challenges here are how to manage all three resources so that our application will run efficiently. If we take the example of fog computation to run without cloud computation, it will be a very difficult and big challenge for the management of resources. The reason is that fog nodes do not have sufficient storage space and computing power for resource management. If we discuss resource management which includes IoT objects and fog nodes. There is a need for communication for exchanging the data between them and processing must assist usage so repeated data will not take up valued space. If various customers are utilizing similar assets, the executives of these assets turn out to be increasingly important to give insignificant delays to the gadgets utilized (Al-Khafajiy *et al.*, 2018; Sarkar *et al.*, 2015).

We have many different techniques for resource management; a few of them are explained in (Mutlag *et al.*, 2019) [12]. One technique stated that the selection of a fog node with the least overall delay inside its present position then that fog node will be used for the given task. Aazam and Huh (2015) proposed a model regarding resource management. That model efficiently manages all the resources. The presented resource-management model identifies the utilization of resources by customer and pre-assigning of devices on the basis of client trends and the possibility of utilizing in the future. This idea is accepted as having better impartiality and effectiveness when the resources are truly utilized.

1.7.3 Programming Platform

The hot research trend in mobile edge computing is a computational offload, having many proposals for offloading workloads on the cloud (Kosta *et al.*, 2012; Chun *et al.*, 2011). As we know always offloading on the cloud may not be conceivable or sensible. Orsini *et al.* (2015) presented an idea called an adaptive Mobile Edge Computing (MEC) programming framework called Cloud Aware, which decreases the load to the edge gadgets, therefore enhancing the designing of adaptable and scalable edge-based mobile applications. The researchers mentioned the kinds of elements that a mobile edge computing application ought to be divided into, therefore the offloading decision is simplified. The model offloads work with the aims to (1) increases the computational speed, (2) reduce energy consumption, (3) utilize bandwidth, or (4) facilitate less delay (Dastjerdi *et al.*, 2016b).

Basic development in the field of fog computing is that it has been developed by mobile fog (Hong *et al.*, 2013), API to develop innovative applications, that influenced the extensive, geo-distribution, and less delay assurance offered by means of the fog-computing architecture. The application that has been designed using the presented API has multiple components, all of them executing on multiple layers in the hierarchy of equipment.

1.8 Conclusion

This chapter concluded that the survey shows that there is widespread processing works, crossing over various arrangement situations and applications, from which benefits will be gained from fog computing. In fact, our survey presents that processing is an important component in mostly ubiquitous medical care applications, therefore these functions frequently require execution somewhere close to sensors and cloud. We have additionally talked about trade-offs while placing computing functions in the network and examined the advantages and difficulties of fog computing identified with unavoidable health applications. Sensor gadgets are frequently not able to do such computation works by themselves, which is the reason why they need to distribute the overloading of computation works.

Furthermore, cloud computation is frequently not realistic to solve such offloading because of limitations concerning reliability, security matters or protocols. Fog computing, along with their adaptability to include computation as a feature of a system framework, shows up consequently as a reasonable idea to achieve the prerequisites of healthcare. The filtering function of fog computing support preserves privacy or decreases overload on the network and resources.

Moreover, with their capability to work nearly to the client-end, fog computing functions include a significant element to become more reliable. The explanation of this chapter is a pointer to the future of healthcare with the association among the IoT technology, fog, edge and pervasive computation. IoT based technologies provide several facilities and revolutions in the medical sector but almost all of the facilities and inventions are still in the initial stages.

Bibliography

Aazam, M., and Huh, E. N. (2015, March). Dynamic resource provisioning through Fog micro datacenter. In 2015 IEEE international conference on pervasive computing and communication workshops (PerCom workshops) (pp. 105–110). IEEE.

Abbasi, B. Z., and Shah, M. A. (2017, September). Fog computing: Security issues, solutions and robust practices. In 2017 23rd International Conference on Automation and Computing (ICAC) (pp. 1–6). IEEE.

AbdElhalim, E., Obayya, M., and Kishk, S. (2019). Distributed Fog-to-Cloud computing system: A minority game approach. *Concurrency and Computation: Practice and Experience*, 31(15), e5162.

Ahmad, M., Amin, M. B., Hussain, S., Kang, B. H., Cheong, T., and Lee, S. (2016). Health fog: a novel framework for health and wellness applications. *The Journal of Supercomputing*, 72(10), 3677–3695.

Al-Fuqaha, A., Guizani, M., Mohammadi, M., Aledhari, M., and Ayyash, M. (2015). Internet of things: A survey on enabling technologies, protocols, and applications. *IEEE communications surveys and tutorials*, 17(4), 2347–2376.

Al-Khafajiy, M., Webster, L., Baker, T., and Waraich, A. (2018, June). Towards fog driven IoT healthcare: challenges and framework of fog computing in healthcare. In Proceedings of the 2nd International Conference on Future Networks and Distributed Systems (p. 9). ACM.

Ansari, N., and Sun, X. (2018). Mobile edge computing empowers Internet of Things. *IEICE Transactions on Communications*, 101(3), 604–619.

Banaee, H., Ahmed, M. U., and Loutfi, A. (2013). Data mining for wearable sensors in health monitoring systems: a review of recent trends and challenges. *Sensors*, 13(12), 17472–17500.

Bangui, H., Rakrak, S., Raghay, S., and Buhnova, B. (2018). Moving to the edge-cloud-of-things: recent advances and future research directions. *Electronics*, 7(11), 309.

Bonomi, F., Milito, R., Zhu, J., and Addepalli, S. (2012, August). Fog computing and its role in the internet of things. In Proceedings of the first edition of the MCC workshop on Mobile cloud computing (pp. 13–16). ACM.

Chakraborty, S., Bhowmick, S., Talaga, P., and Agrawal, D. P. (2016, October). Fog networks in healthcare application. In 2016 IEEE 13th International Conference on Mobile Ad Hoc and Sensor Systems (MASS) (pp. 386–387). IEEE.

Chang, C., Srirama, S. N., and Buyya, R. (2019). Internet of things (IoT) and new computing paradigms. *Fog and Edge Computing: Principles and Paradigms*, 1–23. 133

Chen, M., Li, W., Hao, Y., Qian, Y., and Humar, I. (2018). Edge cognitive computing based smart healthcare system. *Future Generation Computer Systems*, 86, 403–411.

Chiang, M., and Zhang, T. (2016). Fog and IoT: An overview of research opportunities. *IEEE Internet of Things Journal*, 3(6), 854–864.

Chun, B. G., Ihm, S., Maniatis, P., Naik, M., and Patti, A. (2011, April). Clonecloud: elastic execution between mobile device and cloud. In Proceedings of the sixth conference on Computer systems (pp. 301–314). ACM.

Darwish, A., Hassanien, A. E., Elhoseny, M., Sangaiah, A. K., and Muhammad, K. (2019). The impact of the hybrid platform of internet of things and cloud computing on healthcare systems: Opportunities, challenges, and open problems. *Journal of Ambient Intelligence and Humanized Computing*, 10(10), 4151–4166.

Dastjerdi, A. V., and Buyya, R. (2016a). Fog computing: Helping the Internet of Things realize its potential. *Computer*, 49(8), 112–116.

Dastjerdi, A. V., Gupta, H., Calheiros, R. N., Ghosh, S. K., and Buyya, R. (2016b). Fog computing: Principles, architectures, and applications. In Internet of Things (pp. 61–75). Morgan Kaufmann.

De Donno, M., Tange, K., and Dragoni, N. (2019). Foundations and Evolution of Modern Computing Paradigms: Cloud, IoT, Edge, and Fog. *Ieee Access*, 7, 150936–150948.

de Macedo, D. D. J., de Araújo, G. M., Dutra, M. L., Dutra, S. T., and Lezana, Á. G. R. (2019). Toward an efficient healthcare CloudIoT architecture by using a game theory approach. Concurrent Engineering, 1063293X19844548.

Elhayatmy, G., Dey, N., and Ashour, A. S. (2018). *Internet of Things based wireless body area network in healthcare*. In Internet of things and big data analytics toward next-generation intelligence (pp. 3–20). Springer, Cham.

Evans, D. (2011). The internet of things: How the next evolution of the internet is changing everything. *CISCO white paper*, 1(2011), 1–11.

Gia, T. N., Jiang, M., Rahmani, A. M., Westerlund, T., Liljeberg, P., and Tenhunen, H. (2015, October). Fog computing in healthcare internet of things: A case study on ecg feature extraction. In 2015 IEEE International Conference on Computer and Information Technology; Ubiquitous Computing and Communications; Dependable, Autonomic and Secure Computing; Pervasive Intelligence and Computing (pp. 356–363). IEEE.

Happ, D. (2018). *Cloud and Fog Computing in the Internet of Things*. Internet of Things A to Z: Technologies and Applications, 113.

Hassan, M. K., El Desouky, A. I., Elghamrawy, S. M., and Sarhan, A. M. (2019). A Hybrid Real-time remote monitoring framework with NB-WOA algorithm for patients with chronic diseases. *Future Generation Computer Systems*, 93, 77–95.

Hong, K., Lillethun, D., Ramachandran, U., Ottenwälder, B., and Koldehofe, B. (2013, August). Mobile fog: A programming model for large-scale applications on the internet of things. In Proceedings of the second ACM SIGCOMM workshop on Mobile cloud computing (pp. 15–20). ACM.

Hu, P., Dhelim, S., Ning, H., and Qiu, T. (2017). Survey on fog computing: architecture, key technologies, applications and open issues. *Journal of network and computer applications*, 98, 27–42.

Imran, M., and Qadeer, M. A. (2019). Wearable U-HRM device for rural applications. In U-Healthcare Monitoring Systems (pp. 1–14). Academic Press.

Inbaraj, X. (2020). Distributed Intelligence Platform to the Edge Computing. In Architecture and Security Issues in Fog Computing Applications (pp. 108–130). IGI Global.

Iqbal, N., and Islam, M. (2016). From Big Data to Big Hope: An outlook on recent trends and challenges. *Journal of Applied Computing*, 1(1):14–24.

Jara, A. J., Zamora-Izquierdo, M. A., and Skarmeta, A. F. (2013). Interconnection framework for mHealth and remote monitoring based on the internet of things. *IEEE Journal on Selected Areas in Communications*, 31(9), 47–65.

Jia, X., He, D., Kumar, N., and Choo, K. K. R. (2019). Authenticated key agreement scheme for fog-driven IoT healthcare system. *Wireless Networks*, 25(8), 4737–4750.

JoSEP, A. D., KAtz, R., KonWinSKi, A., Gunho, L. E. E., PAttERSon, D., and RABKin, A. (2010). A view of cloud computing. *Communications of the ACM*, 53(4).

Kelly, R. (2016). Internet of things data to top 1.6 zettabytes by 2022. *Campus Technology*, 9, 1536–1233.

Korzun, D., Balandina, E., Kashevnik, A., Balandin, S., and Viola, F. (Eds.). (2019). *Ambient Intelligence Services in IoT Environments: Emerging Research and Opportunities: Emerging Research and Opportunities*. IGI Global.

Kosta, S., Aucinas, A., Hui, P., Mortier, R., and Zhang, X. (2012, March). Thinkair: Dynamic resource allocation and parallel execution in the cloud for mobile code offloading. In 2012 Proceedings IEEE Infocom (pp. 945–953). IEEE.

Kraemer, F. A., Braten, A. E., Tamkittikhun, N., and Palma, D. (2017). Fog computing in healthcare–a review and discussion. *IEEE Access*, 5, 9206–9222.

Kumari, A., Tanwar, S., Tyagi, S., and Kumar, N. (2018). Fog computing for Healthcare 4.0 environment: Opportunities and challenges. *Computers and Electrical Engineering*, 72, 1–13.

Kumar Y and Mahajan M. (2019). Intelligent Behavior of Fog Computing with IOT For Healthcare System. *International Journal of Scientific and Technology Research*. 8(7):674–679.

Liu, Y., Fieldsend, J. E., and Min, G. (2017). A framework of fog computing: Architecture, challenges, and optimization. *IEEE Access*, 5, 25445–25454.

Mehdipour, F., Javadi, B., Mahanti, A., and Ramirez-Prado, G. (2019). *Fog Computing Realization for Big Data Analytics*. Fog and Edge Computing: Principles and Paradigms, 259–290.

Mittal, S., Negi, N., and Chauhan, R. (2017, November). Integration of edge computing with cloud computing. In 2017 International Conference on Emerging Trends in Computing and Communication Technologies (ICETCCT) (pp. 1–6). IEEE.

Mukherjee, M., Matam, R., Shu, L., Maglaras, L., Ferrag, M. A., Choudhury, N., and Kumar, V. (2017). Security and privacy in fog computing: Challenges. *IEEE Access*, 5, 19293–19304.

Mukherjee, M., Shu, L., and Wang, D. (2018). Survey of fog computing: Fundamental, network applications, and research challenges. *IEEE Communications Surveys and Tutorials*, 20(3), 1826–1857.

Mutlag, A. A., Ghani, M. K. A., Arunkumar, N. A., Mohamed, M. A., and Mohd, O. (2019). Enabling technologies for fog computing in healthcare IoT systems. *Future Generation Computer Systems*, 90, 62–78.

Nandyala, C. S., and Kim, H. K. (2016). From cloud to fog and IoT-based real-time U-healthcare monitoring for smart homes and hospitals. *International Journal of Smart Home*, 10(2), 187–196.

Negash, B., Gia, T. N., Anzanpour, A., Azimi, I., Jiang, M., Westerlund, T., .. and Tenhunen, H. (2018). *Leveraging fog computing for healthcare IoT*. In Fog Computing in the Internet of Things (pp. 145–169). Springer, Cham.

Ni, J., Zhang, K., Lin, X., and Shen, X. S. (2017). Securing fog computing for internet of things applications: Challenges and solutions. *IEEE Communications Surveys and Tutorials*, 20(1), 601–628.

Orsini, G., Bade, D., and Lamersdorf, W. (2015, October). Computing at the mobile edge: Designing elastic android applications for computation offloading. In 2015 8th IFIP Wireless and Mobile Networking Conference (WMNC) (pp. 112–119). IEEE.

Paul, A., Pinjari, H., Hong, W. H., Seo, H. C., and Rho, S. (2018). *Fog computing-based IoT for health monitoring system. Journal of Sensors*, 2018.

Peng, L., Dhaini, A. R., and Ho, P. H. (2018). Toward integrated Cloud–Fog networks for efficient IoT provisioning: Key challenges and solutions. *Future Generation Computer Systems*, 88, 606–613.

Peter, N. (2015). Fog computing and its real time applications. *International Journal of Emerging Technology and Advanced Engineering*, 5(6), 266–269.

Prakash, P., Darshaun, K. G., Yaazhlene, P., Ganesh, M. V., and Vasudha, B. (2017). Fog Computing: Issues, Challenges and Future Directions. *International Journal of Electrical and Computer Engineering*, 7(6), 3669.

Priyadarshini, R., Malarvizhi, N., and Neeba, E. A. (2019). A Study on Capabilities and Challenges of Fog Computing. In Novel Practices and Trends in Grid and Cloud Computing (pp. 249–273). IGI Global.

Qi, J., Yang, P., Waraich, A., Deng, Z., Zhao, Y., and Yang, Y. (2018). Examining sensor-based physical activity recognition and monitoring for healthcare using Internet of Things: A systematic review. *Journal of biomedical informatics*, 87, 138–153.

Rahman, G., and Chuah, C. W. (2018). Fog computing, applications, security and challenges, review. *International Journal of Engineering and Technology*, 7(3), 1615–1621.

Rahmani, A. M., Gia, T. N., Negash, B., Anzanpour, A., Azimi, I., Jiang, M., and Liljeberg, P. (2018). Exploiting smart e-Health gateways at the edge of healthcare Internet-of-Things: A fog computing approach. *Future Generation Computer Systems*, 78, 641–658.

Raza, K. (2019). Improving the prediction accuracy of heart disease with ensemble learning and majority voting rule. In U-Healthcare Monitoring Systems (pp. 179–196). Academic Press.

Raza, K., and Qazi, S. (2019). Nanopore sequencing technology and Internet of living things: A big hope for U-healthcare. In Sensors for Health Monitoring (pp. 95–116). Academic Press.

Saad, M. (2018). Fog Computing and Its Role in the Internet of Things: Concept, Security and Privacy Issues. *International Journal of Computer Applications*, 975, 8887.

Sarkar, S., Chatterjee, S., and Misra, S. (2015). Assessment of the Suitability of Fog Computing in the Context of Internet of Things. *IEEE Transactions on Cloud Computing*, 6(1), 46–59.

Shi, Y., Ding, G., Wang, H., Roman, H. E., and Lu, S. (2015, May). The fog computing service for healthcare. In 2015 2nd International Symposium on Future Information and Communication Technologies for Ubiquitous HealthCare (Ubi-HealthTech) (pp. 1–5). IEEE.

Suh, M. K., Chen, C. A., Woodbridge, J., Tu, M. K., Kim, J. I., Nahapetian, A., .. and Sarrafzadeh, M. (2011). A remote patient monitoring system for congestive heart failure. *Journal of medical systems*, 35(5), 1165–1179.

Tabas, I., and Glass, C. K. (2013). Anti-inflammatory therapy in chronic disease: challenges and opportunities. *Science*, 339(6116), 166–172.

Tolentino, R. S., and Park, S. (2010). A study on U-healthcare system for patient information management over ubiquitous medical sensor networks. *International Journal of Advanced Science and Technology*, 18(1), 10.

Uddin, M. Z. (2019). A wearable sensor-based activity prediction system to facilitate edge computing in smart healthcare system. *Journal of Parallel and Distributed Computing*, 123, 46–53.

Varshney, P., andSimmhan, Y. (2017, May). Demystifying fog computing: Characterizing architectures, applications and abstractions. In 2017 IEEE 1st International Conference on Fog and Edge Computing (ICFEC) (pp. 115–124). *IEEE*.

Verma, P., and Sood, S. K. (2018). Fog assisted-IoT enabled patient health monitoring in smart homes. *IEEE Internet of Things Journal*, 5(3), 1789–1796.

Vora, J., Tanwar, S., Tyagi, S., Kumar, N., and Rodrigues, J. J. (2017, October). FAAL: Fog computing-based patient monitoring system for ambient assisted living. In 2017 IEEE 19th international conference on e-health networking, applications and services (Healthcom) (pp. 1–6). IEEE.

Weiner, M., Jorgovanovic, M., Sahai, A., and Nikolié, B. (2014, June). Design of a low-latency, high-reliability wireless communication system for control applications. In 2014 IEEE International conference on communications (ICC) (pp. 3829–3835). IEEE.

Yi, S., Qin, Z., and Li, Q. (2015, August). Security and privacy issues of fog computing: A survey. In International conference on wireless algorithms, systems, and applications (pp. 685–695). Springer, Cham.

Yousefpour, A., Ishigaki, G., Gour, R., and Jue, J. P. (2018). On reducing IoT service delay via fog offloading. *IEEE Internet of Things Journal*, 5(2), 998–1010.

Zhang, P., Liu, J. K., Yu, F. R., Sookhak, M., Au, M. H., and Luo, X. (2018). A survey on access control in fog computing. *IEEE Communications Magazine*, 56(2), 144–149.

Zhang, W., Zhang, Z., and Chao, H. C. (2017). Cooperative fog computing for dealing with big data in the internet of vehicles: Architecture and hierarchical resource management. *IEEE Communications Magazine*, 55(12), 60–67.

Zhou, L. (2016a). On data-driven delay estimation for media cloud. *IEEE Transactions on Multimedia*, 18(5), 905–915.

Zhou, L. (2016b). QoE-driven delay announcement for cloud mobile media. *IEEE Transactions on Circuits and Systems for Video Technology*, 27(1), 84–94.

2

Future Opportunistic Fog/Edge Computational Models and their Limitations

Sonia Singla[1]*, Naveen Kumar Bhati[2], and S. Aswath[3]

[1] *University of Leicester, U.K.*
[2] *Sunder Deep College of Engineering and Technology, Uttar Pradesh Technical University, India*
[3] *Department of Computer Science, PES University, Bangalore-560008, India*

Abstract

The Internet-of-Things (IoT) is the possible future of the Internet, where everything will be connected. The IoT is required to interface with billions of devices and individuals to bring positive benefits to us all. With this advance, cloud computing, close to its edge, prepares perfect models, for instance, multi access edge computing (MEC) and cloudlets, are seen as promising responses for dealing with the large volume of security-important and time-sensitive data that is being created by the IoT. Studies have revealed that Fog/Edge Computing (FEC) based organizations will assume an essential role in expanding the cloud by means of go-between organizations at the edge of the framework. Dimness/Edge Computing-based IoT's (DECIoT) configure service organization provisioning near the Cloud-to-Things continuum, thus making it appropriate for key applications. Moreover, the proximity of fog/edge devices to where the data is created makes it extend the advantage partition, organization transport, and assurance. Edge and fog registering are closely related – both indicate the ability to process information closer to the requester/buyer to reduce idleness cost and improve client experience. Both can channel information before it "hits" a major information lake for further utilization, reducing the amount of information that can be handled. The essential idea of edge and fog processing is to move information rationale (basically around information approval/information sentence structure checks) to an external ring of capacities. In this chapter we review the difficulties and future directions to be investigated in the role of fog, edge and pervasive computing.

Keywords *Fog; edge; Internet of things; Blockchain*

*Corresponding Author: ssoniyaster@gmail.com

Fog, Edge, and Pervasive Computing in Intelligent IoT Driven Applications, First Edition.
Edited by Deepak Gupta and Aditya Khamparia.

2.1 Introduction

The Internet of Things (IoT) brings a sensitive expansion of endpoints. It is difficult in a few different ways. Fog computing, also called edge figuring, has principally been established in the Industrial Internet of Things (IIoT) space. The structure utilizes neighbourhood processing hubs, between the endpoints (for example sensors, cameras, and so on) and cloud server farms, to accumulate, store and process information as opposed to utilizing a remote cloud server farm.

Information is the new fuel for any framework to convey smart and complex management. Information is being promoted as a vital resource for any association to prepare and give cutting edge capabilities clearly and reliably. Regardless of whether information is inside sourced or obtained from various and conveyed sources, it is fundamental for a wide range of information to be consistently and intentionally gathered, transmitted, cleaned, and facilitated on capacity frameworks. There are a few types of scientific technique and machines to do further and definitive investigation on the curated and merged information to separate significant bits of knowledge continuously. Clear and succinct language leads to better leadership and work. We need capable and well-coordinated investigation stages for accelerating, rearranging, and streamlining information examination, which is a difficult problem to solve due to the multi-organized and large amount of information. On the foundation front, we need a profoundly upgraded register, stockpiling and system framework to accomplish easy information examination. Another vital point is that there is group, constant, and intelligent handling of information. The greater part of the individual and expert applications needs on-going experiences so as to create constant applications. That is, continuous problem handling, and basic leadership are being demanded and thus the edge or fog computing idea has turned out to be exceptionally mainstream [1].

Cloud computing is the way towards using remote servers or PCs over the web to perform information tasks, stockpiling and overseeing information as opposed to using a nearby PC or server. The administration given by cloud processing can be of any kind, for example, stockpiling, databases, programming, applications, organization, servers, etc. Fog computing means the expansion of administrations beyond distributed computing to the undertaking's prerequisites. It comprises a decentralized domain for calculating where the foundation gives stockpiling, applications, information, and calculations. Fog computing is also called fog networking or fogging [3].

Cloud computing is an extraordinary arrangement where there is a continuous access to a cloud server equipped to handle and transmit information rapidly to the end device. Mist figuring is chiefly a design of heterogeneous devices in which certain applications and management are overseen at the hub by a smart device, however, the genuine management is by the cloud. Mist registering essentially focuses on versatile clients while the cloud focuses on general web clients. The management type in haze figuring is restricted, while in the cloud it is globalized. Although the capacity is restricted in mist figuring compared to distributed computing, the separation between the clients is considerably less than it might be conveyed through remote associations yet in distributed computing, correspondence is through IP systems. Contrasting the parameters of haze and distributed computing, in haze figuring mobility is upheld whereas in the cloud it is restricted. The number of administration hubs in mist is greater than in haze. Developing corporations are strengthened with the additional material in the past but not maintained further. Giving or assigning security in the cloud is poor while it is possible in fog processing [2].

The term fog computing was coined by Cisco. It is another innovation that gives numerous advantages to various fields, particularly the IoT. Like the cloud, fog computing gives management to the IoT clients, for example, information handling and capacity. Mist registering depends on giving information preparing capacities and capacity locally to mist devices as opposed to sending them to the cloud. Both the cloud and fog give stockpiling, registering and organizing assets [4].

Fog extends out the cloud near the devices which produce or create the information. The devices are called mist hubs. The device with system association, stockpiling, and registering highlight is known as the haze hub. Models incorporate switches, controllers, servers, cameras, *etc.* The mist registering is additionally called edge figuring [5].

Fog computing has insights at the neighbourhood level of system design, preparing information in an IoT door level, while edge can be seen to have the knowledge of the whole distance legitimately on the edge device itself without actually being in the middle [6].

Edge computing (or fog computing) is a strategy for enhancing distributed computing frameworks by performing information handling at the edge of the system, close to the wellspring of the information. Edge figuring is a characteristic following stage after distributed computing. It wouldn't be viable for every device to utilize the cloud in the way that cell phones do. Today telephones send everything to the cloud to be handled, the information is stored in the cloud, and the outcomes are returned to the device.

There are a few models where edge figuring gives IoT a focused edge. For instance, in modern web of things application, for example, aeronautics, smart traffic lights, or assembling, the edge devices catch spilling information that can be utilized to keep a section from falling flat, reroute traffic, advance generation, and anticipate abandoned items. At the point when information investigation is done at the edge of a system, that is known as "edge analytics [7]."

Associated devices send information and get directions to and from a nearby hub, for the most part introduced on premises. The hub could be a passage device, for example, a switch or switch that has the additional handling and capacity abilities. It can get, process, and respond progressively to the approaching information.

As the applications and sellers of mist registering develop, information and interface similarity become an issue. The absence of interoperability between items in the business will block the selection of the innovation.

Open fog consortium was established in 2015. It is sponsored by organizations like Cisco, ARM, Dell and Microsoft. They are driving benchmarks and best practices in fog computing framework configuration. They will likely encourage reception of cross industry measures and systems.

There are even applications and businesses where, just on the degree of sending information, conventional systems do not get the job done, not to mention can be utilized, for example due to their remoteness and the costs it takes to send this information through, for example, satellite communication.

In this way, for a number of reasons (transmission capacity, costs, speed, mechanization, upkeep, prescient investigation, remoteness, and so on) we need a quicker, less expensive and better methodology than the current one which typically would be: accumulate the information, send it through systems to the cloud or different situations where they are handled and utilized, *etc.*

That is where both edge registering and haze processing truly come into their own. On the off-chance that your information is produced at the edge of the IoT, why not bring all your knowledge and investigation as near the edge, the source, as you can, with all the undeniable advantages? Furthermore, it is also where those guaranteed estimates nervous registering and IoT come in.

When contrasted with explanatory calculations that execute on IoT/haze assets in the cloud, those on the edge will:

- produce datasets that are privately obtained, smaller in volume, lower in latencies, progressively specific, and primarily persevered in memory, as befits calculations that are designed with sensors;
- develop more easily and progressively particularly with regards to the assignment explicit nature of most IoT endpoints;
- require less processor parallelism, reliable with the smaller remaining tasks near the edge devices;
- execute at higher PC per-datum rates, attributable to the various levelled handling necessities of profound learning for undertakings, for example, picture, video, sound, and other complex example acknowledgment assignments;
- be designed into less complex gathering models, considering that machine/ profound learning calculations will demonstrate streamlining for the expected necessities of specific edge devices;

- be bundled into commoditized equipment and be streak upgradeable with changed calculations over remote associations, with regards to their prerequisite to run the most recent generation calculations that have been prepared on shared IoT/haze bunches;
- express less complex element spaces, predictable with the discrete, repeatable, organized, and concentrated nature of the machine information handled by an edge device; and
- participate in less processor correspondence and framework roundtripping, with regards to the requirement for edge devices to work in irregularly associated, low-transmission capacity, independent decisioning situations.
- Edge and haze figuring are firmly related – both allude to the capacity to process information closer to the requester/purchaser to diminish dormancy cost and improve client experience. Both can channel information before it "hits" a major information lake for further utilization, decreasing the measure of information that should be prepared. The fundamental idea of edge and mist figuring is to move information rationale (predominantly around information approval/information sentence structure checks) to an external ring of abilities.

Edge and mist figuring ideas have been created to react to the increase in information transmission capacity required by end devices and has been fuelled by the expansion of the IoT which in this way has increased the need to progressively process the produced information closer to the source. Edge and haze processing push the cloud (read server farm) closer to the requester to limit inertness, limit cost and increment quality.

For instance, a Boeing 787 creates 40 TB for each hour of flight, however, just a portion of a TB of this is finally transmitted to a server farm for examination and capacity. Similarly, an enormous retail location may gather roughly 10 GB of information each hour, however only 1 GB of that is transmitted to a server farm. As it is not reasonable, nor conceivable to introduce a full server farm either on a plane or inside a store, edge or mist figuring steps in to approve and pre-process this information either inside a nearby system (mist) or a portal device (edge).

What is more, this is the principle difference between edge and mist figuring – the area of the devices. Haze registering pushes the information approval insight further into the neighbourhood organization, while edge figuring places the information approval and knowledge preparation onto focal edge devices like switches.

Cloud and edge/fog processing provide food for an alternate set of prerequisites as set out in the correlation table below.

IoT devices are "loquacious", they produce a steady stream of information that must be approved, investigated and prepared. Conventional value-based frameworks had the requester (state a customer application) sending unapproved information to an information processor that was commonly introduced

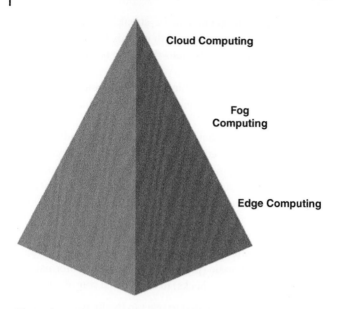

Figure 2.1 Cloud, fog and edge computing.

in a focal server farm. With the increase in IoT devices the information approval/information punctuation checks need to happen nearer to the request.

2.2 What are the Benefits of Edge and Fog Computing for the Mechanical Web of Things (IoT)?

Edge registering is an immediate reaction to the grand increment of data transmission required by the end devices that support the IoT. These devices are "glib" and produce a steady stream of information that must be approved, dissected and handled continuously to give a good end-client experience. As edge and haze processing pushes the information approval closer to the requester, it can smash through information at a quicker pace than if it were held in a focal area. It likewise takes into consideration disconnected or detached approval of information, which decreases the aggregate sum of start to finish transfer speed required, at the same time also bringing down data transmission costs.

Edge figuring can also reinforce digital safeguards, as different safety efforts, for example, encryption can be actualized in the nearby system before the information crosses through unprotected pieces of the web.

Open administrations carry us to foundation. Once more, this is a general classification which can be created by a few accomplices in the administration environment. Savvy framework is a model, Brilliant Streets another (in situations where street foundation is a national or "shared" matter). Yet additionally consider applications, for example, toll accumulation.

Next there is wellbeing and security. On a national level this positively also incorporates protection and the mechanical military complex. On progressively local levels we see applications, for example, bright lighting (there is a connection between lighting of open spaces and criminal activity), different types of character control, observation, *etc.* To summarize, there is a role for the IoT in security alarms, battling cataclysmic events and so on.

We now consider medical services, another area experiencing advanced change, and firmly related to government. Medical services are composed distinctively over the globe, from funding to human services protection and real care. However, there is a constant requirement for a management segment. Medical services is a key role in the IoT market.

In addition, governments have a role in general wellbeing which can be improved by taking activities utilizing the IoT and in a joint effort with a private state-supported partner. Coincidentally, the same also applies to open wellbeing. A model: joint efforts among governments and protection firms, utilizing telematics [8].

Fog computing, also called fogging, has a wide application in video streaming, gaming, health monitoring systems and its structure is well designed which makes it very suitable in applications related to 5G and big data analysis. It is considered to be the best gateway to handle the large amount of data created by the IoT and is found to be in close proximity with end users; however, it should not be considered as a replacement for cloud computing but rather should be taken as an additional support or supplement. Despite having many useful applications with cost reduction and expanded data storage, it also faces industrial challenges. This work has addressed some key qualities and considered general ideas regarding potential difficulties such as how mist figuring can grow CC administrations at the edge. Subsequently, it explains some utilization cases that provoked the need of mist registering particularly regarding the ongoing information examination significance for IIoT, commonly in medical services, STLS, and keen framework. Our work stresses haze's results and disturbance in three primary perspectives, IoT, huge information examination, and capacity. With a focus on the future capability of the rising time of mist, this review gave some information about up and coming use cases which describe some close to time inquiries about headings and circulated application work for adaptability in QoS of haze processing systems. Mist figuring, having some reliance on existing key innovation, can be classified as processing (*e.g.*, calculation offloading and inactivity the executives), correspondence (*e.g.*, SDN, NFV, and 5G), and capacity (precache framework and capacity development). Haze systems administration could be progressively smart with the organization of existing key advancements. SDN and NFV together can expand the system versatility just as cost decrease and live VM relocation (calculation offloading) can address the asset imperative difficulties tense devices.

Many research works considering IIoT applications and others such as manufacturing oil or gas have been done and are ongoing; however, there are certain research studies which are still pending such as the development of smart cities. Cloud computing mostly provides us with networking, databases, and storage while fog computing will provide fast development and constant handling by reducing the bandwidth and communication between sensors and clouds. It will be interesting and amazing to know see where fog will be in the coming ten years and what further changes will come into existence in virtual computing [9].

2.3 Disadvantages

Encryption calculations and security arrangements make it increasingly hard for self-assertive devices to trade information. Any errors in security calculations lead to presentation of information to the programmers. Other security issues are IP address parodying, man in the centre assaults, remote system security and so on.

To accomplish high information consistency in the mist figuring is testing and requires more work. Mist figuring will see worldwide stockpiling with limitless size and speed of nearby stockpiling yet information the board is a test. Trust and verification are significant concerns.

Planning is difficult as undertakings can be moved between customer devices, haze hubs and back end cloud servers.

Power utilization is high in mist hubs in contrast with concentrated cloud engineering.

2.4 Challenges

An underlying test is to decide on your business targets. To do this, distinguish the undertakings the association and its clients need to accomplish at the edge. Human services specialists taking patient vitals in remote regions may require certain abilities. Software engineers creating and testing computer generated reality highlights for another videogame may require others. Keen urban areas may have huge limit and inactivity prerequisites. However, in all cases, execution and speed of information transport are basic.

Next, consider the utilization cases for these verticals, including reconnaissance, self-driving vehicles or prescient support for machines. All these true applications require quick, ongoing, high-volume information. At that point, select explicit applications inside these utilization cases. An application may distinguish deserted bundles, screen sustainable power sources on a power lattice or find approaching bearing disappointments. Knowing the scientific categorization of your chosen verticals, use cases and applications will enable you to create basic prerequisites.

The following test is to create arrange engineering and component apportioning to meet the necessities of clients and applications. It is essential to comprehend which bits of the framework will keep running in the cloud and which segments will execute at the edge.

As a rule, a given amount of CPU power or capacity is less expensive the closer it lives to the cloud, however its presentation and idleness, transfer speed, unwavering quality and security properties improve as procedures draw nearer to the edge.

Edge processing design includes a progressive system of levels (for instance, local, neighbourhood, road and building-level hubs in a brilliant city), and each level may have various companion hubs sharing the heap. Apportioning can delineate all over the chain of importance and disperse them over the various haze hubs on each level.

A few models' bodies are grinding away consummating haze and edge processing engineering, including the Open Fog Consortium. The ETSI Multi-get to Edge Computing activity gives a superb edge forthcoming. At last, the Industrial Internet Consortium edge figuring errand gathering has contemplated these zones broadly. It's beneficial to counsel these assets just as specialists as you refine your IoT organize building models.

2.5 Role in Health Care

Before the beginning of IoT and distributed computing times, doctor understanding corporations were restricted to face to face visits, broadcast communications, and content correspondence. It was inconceivable for specialists to screen patients' wellbeing condition remotely to make an opportune treatment. In any case, as of late, IoT and distributed computing based social insurance frameworks make continuous applications in the human services division conceivable, release the maximum capacity of IoT and distributed computing in the medical services, and bolster doctors in conveying astounding social insurance administrations. IoT and distributed computing have expanded patient commitment and fulfilment since correspondences among patients and specialists have turned out to be increasingly available and more efficient. Besides, remote checking decreases the length of clinic stays and maintains a strategic distance from emergency clinic readmissions. Accordingly, these new advancements have significant impacts on decreasing human services costs and improving treatment results. IoT and distributed computing advancements are improving the social insurance industry by adding to the development of another variety of IoT-associated restorative devices and improving individual's cooperation in human services frameworks. Increasingly more IoT and distributed computing based medical services applications have been created to serve patients, families, doctors, emergency clinics and insurance agencies.

Restricted time and capacity: Increase in population has limited the capacity for doctors to check and talk to patients about diet, exercise, routine check-ups and their daily life as it is not possible to give more time to each patient because of the increased number of patients with disabilities and suffering from various diseases. Doctors find it difficult to complete this essential and necessary part of the proper treatment.

Strictness: It is difficult for doctors to maintain records of whether the patient is taking the prescribed medication daily and eating correctly which increases the risk of them being admitted to hospital and also proves costly to the patients and their family.

Adult Population above 60: With the increase in population it is natural that the adults aged over 60 will require more care and medical facilities, these will almost double from 831 million in 2013 to two billion in 2050.

Increase in Urban population: According to the WHO about 70% of the population will be living in cities rather than in rural areas which will demand more facilities and could be the cause of widespread of disease, such as COVID, which will spread quickly areas of high population.

Lack of workers: Rise in demand for medical facilities and increase in patients will increase the need for staff such as nurses, doctors, and surgeons.

Increase of cost: There is high increase in cost for medical facilities, in countries like USA the cost for diabetes increased by 21% from 2007 to 2016 which is approximately $250 billion [10].

The worldwide market for IoT medical devices is expected to surpass $500 billion by 2025, which will probably cause a significant change in outlook in human services IT. That is on the grounds that most processing presently occurs in on-premises server farms or, increasingly, in the cloud.

Analysing data from a distance results in delay, insufficient bandwidth, data traffic and long waits in queues which can be overcome by moving to fog edge computing which is indeed the need of time and response to act just on data which is required instead of the whole data and by maintaining the same quality of data. Its most prominent effect will be in the management of long term illness. The blend of IoT and quick 5G cell associations will improve the provision of at-home care and allow regular checking of patients for illness, for example, diabetes and congestive cardiovascular breakdown.

Edge registering is most helpful for devices whose information must be followed up on quickly because there isn't the ideal opportunity for it to be transferred to the cloud. A model would be emergency unit that required quick examination of information and execution of directions, for example, closed circle frameworks that keep up physiologic homeostasis. As sensors become progressively more up to date, we'll see comparable closed circle control of devices that screen insulin levels, breath, neurological action, cardiovascular rhythms and GI capacities.

Prescient investigations will run, and preventive considerations will be the standard. New markets will be created to allow expanding interest for information driven consideration, and new advancements — for example, exoskeletons for frailer patients — will be required as future increments. This will likewise raise new social issues, as we address the expanding gap between those who are well off and the poor [11].

Health care associations are quickly embracing examination arrangements as a component of their HIT frameworks to offer better patient consideration. As the quantity of investigation arrangements and IoT devices brought into human services systems develop, further developed methods for taking care of information are expected to guarantee that clinicians get information rapidly.

Edge figuring is becoming progressively prominent in medical services as associations bring increasingly linked restorative devices into their wellbeing IT biological system.

IoT and associated medical devices produce information at the edge of the system. Customarily, information is created at the edge of the system and moved back to the server farm. Edge figuring forms the information at the source or the edge of the system.

As per a report distributed by the *IEEE Internet of Things Journal*, distributed computing isn't a proficient method to process information when the information is created at the edge of the system.

The report creators characterized edge processing as "empowering advances enabling calculation to be performed at the edge of the system, on downstream information in the interest of cloud administrations and upstream information for the benefit of IoT administrations. The 'edge' is as any processing and system assets along the way between information sources and cloud server farms."

Social insurance associations are presently managing high volumes of IoT devices as increasingly associated therapeutic devices are used in ordinary medical clinic settings.

"An average hospital room will have between 15 and 20 medical devices, and almost all of them will be networked, either wired or wireless," Aruba Networks Product Marketing Manager Rick Reid told HITInfrastructure .com. "That's a pretty high density if you think about the size of an ICU room which is usually about 15'x15' with 20 devices in it and the room next door has 20 devices in it. A ward is typically 20 beds, so that's quite a lot of devices in a relatively small area."

An enormous clinic can have upwards of 85 000 associated medical and IoT devices putting great strain on the human services organization.

Huge social insurance associations with present or future great information investigation prospects need to investigate edge processing to guarantee that clinicians and patients will get the best reaction times from the information they produce. As the quantity of IoT devices continues to develop, edge processing may turn into a standard in wellbeing IT foundation [12].

We are living within the era of the IoT where we deal with large data every day from billions of devices. The present technique which is used to manage the data is cloud computing but this raises many problems in networking issues leading to delay and the need for techniques which can deal with quality of data without any delay. Thus, to manage these issues another framework viewpoint, closer to the IoT end devices is exhibited called fog computing. At whatever point displayed effectively then cloudiness calculating can lead to improvements in the organization (QoS) offered to systems that require preparation to concede difficult data like social insurance structures that could benefit from the fast handling of data from sensors to allow patients to be seen [13].

2.6 Blockchain and Fog, Edge Computing

The blockchain is certainly a smart innovation – the brainchild of an individual or gathering of individuals known by the pen name, Nakamoto. However, from that point forward, it has advanced into something more prominent, and the principle question everyone is asking is: what is Blockchain?

By enabling computerized data to be circulated yet not duplicated, blockchain innovation allowed the foundation of another kind of web. Initially conceived for the advanced money, Bitcoin, (Buy Bitcoin) the tech network has now discovered other potential use for the innovation.

A blockchain (initially two words: square chain) is a consistently developing rundown of computerized records in bundles (called squares) which are connected and verified utilizing cryptography.

These carefully recorded "obstructs" of information are stored in a direct chain. Each square in the chain contains information (for example bitcoin exchange), is cryptographically hashed, and time stepped. The squares of hashed information draw upon the previous square (which preceded it) in the chain, guaranteeing all information in the generally "blockchain" has not been messed with and has not been modified.

Edge computing is an approach to bring the handling focus nearer to the wellspring of information, or the "edge," essentially reducing expenses and saving time by tapping on a system of PCs who are offering their stockpiling and preparing capacity to the system's customers in return for compensation. Edge figuring doesn't really need to be blockchain-based, yet in a few different ways, the two advancements overlap.

Generally, they're like blockchain diggers, anybody can utilize their handling power for any procedure at some random time—it could be mining, logical computations, video spilling, or whatever else. Not at all like blockchains, edge figuring administrations are not restricted to a particular use case.

The most concise differentiator I've seen about edge and mist is from Cisco: "Haze processing is a standard that characterizes how edge figuring should function, and it encourages the activity of register, stockpiling and systems administration benefits between end devices and distributed computing server farms. Also, many use fogs as a hopping off point for edge figuring."

Fog computing is another developing innovation that can make blockchains considerably more dominant than they are now. Haze processing, which Cisco depicts as an approach to "stretch out the Cloud to where the things are," is entirely moving into the blockchain space. Now, some may now have a suspicion about how this identifies with blockchain and handling. However, aside from the undeniable beginning suspicions, one of its greater ramifications brings blockchain applications to a significantly higher universality level: versatile and IoT (IoT) devices.

Bieler (2016) contended that both the web of things and blockchain innovation depend on "decentralized, disseminated approaches". As indicated by him, the decentralized and self-sufficiently working frameworks in the (modern) web of things need direct shared correspondence and collaboration "instead of through existing concentrated models". Kranz (2017) called attention to that, in the advancement of the (mechanical) web of things, a blockchain can help. Given the essential highlights of blockchain innovation, for example shortcoming tolerant correspondence, a disseminated record, casting a ballot and agreement joined with execution conventions, blockchain innovation could be utilized to lay a protected and dependable establishment for the guideline of information and data exchanges between independently working devices on the edge and decentralized mist units of the focal cloud foundation. Blockchain innovation would thus be able to make decisions that empower decentralized, self-governing and mutually working frameworks to choose and manage for themselves on what conditions they can give their information and data, just as where, how and to whom [14].

One of the difficulties in an appropriated figuring condition, for example, fog computing is the manner by which to defend arrange assets and exchanges with a similarly disseminated security structure. Haze processing makes a kind of work framework where all hubs have equivalent jobs, in view of their abilities, and onto which we disperse computational burdens. Alongside dispersed figuring, we need conveyed trust and security. This is particularly significant for situations where the framework and layers of the haze hub stack are possessed and overseen by a wide range of elements.

How would you oversee trust in a decentralized, dispersed way among on-screen characters that don't really confide in one another? Blockchain innovation is made for this sort of test.

There is a characteristic match between mist processing and blockchain—or dispersed record innovation. This isn't just valid in situations where various entertainers need to cooperate but don't confide in one another, it's additionally valid for disengaged or self-ruling frameworks. The self-sufficient highlights of haze registering require believed exchanges in circumstances where a framework needs to work while disengaged from the cloud or server farm. As haze innovation turns into a basic piece of more IoT (IoT) and 5G applications, there is a developing need to use appropriated record innovation to set up accord for every exchange.

The frameworks for dispersing trust among various on-screen characters in a haze situation must be established in the Open Fog design system to guarantee interoperability. This will empower even use situations where circulated framework suppliers can commoditize and sell mist processing assets, with the affirmation of being paid for these administrations. Any asset associated with an Open Fog framework could then turn into an item of such a commercial centre.

2.7 How Blockchain will Illuminate Human Services Issues

Blockchain can totally change the medical services framework. It can help the medical services industry reverse the difficulties that it is experiencing. For instance, it can help improve general access, uprightness, security, discernibility, and interoperability. Blockchain medical services applications hold the key to improving the present social insurance condition.

With blockchain, various human services systems (HSS) can meet up and trade information with one another, because of the conveyed structure that it brings to the table. All in all, which difficulties can blockchain tackle? We will describe them below.

Perhaps the greatest leeway for utilizing blockchain is the interoperability that it brings to the table.

Blockchain's reliability likewise takes care of many issues that the present medical services industry is experiencing. Blockchain in human services today agrees completely on how the information is transmitted between frameworks. This prompts errors invalidating the significance of the information stored.

With blockchain, the information reliability can be constantly maintained at all levels. It likewise empowers various cases to expel out of date understanding information. Likewise, when the information is transferred on the blockchain, it can't be changed by malignant on-screen characters, safeguarding the reliability of the information. Just the patients can change the details when working with a doctor. There are many qcontextual investigations for blockchain in social insurance indicating that blockchain can carry the essential reliability to shield information from being taken or abused.

The current conventional social insurance industry experiences information releases that can be costly. With no appropriate arrangement, they are intensely dependent on experimentation, and that is too neglecting to even consider securing every one of their foundation including information honesty. Information altering and theft is becoming a major concern and needs to be appropriately managed. Blockchain uses information encryption utilizing private keys, and just the collector can unscramble the substance utilizing his key.

Upkeep cost is also an essential issue within the present human services frameworks. The present framework requires support across various activities and necessities a group to guarantee that every one of the capacities are running the same and are in a state of harmony. Blockchain doesn't experience the ill effects of this issue as it is a conveyed decentralized system.

Information is appropriated over the system which implies that there is no single purpose for disappointment. If a hub goes down, information can be brought from different hubs as there are numerous duplicates on the system. Every hub has its duplicate of the database. The reinforcement system is astounding as it will assist clinics to cope with crises better. Another advantage of having a reinforcement on every hub implies that there is less exchange cost with regards to putting away or recovering data.

Blockchain gives general access to every one of its clients. It isn't reliant on a focal position which makes widespread access conceivable. Approved substances can, without much of a stretch, access information at whatever point required, and the entire procedure can be robotized with various instruments like brilliant agreements. To put it plainly, blockchain for therapeutics is promising and ought to likewise be supported in the workplace.

2.8   Uses of Blockchain in the Future

Blockchain is at the cutting edge of numerous exchanges in view of its job in the dissemination of digital currencies, for example, bitcoin. Over the long haul, these computerized money exchanges may turn into a small piece of blockchain innovation's general impression and the way resources are moved on the web.

The conceivable outcomes for blockchain execution appear to be unending, as its hidden innovation can be utilized in numerous fields to play out these significant undertakings and that's just the beginning:

- execute contracts
- purchase and sell licensed innovation
- convey medical data
- guarantee that casting of a ballot in elections is morally sound.

Private blockchains enable organizations to reform inner procedures. Open, open-source varieties change the way individuals handle business in their day to day lives. World society has quite recently started to start to reveal blockchain applications. New uses for blockchain are found routinely [16].

2.9 Uses of Blockchain in Health Care

Issue: the fake medication issue is one of the most serious issues in the human services industry. With around 10 to 30% of medications being fake, the opportunity has already come and gone for medical services foundations to fix the issue for good. Not only do they lose a large amount of income but additionally there are impacts on the patients. At the present time, the size of the fake medication market is currently worth $200 billion annually. The two primary nations which are the biggest producers of fake medication are China and India. That is a significant issue, and should be fixed using blockchain. Fake medication is increasingly unsafe as it either doesn't contain the essential ingredients or is comprised of various ingredients.

Arrangement: Blockchain can comprehend every one of these issues by giving security and recognizing medication. Blockchain works by adding exchanges to the square. These exchanges are changeless and are also timestamped for later confirmation. In this way, if the entire store network is moved to the blockchain, most of the issues can be fixed. This is an ideal case of the use of blockchain in pharma.

2.10 Edge Computing Segmental Analysis:

MRFR's report offers a basic segmental investigation of the market dependent on innovation, organization, segment, end-client, applications, and locale. By innovation, the market is portioned into versatile edge processing, haze figuring. Mist figuring is relied upon to hold the biggest piece of the overall industry during the gauge time frame. By organization, the market is sectioned into on-premises and on-cloud. The on-cloud portion is relied upon to overwhelm the market during the figure time frame.

The market has been divided into equipment and programming. Among these, the equipment portion now holds the post position and is probably going to see sound development over the next couple of years. By end client, the market is sub-divided into reconnaissance, car, gaming, and media, medical services and training. Reconnaissance held the predominant piece of the pie in 2017. In the meantime, the car fragment is required to show the most elevated rate during the estimate time frame. By application, the market has been fragmented into IoT, information storing, investigation, and condition checking. The IoT section is required to hold its top situation beyond 2023, mirroring a CAGR of 25.6% [17].

2.11 Uses of Fog Computing

Remote checking for Oil and Gas tasks: Edge processing for Oil and Gas activities can mean the difference between normal activities and a catastrophe. Conventional combined information examination frameworks can reveal what took up personal time and can foresee dissatisfaction dependent on managed learning applied to various types of tasks. However, setting up of standards that can do close moment examination at the site where the information is being made can see the indications of a failure and take measures to avert a disaster before it even starts.

AI Models: Anomaly Detection models (*e.g.* Kalman Anomaly), Predictive models (*e.g.* Bayesian change discovery), Optimization Methods (*e.g.* Straight improvement)

Retail client conduct investigations: Retail Analytics to diminish truck relinquishment and improve client commitment utilizing close moment edge examination — where deal information, pictures, coupons utilized, traffic examples, and recordings are made — gives extraordinary knowledge into buyer conduct. This insight can enable retailers to target product, deals, and advancements and help overhaul store formats and item position to improve the client experience. For instance, use reference points to gather data, for example, exchange history from a client's cell phone, at that point target advancements and deals as clients walk through the store.

AI Models: Statistical Methods (*e.g.* advertise container examination, Apriori), Time arrangement bunching, and so forth.

Self-Driving Cars: With cutting edge advanced driver assistance systems (ADAS), vehicles will become more secure and progressively effective as they become more aware of and respond to the surrounding driving conditions. Genuine progress will mean the democratization of ADAS where the innovation is accessible at section level to premium vehicles, for young to seniors drivers, in travel and business vehicles, and everywhere in the middle.

Autos must expect to contend with the advancement and arrangement of cutting edge innovations — and self-driving vehicles soon — it is essential to look at how the aggregate arrangement of frameworks inside the vehicle can convey a superior encounter as opposed to moving toward the vehicle as a bunch of autonomous advances [7].

2.12 Analytics in Fog Computing

Today, pictures and picture successions (recordings) make up about 80% of all corporate and open unstructured huge information. As development of unstructured information increases, expository frameworks must acclimatize and decipher pictures and recordings just as they translate organized information, for example, content and numbers. A picture is a lot of signs detected by the human eye and prepared by the visual cortex in the cerebrum making a distinctive encounter of a scene that is in a split second connected with ideas and articles recently seen and recorded in one's memory. To a PC, pictures are either a raster picture or a vector picture. Basically, raster pictures are an arrangement of pixels with careful numerical qualities for shading; vector pictures are a lot of shading commented on polygons. To perform investigation on pictures or recordings, the geometric encoding must be changed into shapes portraying physical highlights, articles and development spoken to by the picture or video [17].

2.13 Conclusion

With edge computing, the IoT information is gathered and investigated legitimately by controllers, sensors, and other associated devices, or the information is transmitted to a close by a processing device for examination.

While fog computing is like edge processing and they are regularly confused with one another, there is a slight distinction between them. In fog computing, there is just one brought together registering device liable for handling information from various endpoints in the systems. In edge computing, each system takes an interest in handling information. Cost: there are still many areas of research work pending to improve and utilized fog computing along with overcoming the challenges of massive data to maintain the quality of data analysed within a time.

Bibliography

1 P. Raj and P. J., "Expounding the edge/fog computing infrastructures for data science," in *Handbook of research on cloud and fog computing infrastructures for data science*, P. Raj and A. Raman, Eds. IGI Global, 2018, pp. 1–32.

2 "ResearchGate." [Online]. Available: https://www.researchgate.net/publication/320855949_Fog_Computing_Issues_Challenges_and_Future_Directions. [Accessed: 18-Oct-2019].

3 "Cloud Computing vs Fog Computing - 7 Amazing Comparison." [Online]. Available: https://www.educba.com/cloud-computing-vs-fog-computing. [Accessed: 18-Oct-2019].

4 Atlam, H., Walters, R., and Wills, G. (Apr. 2018). Fog computing and the internet of things: A review. *BDCC* 2 (2): 10.

5 "RF Wireless Vendors and Resources | RF Wireless World." [Online]. Available: https://www.rfwireless-world.com/Terminology/Advantages-and-Disadvantages-of-Fog-Computing.html/. [Accessed: 18-Oct-2019].

6 "IoT Edge Computing | 2019 Overview of Software, Devices and Use Cases." [Online]. Available: https://www.postscapes.com/iot-edge-computing-software. [Accessed: 18-Oct-2019].

7 "Fog Computing: Outcomes at the Edge with Machine Learning." [Online]. Available: https://towardsdatascience.com/fog-computing-outcomes-at-the-edge-using-machine-learning-7c1380ee5a5e. [Accessed: 18-Oct-2019].

8 "The Internet of Things (IoT) - essential IoT business guide." [Online]. Available: https://www.i-scoop.eu/internet-of-things-guide/#The_growing_role_of_fog_and_edge_computing_in_IoT. [Accessed: 18-Oct-2019].

9 Anawar, M.R., Wang, S., Azam Zia, M., Jadoon, A. K., Akram, U., and Raza, S., "Fog computing: an overview of big iot data analytics". *Wirel. Commun. Mob. Comput.* 2018: 1–22.

10 [Online]. Available: https://par.nsf.gov/servlets/purl/10092283. [Accessed: 18-Oct-2019].

11 "Edge Computing in Healthcare: Experts Explain How IoT Will Change the Landscape." [Online]. Available: https://healthtechmagazine.net/article/2019/08/will-edge-computing-transform-healthcare. [Accessed: 18-Oct-2019].

12 "Edge Computing Uses IoT Devices for Fast Health IT Analytics." [Online]. Available: https://hitinfrastructure.com/news/edge-computing-uses-iot-devices-for-fast-health-it-analytics. [Accessed: 18-Oct-2019].

13 "ResearchGate." [Online]. Available: https://www.researchgate.net/publication/327324905_Towards_fog_driven_IoT_healthcare_challenges_and_framework_of_fog_computing_in_healthcare. [Accessed: 18-Oct-2019].

14 "Blockchain between edge and fog computing." [Online]. Available: https://www.centric.eu/NL/Default/Themas/Blogs/2018/02/13/Blockchain-between-edge-and-fog-computing. [Accessed: 18-Oct-2019].

15 "Blockchain Technology Explained." [Online]. Available: https://www.lifewire.com/blockchain-explained-4150034. [Accessed: 18-Oct-2019].

16 "Edge Computing Market 2019 Global Trends, Size, Share, Industry Growth by Forecast to 2023 - MarketWatch." [Online]. Available: https://www

.marketwatch.com/press-release/edge-computing-market-2019-global-trends-size-share-industry-growth-by-forecast-to-2023-2019-02-14. [Accessed: 18-Oct-2019].

17 A. Jayanthiladevi, S. Murugan, and K. Manivel, "Text, images, and video analytics for fog computing," in *Handbook of research on cloud and fog computing infrastructures for data science*, P. Raj and A. Raman, Eds. IGI Global, 2018, pp. 390–410.

3

Automating Elicitation Technique Selection using Machine Learning

Hatim M. Elhassan Ibrahim Dafallaa[1], Nazir Ahmad[1,*], Mohammed Burhanur Rehman[1], Iqrar Ahmad[1], and Rizwan khan[2]

[1]Department of information system Community College, King Khalid University Muhayel, Kingdom of Saudi Arabia
[2]Al-Barkaat College of Graduate Studies, Aligarh, India

Abstract

Technique selection is one of the most important issues in the field of requirement engineering, specifically requirement elicitation, it's impact on IS projects range from minor to catastrophic, the current primitive practice of selecting elicitation technique has a high risk of falling into inappropriate and improper classification in the context of the requirement design, which eventually leads to an increase of the risk factor to project failure. This chapter is an attempt to mechanize the requirement elicitation technique selection according the characteristics of the individual elicitation scenarios to produce a higher accuracy percentage for the selection process.

Keywords — *Requirements elicitation; Elicitation Technique; Technique selection; Machine learning*

3.1   Introduction

The emphasis of the requirement elicitation phase accuracy is drawn by many researchers in the field. The degree of error at this stage of the process has an immediate effect on the final product, in other words it's the make or break factor for the success of any IS project.

The requirement elicitation process for any system development project is an intensive work, that requires several tedious phases such as searching, learning and acquiring the requirements along with the validation process of all the proposed system components. These mentioned phases are initialized through the

Fog, Edge, and Pervasive Computing in Intelligent IoT Driven Applications, First Edition.
Edited by Deepak Gupta and Aditya Khamparia.

elicitation technique selection process in collaborative form that involves both the requirement engineer and the stakeholder.

We cannot discuss the elicitation technique selection without reflecting on the limitations of the current applied methodologies and approaches, these limitations manifest in the application frame, specifically the scenarios and the condition that pertain the optimal and worst cases, the elicitation techniques differs in the characteristics composition and it needs a specific election case characteristics to produce a higher accuracy percentage that reflects the actual desired design, so relying on the requirement of engineer premonition and judgement does put the developing process in the zone of failure and uncertainty. Which was perfectly depicted by the Standish report that considered that 51 % of the software development projects are considered late or challenged, the survey depicted an alarming failure percentage of 31% [1]. The study considered classified the prime reasons influencing the requirement elicitation process into Incompleteness of Requirements & Specifications which is credited to the inappropriate mechanism of selecting the requirement technique that fits the elicitation case characteristics. In the light of the previous elaboration, we propose a machine-learning model to automate the requirement technique selection process, to improve the current records of successful projects.

This chapter is composed as follows. In Section 2, we discuss the existing approaches for the elicitation technique. In Section 3, we introduce the machine learning model and the selection attributes influencing the selection process. In Section 4, we analyze the produced results. In Section 5, we evaluate the error rate of the proposed model. In Section 6, we validate the produced result by conducting an experiment. Section 5 concludes this chapter.

3.2 Related Work

The most related research in the area of supporting requirement techniques, selection is a framework called "A Model for Two Knowledge-Intensive Software Development Processes" by Ann M. Hickey and Alan M. Davis. According to the authors conducted study the analysts select a particular elicitation technique for a combination of any of the following four reasons: (1) it's by miles the simplest method that the analyst is aware of, (2) it is by miles the analyst's favorite approach for all situations, (3) the analyst is following a few explicit techniques, and that method prescribes a specific approach on the present day time, and (4) the analyst understands intuitively that the approach is powerful in the modern-day situation. Really the fourth purpose demonstrates the most "maturity" by the analyst. Hickey and Davis hypothesize that such maturity ends in progressed know-how of stakeholders' needs, and accordingly a higher chance that a resulting gadget will satisfy ones requirements. Unfortunately, most practicing analysts no longer have the perception vital to make such a knowledgeable selection, and therefore depend on one of the first three motives [1]. Their look at "new unified model of necessities elicitation" can also be said to be a pioneering observation within the area of requirement elicitation, to improve our knowledge of the elicitation process, elicitation methodologies and techniques, and the elicitation technique selection process. However, the modeling approach provided a mathematical, logical illustration of best case scenario, lacking the in-depth details of complexity, analyst and technique characteristics of the situational case. Moreover, the design of Hickey and Davis was a more theoretical than a practical approach to address the problem.

Work in 2003 by Hickey and Davis after the unsolved Problem Software Development Software Solutions in 2002 aimed to quantify how the experts select an Elicitation Technique Selection [2]. Their study considered the previous research limitation and theory regarding elicitation technique selection, by choosing a qualitative research approach, using three primary qualitative information-gathering methods (participation in the setting, document analysis, and in-depth interviews) to provide the data needed to discover a situational technique selection theory. The study by Hickey and Davis has produced new theories and consideration of elicitation technique selection that for each elicitation technique, there exists a specific, unique, small set of predicates concerning situational characteristics that drive experts to seriously consider the technique. For each elicitation technique, there exists a set of basic analyst skills that must be present or the technique will not be effective, "Prerequisite Skills." Hickey and Davis suggest the creation of models (perhaps multiple models) to aid analysts in fully comprehending a situation and in communicating with stakeholders. However, the state of the research has not reached a definitive conclusion on all the situations and techniques. Moreover, this study lacks state of the practice in requirements elicitation.

The first attempt to systematically apply information-processing theory to requirements elicitation technique selection was proposed by Marakas *et al.* [3], they applied models borrowed from the organizational literature to systematically understand the effects these issues have on uncertainty and equivocality, and thus on elicitation technique selection. Although some prior work was implicitly organized along these lines. It should be noted that many of the contingencies in the extant literature have not been included in the categorizations such as those concerning the proficiency of the analysts and developers, and those imposing constraints on the methodologies that could be employed. From a normative point of view, neither of these should influence the ideal pattern of alignment between contingencies and elicitation techniques. However, the efforts have been limited to grouping characteristics or examples around a common label (*e.g.* characteristics of users, of the utilizing system, of the analysts, *etc.*). Moreover, an organizing framework, or a set of them, tying these lists of characteristics to their impact on the selection of techniques has not yet emerged.

Li Jiang *et al.* [4] also proposed the Knowledge-based Approach for the Selection of Requirements Engineering Techniques (KASRET) throughout a qualitative and quantitative data collection . This approach has three major features. First, a library of requirements techniques was developed which includes detailed knowledge about techniques. Second, KASRET integrates the advantages of different knowledge representation schemata and reasoning mechanisms. Thus, KASRET provides mechanisms for the management of knowledge about requirements techniques and support for RE process development. Third, as a major decision support mechanism, an objective function evaluates the overall ability and cost of RE techniques, which is helpful for the selection of RE techniques. However, the following approach cannot be used as formal proof that the KASRET approach will always provide the best solution for a software project. Another limitation of the study is the accuracy of the data derived from the case study. It is not realistic to assume that all the data derived have equal levels of accuracy. Moreover, the depicted approach still depends on the expert knowledge to select and weight the technique and didn't consider the situational characteristics of the elicitation case scenario. These illustrated factors reduce the validity of the case study.

Kausar *et al.* [5] provided a guideline for the practitioners for selecting the elicitation technique that suits their needs, based on the study conducted by Jiang *et al.* [4]. They modified the techniques by associating weights with the different parameters for the selection, according to the needs and priorities of the project. However, this proposal was merely a requirement technique, classification based on the project characteristics as selection parameters. Moreover, the study does not consider the situational characteristic elicitation case scenario nor the idea of automating the selection of the requirement elicitation techniques.

Muqeem and Beg [6], proposed an effective model for requirement elicitation technique selection. In their study sought to classify the stakeholders by observing non-verbal communication (NVC) and use it as a base for elicitation technique selection. Whereas the study focused only the area of NVC, by intensifying the pre-domain analysis using the interviewing technique for the sake of classifying the stakeholder maturity level, there is no novel contribution regarding the elicitation technique selection that may help in the selection of the appropriate technique according the situational complexity characteristics. Moreover, the illustrated model was never tested for validation.

Anwar [7] presented a practical guide to requirements elicitation technique selection based on technique features, stakeholder characteristics, requirements sources and project environment, based on an empirical study involving five practitioners from the software industry. However, the study was merely a guideline for elicitation technique selection, which didn't provide any characteristic mapping of the selections attributes.

Tiwari *et al.* [8], developed a framework to select elicitation techniques for a given software project based on the alignment of the project's contextual information and the elicitation techniques. However, the proposed framework has no novel contribution regarding the elicitation technique selection that may help in the selection of an appropriate technique according to the situational complexity characteristics. Moreover, the illustrated model was never tested for validation.

Carrizo *et al.* [9], introduced a new approach to systematize the requirements elicitation technique selection. Carrizo *et al.* validated the proposed framework by running two experiments on software engineering master students. However, their framework is at an early stage: it contains relevant attributes and adequacy values for techniques established on the basis of information available. Moreover, their methodology does not have the top to bottom subtleties of the attributes mapping and classification. Furthermore, the implementation is impractical, their main focus is on choosing the technique that suits the whole project, rather than choosing a technique for the situational cases complexity characteristic.

A research study conducted by Roy Egas [10] has some similarity with Carrizo *et al.* [9] "systematize the requirements elicitation technique selection" in terms of the attributes selected for comparison. This research focused on the theory of requirements elicitation selection and how requirements elicitation selection is executed in practice. However, this research did not take into account the elicitors' familiarity with the various methods of the requirements elicitation, nor help the elicitors to identify their position and learn which contextual aspects have a negative impact on the commonly used elicitation methods. Moreover, the research needs to investigate the influence of certain contextual factors on the efficiency of the elicitation method for new additional attributes to determine the context.

Kim *et al.* [11] proposed a study of Selection of Requirement Elicitation Techniques Using Laddering, their study analyzed the cognitive structure of IT professionals based on the factors that actually affect the use of the requirement elicitation techniques using laddering. They also use various statistical techniques to examine the relationship between the attributes of the actual project and the requirement elicitation techniques. However, it was too complicated to determine the dominant cognitive structure of respondents when they tried showing all the relationships between factors. Moreover, the approach mainly relied on the quantitative reports of the experience of previous projects to build their data model regardless of the authenticity factor, to focus only on the project characteristics as a sole parameter for selecting the requirement elicitation techniques.

In recent attempts Mishra *et al.* [12] created a tool to perform a continuous process improvement, used by requirement engineers to create and use effective requirement elicitation methods according to the current project situation, rather than choosing a technique for the situational cases complexity characteristic, the selection of technique was based on database system accumulative experience on conducted projects.

As can be seen from the above summaries, existing research has presented different approaches to assist the requirement engineer in selecting the appropriate technique and most of these approaches work as a theoretical guideline concept to show when they should or should not apply the technique as stated in Table 3.1. We believe that requirements analysts who are very experienced (and are believed to be masters of elicitation) appear to have the capacity to pick out appropriate elicitation strategies on an everyday basis. Since most working analysts have less experience and are journeyman rather than masters, it is no surprise that over half the goods created by means of the software industry fail to meet customers' needs [13]. However, these researched approaches were helpful only in classifying the elicitation techniques for the project as a whole, rather than the internal situational complexity of the elicitation cases. Moreover, the existing researches do not have the top to bottom subtleties of the attributes mapping and validation. Currently, there's no elicitation necessities choice technique that holds the minimum properties that we consider to be vital to warrant consideration as a importance and systematic aid for requirements engineering practitioners based on the elicitation scenario characteristics of the requirement, rather than selecting an elicitation technique or two to be used on the whole project. There is an obligation to increase the level of reliability in the requirement elicitation process to make the technique an efficient selecting mechanism using machine language and algorithms. Thus, automating and systemizing the process to produce a higher accuracy percentage to improve the industry records. This is the gap that our model sets out to fill.

3.3 Model: Requirement Elicitation Technique Selection Model

Figure 3.1 shows the systemization model components and attributes, the model application of k-nearest neighbors algorithm and machine learning to produce the most accurate elicitation technique according to the requirement scenario characteristics. The accuracy of the model can be tested through the produced percentages and predication. The model tackles the core issues of:

- absence of systematic approaches
- absence of the automation in requirement technique selection
- project failure and disuse
- incomplete requirement as result of improper application of the elicitation technique

Table 3.1 Comparison of various Requirement Elicitation Selections Techniques.

Reference	Model	Approach	Systematic	Automated
Hickey & Davis 2002	"New unified model of requirements elicitation"	Theoretical		×
Hickey & Davis 2003	"Requirement Elicitation Technique Selection" How Do Experts Do It?	Theoretical	×	×
Marakas, Aguirre-Urreta, M. I., & Marakas, G. M. 2007	"Systematically understand the effects these issues have on uncertainty and equivocality"	Contingency theory	✓	×
Li Jiang, Armin Eberlein, Behrouz H. Far, 2008	"A case study validation of a knowledge-based approach for the selection of requirements engineering techniques"	Case-based reasoning (CBR)	✓	×
Sumaira Kausar, Saima Tariq, Saba Riaz, Aasia Khanum 2010	"Guidelines for the Selection of Elicitation Techniques"	Classification	×	×
Md. Muqeem, Md. Rizwan Beg 2012	"Nonverbal communication (NVC) based model for selecting effective requirement elicitation technique"	Observing non-verbal communication	×	×
Razali, Anwar. 2012	"Practical guide to requirements elicitation technique selection"	Theoretical	×	×
Tiwari, S., Rathore, S. S., & Gupta, A. 2012	"Framework to select elicitation techniques"	Contextual information	×	×
Carrizo, Dieste, & Juristo. 2014	"Systematize the requirements elicitation technique selection"	Logical design	✓	×
Roy Egas 2015	"Research and validation of elicitation selection models"	Contextual factors	×	×
Ja-Hee Kim, Seung Mo Ham, Hwan-Ju Cha, Kwang Kook Kim 2017	"Selection of Requirement Elicitation Techniques Using Laddering"	Laddering / cognitive	✓	×
Mishra, Aydin & Ostrovska 2018	"Knowledge management in requirement elicitation: Situational methods view"	Database tool	✓	×

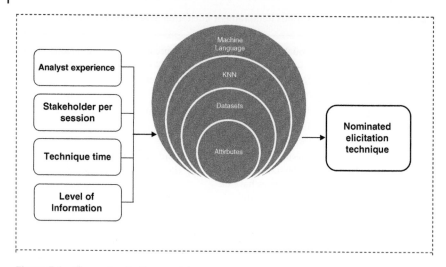

Figure 3.1 The systemization model.

As such, the proposed systemization model will enhance the average requirement engineering ability to select a proper elicitation technique, it will most likely improve our record of successful product.

3.3.1 Determining Key Attributes

Before deciding on the most suitable elicitation approach to apply at any time through an undertaking, it is first important to determine most influential attributes that are required in the decison making process of the selection of the elicitation approach. We have reviewed the related literature in search of theoretical proposals and empirical studies directly reporting or inferring the technique selection attributes that possibly influence the elicitation technique selection process. We have defined four attributes or parameters (analyst experience, session, technique duration and information rate).

3.3.2 Selection Attributes

These selected attributes help the system analyst to know exactly which techniques are best suited for his or her experience, the capacity of the technique, the level of the information about the technique and the timeframe of the technique. As such these attributes will represent the given elicitation case scenario

- session
- analyst experience

- technique duration
- information rate

The stakeholder and the system analysts are the main entities in the elicitation process, each having a role in building the system; the identified attributes are directly related to both the stakeholders and the system analyst characteristics in the domain of the decision making process as main influential factors in selecting the elicitation technique.

The technique selection attributes were explored in the research survey; these attributes have a major influence on the decision-making process of the elicitation technique selection. The surveyed selection parameters will help the systematic transformation and automation of the selection process.

3.3.2.1 Analyst Experience

System analysts are integral to the elicitation process and their role is not confined only to selecting an elicitation technique, they also play the role of the mediator and the creator of the system. Assuming the role of a system analyst is not an easy job, it requires tremendous skills such as technical, communication, elaborations, problem solving and lastly patience.

Most of these skills are accumulative experience gained through involvement in different software development projects, different stakeholders as well as techniques.

Hickey and Davis [1] discussed the elicitation selection dilemma reflecting on the system analysts lack of experience and mechanism on that field. Of course not all system analysts possess the same level of experience and knowledge, some basically don't have it, others are only accustomed to or familiar with one or two techniques, as such, they will try to impose them in every elicitation case scenario where they will not give the best performance and result. There is a need to classify the system analyst according to their level of involvement in software development projects and their familiarity with the specific technique as the parameter, during the course of deploying a software development project to ensure the suitability of system analyst to the elicitation case complexity for a better outcome. Thus, the system analysts background, skill, and experience were selected as an attribute and a parameter in the technique selection process of the proposed model.

3.3.2.2 Number of Stakeholders

This is the range of individuals that may participate at the equal time in the elicitation consultation. A few techniques are confined only to one or stakeholders in line with consultation. While, others are allowed extra stakeholders in keeping with consultation. Identifying the ratio for each technique will help the model narrow down the possibilities to a unique technique which fits the description of the present elicitation case scenario complexity. Thus, the number of stakeholders was selected as an attribute and a parameter in the technique selection process of the proposed model.

3.3.2.3 Technique Time

This is the time spent on the sessions for each on the techniques before deployment, the factor of time is very important when working on software development projects. Schedule constraints have their own impact and significance in the mission; some have strict and precise closing dates whilst others are more flexible. The elicitation techniques chosen for deadline specific projects ought to be quick and effective, whilst others with a flexible closing date has a few margins. Depending on the task in hand and time restriction the system analyst can identify which technique will be more suitable for the elicitation case situation. As is already known, some techniques consume more time than others. This attribute will help the analyst to select the proper technique and keep the project on track within the time frame to avoid late delivery. Thus, the technique time was selected as an attribute and a parameter in the technique selection process of the proposed model.

3.3.2.4 Level of Information

Information is the key and the foundation of elicitation; it is requirement elicitation's prime objective to elicit information for the purpose of building the desired project or system. The availability of information is most crucial to the elicitation process. Although the elicitation technique's sole purpose is to elicit information, they all vary in the level and depth of information they can elicit, for some elicitation techniques have the ability to extract more information than others. Therefore, for the sake of building the systematic selection model, the system analyst needs to be aware of the level of information required for the elicitation case scenario.

Thus, the level of information about the elicitation technique was selected as an attribute and a parameter in the technique selection process of the proposed model.

3.3.3 Selection Attributes Dataset

The extracted selections attributes and parameters were explored in the research survey to build the machine learning datasets, the goal of the survey is to collect data from the analyst in the telecommunication companies, thus the researcher sought help from a well-known worldwide survey platform, such as the Survey Monkey services, to locate as many participants as needed, in the efforts to verify the requirement technique attributes.

The dataset consists of 1149 observations and four attributes: session, analyst experience, technique duration and information rate. These are the attributes of

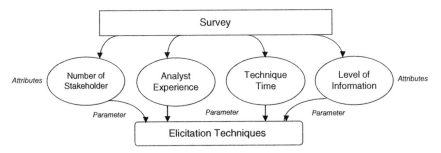

Figure 3.2 Mapping the selection Attributes.

requirements elicitation techniques. There are fourteen classes in the dataset protocol analyses, interview (structured /unstructured), focus groups, task analysis, card sorting, observation, surveys, protocol analyses, repertory grid, brainstorm, prototyping, JAD workshop, scenario analysis, and nominal group technique. These present all the requirements elicitation techniques.

Based on the information gathered from these sources, we plan to develop tentative relationships among the extracted attributes to build the machine learning dataset for the sake of fulfilling the elicitation case scenario situational characteristics. We then plan to validate this relationship through further interviews and surveys, which can be used by less-experienced analysts who wish to develop more expertise.

3.3.3.1 Mapping the Selection Attributes

Mapping the extracted selection attributes and parameters to the elicitation technique will allow us to identify the appropriate analyst to perform the technique, along with which technique will serve the potential number of stakeholders.

Every technique has at time-frame and a level of information which it can elicit, with the given parameters, the time required to accomplish the task and the amount of information needed, which will enable us to identify which technique will best fit the given situation. Furthermore, it will also allow us to apply the situational characteristics for the current elicitation case problem in hand in terms of time and the level of information needed as can be seen in Figure 3.2.

3.3.4 *k*-nearest Neighbor Algorithm Application

The *k*-nearest neighbor algorithm (KNN) is a type of instance-based learning, or lazy learning, where the function is only approximated locally and all computation is deferred until classification. An object is classified by a majority vote of its peers, with the object being assigned to the class most common among its *k*-nearest

Table 3.2 Selection Attributes Dataset Sample.

Technique	Analyst Experience	Session	Technique duration	Information rate
Interview (unstructured)	4	1	1	8
Interview (structured)	9	1	9	1
Task analysis	6	5	9	6
Card Sorting	6	9	8	7
Surveys	3	-	8	7
Protocol analysis	4	1	3	3
Repertory grid	8	9	4	3
Brainstorm	3	9	5	6
Nominal Group Technique	5	9	5	3
Observation	4	2	6	5
Prototyping	5	8	3	1
Focus Groups	4	6	6	4
JAD workshop	7	5	5	6
Scenario analysis	7	7	2	2

neighbors (k is a positive integer, typically small). If $k = 1$, then the object is simply assigned to the class of that single nearest neighbor.

The peers are extracted from a set of objects for that class (for k-nearest neighbor algorithm classification) or the object property value (for k-nearest neighbor algorithm regression) is known. This is considered to be as the training set process for the KNN algorithm, using the Euclidean distance function.

$$d(p, q) = d(q, p) = \sqrt{\left(q_1 - p_1\right)^2 + \left(q_2 - p_2\right)^2 + \ldots + \left(q_n - p_n\right)^2}$$

$$= \sqrt{\sum_{i=1}^{n} \left(q_i - p_i\right)^2}$$

Equation 1: The Euclidean Metric Equation

In this study, a systematization model should be selected as the elicitation approach given the requirement scenario characteristics parameters. Python's scikit-study library may be used to implement the KNN algorithm.

The systematization model implementation starts by loading the elicitation technique dataset function form scikit-learn dataset module. Table 3.2 illustrates the first fourteen observations of the dataset.

Table 3.3   Confusing Matrix Report.

	Precision	Recall	f1-score	Support
Brainstorm	0.92	0.96	0.94	24
Card_Sorting	0.82	1.00	0.90	9
Focus_Groups	0.93	0.93	0.93	14
Interview (structured)	0.93	1.00	0.96	13
Interview (unstructured)	1.00	0.95	0.97	19
JAD_workshop	1.00	1.00	1.00	20
Nominal_Group_Technique	0.94	1.00	0.97	16
Observation	1.00	1.00	1.00	11
Protocol_analyse	1.00	1.00	1.00	19
Prototyping	1.00	1.00	1.00	8
Repertory_grid	1.00	0.96	0.98	23
Scenario_analysis	1.00	0.94	0.97	18
Surveys	1.00	0.95	0.98	21
Task_analysis	1.00	0.93	0.97	15
micro avg	0.97	0.97	0.97	230
macro avg	0.97	0.97	0.97	230
weighted avg	0.97	0.97	0.97	230

The dataset observation portrays the illustrated elicitation technique parameters, in other words their optimal requirement elicitation environment scenarios.

Training and testing operation on the same data is not a recommended approach, so we split the data into two parts, [training set] and [testing set]. We use the Python functionality ['train_test_split'] to split the data. The optional parameter ['test-size'] specifies the split ratio. ['random_state'] parameter makes the data split the same way every time you run. The result of the process will reflect on the testing accuracy while performing on unseen data. In this model we split the dataset into 80% train data and 20% test data. This means that out of total 1149 records, the training set will contain 919 records and the test set contains 230 of those records.

The training session of the systemization model begins by importing the KNeighbors Classifier class from the sklearn, which in turn utilizes the Euclidean distance equation (1) measures the distance between the parameters. The classified elicitation technique with the majority votes of its peers, this model will employ the top five most common among its k nearest neighbors $k=5$, for the sake of the diversity of nominated techniques. Therefore, the requirement engineers intuitive judgment will be subject to similarity reading in the compared results.

3.4 Analysis and Results

In this part of the chapter, we reason to evaluate the systemization model using, the confusion matrix, and accuracy, precision, and f1 score. The type table shows the confusion matrix and class file, that is a technique of the sklearn.

Accuracy is the most intuitive performance measure and it is clearly a ratio of the effectively predicted statement of the entire observations, which may be calculated as follows:

$$Accuracy = (TP + TN/)(TP + FP + FN + TN)$$

For this model, the produced accuracy was (0.969) which means our model is approx. 96% accurate.

The produced results show the systemization model was able to classify all the (230) observation in the test set with a total average percentage of 0. 96% accuracy.

Regarding precision, the ratio of successfully expected advantageous observations of the whole expected fine observations may be calculated as follows:

$$Precision = (TP/TP) + FP$$

The results show the systemization model become efficaciously able to score 0.97 precision and classifying all of the 230 records in the test set with the total average of 0.97 precision, which reflects a low false positive rate.

Recall (sensitivity): the ratio of correctly predicted positive observations compared to all observations in the actual class, can be calculated as follows:

$$Recall = (TP/TP) + FN$$

The results show the systemization model was successfully able to score a total average of 0.97 percentage sensitivity by classifying all the 230 records in the test set.

F1 score, which is the weighted average of precision and recall. This score takes both false positives and false negatives into account, which can be calculated as follows:

$$F1\ Score = 2 * Precision * Recall/Precision + Recall$$

The results show the systemization model was successfully able to score 0.97 F1 scores by classifying all the 230 records in the test set.

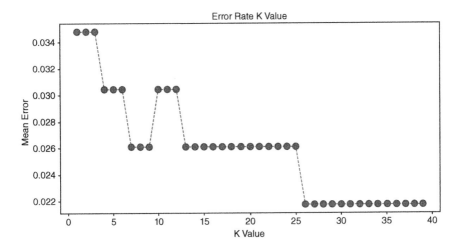

Figure 3.3 Error Rate with the *K* value.

3.5 The Error Rate

In our efforts to evaluate the machine learning model prediction accuracy in contrast to the error rate to the (*K*) value of the KNN algorithm and to foresee whichever rate of (*K*) yields the optimal results. For the sake of evaluation, we designed an experiment to measure the mean of error for all the foreseen objects where (*K*) ranges from (1 to 40), as can be seen in Figure 3.3.

The experiment has was successful in identifying the errors rates value in contracts with the *K* value, between the *k* values of 1 and 4 the error rate scored its higher error rate with an average of 0.035. Nevertheless, it continued to decrease with an average score of 0.022. From this experiment, we can deduce that the produced result was satisfactory.

3.6 Validation

In this section, we will validate the proposed systemization model produced results, by testing the nomination accuracy against the requirement engineer expert's judgement and decision making process. This section will also help in evaluating the effectiveness and the efficiency of the currently employed methodologies and approaches in contrast of the systemization model and whether the automation and systemization will actually assist this field in producing a better requirement design with higher accuracy rates.

To proceed with the validation phase, we sought help from five expert requirement engineers along with five novice requirement engineers who worked in system development projects and were involved in the decision making process of selecting an elicitation technique based on their intuition and experience.

The experimental process turned out as follows: subjects were given a specific elicitation case scenario parameter at five distinctive instances during the course of the test. At each time, they were asked to pick which elicitation techniques they could use in line with the given elicitation requirement characteristics.

We compared the selections with the aid of participants against the choice made by using the systemization model. The information accumulated in the validation segment is given in Table 3.4.

3.6.1 Discussion of the Results of the Experiment

The validation section Table 3.4, reflected the expected alignment and confirmed our suggested hypothesis, where our participants who were expert requirement engineers with extended experience and knowledge of IS project prediction have conformed with an average of 70% similarity to the systemization model prediction. This satisfying result portrayed how this model was able to emulate the decision-making method of the requirement engineer (professional) in deciding on a selected elicitation approach in keeping with the given requirement situation traits.

Moreover, the validation phase has also confirmed the speculation of Hickey and Davis [1] that the absence of systematic techniques in requirement choice method in conjunction with its generalized difficulty in terms of the involvement of the requirement engineer (novice) and their dependence on less accumulated knowledge and experience in selecting an elicitation technique, where the intuition and their personal preference plays a bigger part in the process, might lead the project towards failure. This can be justified by the lower confirmation rate of 20 % with the proposed systemization model.

3.7 Conclusion

Requirement elicitation is a crucial phase of any system development or IS project, the phase emphasis is manifested in how well the proposed design will satisfy the intended stakeholders, or whether it will fail, which is a catastrophic ending that no one desires.

The IT Business Edge report has given us a glimpse of what is really happening in this phase and its effects on the world of developing business, the report estimates the annual worldwide cost of failures to be "about $500 billion per month"

Table 3.4 Validation Experiment.

	Selections values				Interview (unstructured)	Interview (structured)	Task analysis	Card Sorting	Surveys	Protocol analysis	Repertory grid	Brainstorm	Nominal Group Technique	Observation	Prototyping	Focus Groups	JAD workshop	Scenario analysis
	Analyst Experience	Number of Stakeholders	Technique Time	Level of Information														
Model					✓													
Expert 1	9	1	9	9		✓												
Analyst 1															✓			
Model																	✓	
Expert 2	3	15	8	7					✓									
Analyst 2									✓				✓					
Model							✓					✓						
Expert 3	7	5	5	6														
Analyst 3																		
Model														✓				
Expert 4	7	2	6	5										✓				
Analyst 4											✓			✓				
Model												✓						
Expert 5	8	9	4	3											✓			
Analyst 5																		

[14], this huge assessment was our motive to try to boost the success rate with the new technology such as machine learning and AI, to assist us as requirement engineers in building a proper design that satisfies all needs, to clear the ambiguity and uncertainty hovering around the elicitation process.

The data analysis reports in this field have projected unsatisfactory results, the failure percentage ratio in IS projects are actually increasing over the last five years and this continuing issue is attributed to the limitation of the traditional approaches, they are lacking the profundity of the systematic details and mapping coordinates which create a state of ambiguity in the requirement elicitation process. This state is manifest in the nonexistence of a proper elicitation technique selection model that can satisfy the minimal properties and functions and operates at the level of the requirement elicitation scenario characteristics. Therefore, we ought to create a systemization model that takes advantage of machine learning and AI ability to learn and improve automatically from experience and to dispense the dependency on requirement engineering novice experience and intuition to selection an improper elicitation technique that misfits the actual scenario and produces the wrong design requirement which ultimately leads to failure or disuse.

In this systemization model, we have traced an inner detailed operation of the decision-making process involved in the selection of the elicitation technique, we started by classifying attributes influencing the process as steps towards creating a fully satisfactory dataset to run as a backbone of the proposed systemization model. The surveyed dataset parameters were weighted by range of requirement engineering from novices to experts. The systemization model was implemented in creating a detailed mapping process that clearly interprets elicitation environment of the elicitation cases.

The implementation of the proposed systemization was created using Python's scikit-learn library, specifically the k-nearest neighbors algorithm (KNN). The resulting reading was satisfying by total average of 0.96% accuracy and 0.97% sensitivity giving a clear declaration of the reliability and applicability to minimize the reliance on the requirement engineers' unreliable implicit knowledge.

The systemization model was validated by the conducting experiment that included ten requirement engineers ranging from novices to experts in the field against the model predication and nominations. The results were all in the average of 70% similarity rate of the experts judgement, which portrays that the model ability to mimic the decision making process of those who are experts in field. Although, the result is at a satisfactory rate, we still need to repeat more experiments on different platforms to generalize it as a standard.

We validated the proposed system gaining knowledge of the model by running an experiment using five specialist requirement engineers and five novice requirement engineers towards the proposed version. The test has proven that our notion had 70% similarity rate with the expert opinion, which reflects how useful our version is in moving the information of the professional to the proposed device. Although the outcomes are favorable for our proposed version, we nonetheless want to repeat the experiment with other larger samples to generalize the effects. The systematization version is considered to be a new mechanism to that takes the advances of device learning to automatize the selection method and to help requirement engineers to effortlessly tailor current methodologies for specific situations.

Bibliography

1 Hickey, A. M., & Davis, A. M. (2002). Requirements Elicitation and Elicitation Technique Selection : A Model for Two Knowledge-Intensive Software Development Processes Unsolved Problem Software Development Software Solutions, (C), 2005–2010.

2 'Elicitation technique selection: how do experts do it?' (2003) Proceedings. 11th IEEE International Requirements Engineering Conference, 2003., Requirements Engineering Conference, 2003. Proceedings. 11th IEEE International, Requirements engineering conference, p. 169. doi: 10.1109/ICRE.2003.1232748.

3 Marakas, G., Aguirre-Urreta, M., Aguirre-Urreta, M. I., & Marakas, G. M. (2007). Requirements Elicitation Technique Selection : A Theory-Based Contingency Model R EQUIREMENTS E LICITATION T ECHNIQUE S ELECTION : A T HEORY -B ASED C ONTINGENCY M ODEL.

4 Jiang, L. jiang.li@adelaide.edu.a., Eberlein, A. eberlein@ucalgary.c. and Far, B. far@ucalgary.c. (2008) 'A case study validation of a knowledge-based approach for the selection of requirements engineering techniques', Requirements Engineering, 13(2), pp. 117–146. doi: 10.1007/s00766-007-0060-2.

5 'Guidelines for the selection of elicitation techniques' (2010) 2010 6th International Conference on Emerging Technologies (ICET), Emerging Technologies (ICET), 2010 6th International Conference on, p. 265. doi: 10.1109/ICET.2010.5638476.

6 Md. Muqeem, Md. Rizwan Beg 2012 & Effective. NVC BASED MODEL FOR S ELECTING EFFECTIVE, 3(5), 157–165.

7 Razali, F. A. and R. (2012). A Practical Guide to Requirements Elicitation Techniques Selection - An Empirical Study. Middle-East Journal of Scientific Research, 11(8), 9.

8 Tiwari, S., Rathore, S. S., & Gupta, A. (2012). Selecting requirement elicitation techniques for software projects Selecting Requirement Elicitation Techniques for Software Projects, (September). https://doi.org/10.1109/CONSEG.2012 .6349486

9 Carrizo, D., Dieste, O., & Juristo, N. (2014). Systematizing requirements elicitation technique selection. Information and Software Technology, 56(6), 644–669. https://doi.org/10.1016/j.infsof.2014.01.009

10 Egas, R. (2015). Requirements elicitation, which method in which situation.

11 'Selection of requirement elicitation techniques using laddering' (2017) 2017 4th International Conference on Systems and Informatics (ICSAI), Systems and Informatics (ICSAI), 2017 4th International Conference on, p. 1604. doi: 10.1109/ICSAI.2017.8248540.

12 Mishra, D., Aydin, S., Mishra, A., & Ostrovska, S. (2018). Computer Standards & Interfaces Knowledge management in requirement elicitation : Situational methods view. Computer Standards & Interfaces, 56(February 2017), 49–61. https://doi.org/10.1016/j.csi.2017.09.004

13 The Standish Group, "The Chaos Report," 2014, and "Chaos: A Recipe for Success," 2014, www.standishgroup.com.

14 All, A. (2009, December 22). Failed IT Projects Cost Way Too Much, No Matter How You Crunch the Numbers. Retrieved November 10, 2010, from IT Business Edge.

4

Machine Learning Frameworks and Algorithms for Fog and Edge Computing

Murali Mallikarjuna Rao Perumalla[1], Sanjay Kumar Singh[1,],
Aditya Khamparia[1], Anjali Goyal[2], and Ashish Mishra[3]*

[1] School of Computer Science and Engineering, Lovely Professional University, Phagwara, Punjab
[2] Department of Computer Applications, GNIMT, Ludhiana, Punjab, India
[3] Gyan Ganga Institute of Technology, Jabalpur, Madhya Pradesh, India

Abstract

Fog, edge and pervasive computing are technologies developed to overcome the limitations of cloud computing. In this chapter we will cover the role of various machine learning, deep learning frameworks, techniques and algorithms in fog, edge and pervasive computing. Latency, privacy, and bandwidth are some of the limitations or problems with the cloud computing and in this chapter, we will discuss how machine learning combined with these computing technologies can help to overcome the limitations of cloud computing. Inferencing is the main challenge in using machine learning or deep learning models, this chapter covers the various frameworks that help to provide inference and quantize the models. Even machine learning has some advantages and disadvantages, in this chapter we will cover the advantages and disadvantages of using machine learning in edge/fog/pervasive computing. We will also cover the various studies done by the researchers. Every field has numerous applications, in this chapter we will discuss the few possible applications in this fog era using machine learning techniques. By the end of the chapter you will know about the ML frameworks and the various machine learning algorithms used for fog/edge computing.

Keywords *Machine Learning (ML); Deep Learning (DL); Reinforcement Learning; Edge Computing; Fog Computing; Transfer Learning; Ensemble Learning; Multi-Task Learning (MTL); Inferencing; Quantization*

Fog, Edge, and Pervasive Computing in Intelligent IoT Driven Applications, First Edition.
Edited by Deepak Gupta and Aditya Khamparia.
© 2021 John Wiley & Sons, Inc. Published 2021 by John Wiley & Sons, Inc.

4.1 Introduction

In in this section we cover a brief introduction about what fog/edge/pervasive computing is and the role of machine learning in it.

4.1.1 Fog Computing and Edge Computing

Fog computing is also known as fogging. Both technologies promote the computing, storage and networking services between end devices and cloud computing data centers. Fog computing concerns the local network performing computational task that would normally have been performed in the cloud. Organizations are helped to reduce their dependence on cloud-based platforms to analyze data, which often leads to latency issues. Both technologies can take decisions driven by data faster. Fog is another layer of the distributed network environment that is strongly linked to cloud computing and the internet. It moves the edge computing activities to the processors that are connected to a LAN. with fog computing, data is processed through fog nodes or IoT gateways present within the LAN.

Edge computing usually occurs directly on the devices to which sensors are attached or at gateway device close to sensor. In edge computing the data processed resides on the sensor itself, it is not transferred to any hardware. So, prediction, image classification, acoustic detection, gesture recognition and motion analysis can be done on the edge devices using machine learning and deep learning techniques so that only the useful and necessary data is sent through the gateways, which improves the latency issues and increase the security.

4.1.2 Pervasive Computing

Pervasive computing is also known as ubiquitous computing. Pervasive computing makes computing available where it's needed. It is the idea that almost any device can be embedded with chips to connect to an infinite network of other devices. It integrates the current technologies with wireless computing, voice recognition, internet capability and artificial intelligence. It aims to build an environment in which connectivity of devices is incorporated in a manner that is discreet and always accessible.

So, in every computing the main aim is to bring the computing services near to the user and reduce latency, bandwidth and other issues. The role of machine learning within those services is improving efficiency and accuracy. With recent advances it is possible to perform object detection, classification, prediction, monitoring *etc.* in small and low resource devices.

4.2 Overview of Machine Learning Frameworks for Fog and Edge Computing

In this section, we will cover the various machine learning frameworks that are mostly used. Machine learning has many frameworks but the most popular are TensorFlow, Keras, PyTorch. These frameworks provide many inbuilt features and libraries that have made machine learning easy.

4.2.1 TensorFlow

TensorFlow is an open source library made for numerical computation. It makes machine learning faster and easier. It is Python friendly and provides front-end API for building applications with the framework. TensorFlow works on Graphs. It allows the developers to create dataflow graph structures and it runs on both CPU and GPU.

In TensorFlow all computations are performed with the help of tensors. Tensors are used to represent all types of data. Tensor is an n-dimensional vector or a matrix. As we know that TensorFlow makes use of graph, graph is useful to examine the working of the model. A graph describes all the computations. You can visualize the working of computations with the help of TensorBoard. TensorFlow contains all the machine learning algorithms.

TensorFlow has several benefits. They are:

A. **Easy Model Building**

TensorFlow makes model building easy by providing abstractions. In programming abstraction means layers of code on top of existing code. Abstraction provides simplified code which is easier to write, read and debug. So, we don't need to code the whole algorithms from scratch. TensorFlow abstraction helps in writing cleaner code. It also reduces the development time and code length.

B. **Robust ML production**

TensorFlow models can be deployed with ease no matter what language or platform you use. We can make use of TensorFlow extended (TFX) for full production of a ML pipeline. For edge computing we use TensorFlow Lite to make inference.

C. **Eager execution**

Eager execution allows us to perform numerical computations with the support of GPR acceleration. It is flexible for research and experimentation in machine learning. Eager execution makes code simple. We can easily debug using native Python debugging tools. It supports Python data structures to use and structure your code.

4.2.2 Keras

Keras is also an open source library and high-level neural network API. It runs on top of TensorFlow, CNTK or Theano. Keras contains many inbuilt functions, libraries and algorithms for machine learning. It was developed with a focus on enabling fast experimentation. Keras supports deep learning *i.e.* neural networks, CNN and RNN and their combinations. Keras is a user-friendly API and it is compatible with Python.

4.2.3 PyTorch

PyTorch is an open source machine learning library. It was developed by the Facebook artificial intelligence research group. It is used for computer vision, Natural Language Processing applications. It is a replacement for NumPy to use the power of GPUs. It is a research platform that provides maximum flexibility and speed.

4.2.4 TensorFlow Lite

TensorFlow Lite is a deep learning framework which provides inference for on-device. It is an open-source platform. It is used for edge/fog computing optimized but efficient model to run in low computing devices. Using TensorFlow Lite we can convert TensorFlow models and can run in edge computing devices. In TensorFlow Lite we cannot create a model. So, with TensorFlow Lite we can train, re-train, convert, and quantize the TensorFlow models.

4.2.4.1 Use Pre-train Models

TensorFlow Lite provides a set of ready to use models. These models can be used for various machine learning problems. These pre-trained models include:

- image classification
- object detection
- smart reply
- pose estimation
- segmentation

There are many other sources to get the pre-trained models, these models needed to be converted into TensorFlow Lite format.

4.2.4.2 Convert the Model

In order to execute models on low computing resources and edge devices, we need to convert and optimize the model so that model runs efficiently. To achieve the efficiency, we need to convert a model into some special format and store them.

The TensorFlow Lite converter can be used to convert models. When you convert a model the file size of the model is reduced, and some optimizations are added that do not affect the accuracy of a model. The TensorFlow Lite Converter is a tool that converts the models into TensorFlow Lite models

4.2.4.3 On-device Inference
To run inference with the model, we need a model, interpreter, and data. We run the model on the input data to get predictions which is known as inference. We can use the TensorFlow Lite Interpreter. It is supported on various platforms. Interpreter executes the operations on input data and give access to output data.

4.2.4.4 Model Optimization
In order to run the model on low computing devices we need to optimize the model. Optimization should not reduce the accuracy. Using TensorFlow Lite tools we can optimize the size and performance of the model. There may be a minimal impact on the accuracy. Complex training, conversion is required for an optimized model. We can make use of TensorFlow Lite model optimization toolkit.

Quantization: Quantization is a technique to reduce the size of model and time for inference without loss of accuracy. In quantization we reduce the precision of the model *i.e.* a 32-bit model is converted into a 16-bit or 8-bit model. There are various options for quantization namely weight quantization, full integer quantization of weights and activation and Float16 quantization.

4.2.5   Machine Learning and Deep Learning Techniques

In this section we will cover the various algorithms for fog/edge/pervasive computing. Here we give a brief explanation of what supervised, unsupervised and reinforcement models mean and explain various algorithms that are mostly used for fog/edge/pervasive computing.

4.2.5.1   Supervised, Unsupervised and Reinforcement Learning
ML is a part of artificial intelligence. With ML we teach a computer to perform a task without using any explicit knowledge. In ML, we study different algorithms, statistical models to teach a machine. There are different types of learning namely, supervised learning, unsupervised learning and reinforcement learning. We need a large amount of data to teach a machine to perform a task. These data are further divide into training data and testing data. We use training data to teach the machine. We validate the machine using test data. There many loss functions and error minimization techniques to calculate the error and minimize it.

Supervised Learning This is a type of learning where previous experience is used for learning. So, when we train a machine, we feed it with a labeled data so that it can learn from that. These label data work as knowledge to the machine. Validation is performed using testing data to assure that machine has learnt well. These trained ML models are then used to predict the new data. Supervised learning can be further divided into two groups of classification and regression.

Classification. Classification is done on categorical values *i.e.* output is a categorical variable for example, Yes or NO, a disease is malignant or not- malignant. SVM, KNN, Random forests are a few algorithms used for classification tasks.

Regression. Regression deals with the real values. In regression the output variable is a real value *i.e.* price, dollars, temperature *etc.* Linear regression, logistic regression, random forests are a few algorithms for regression.

Unsupervised Learning This is a type of learning in ML where we don't have any knowledge about the task. All we have is unlabeled data. The model is trained to learn the hidden patterns in the data. Unsupervised learning discovers the similarities in the data and groups the data into different clusters. Various assumptions are made on the data defined by some metrics. There are various algorithms namely KNN, K-means clustering, self-organizing maps, *etc.* Unsupervised learning can be further divided into the two groups of clustering and association.

Clustering: Clustering is dividing the data into different groups based on the similarities in the data. For example, grouping the data collected from different edge nodes, IOT sensors and then grouping the collected data.

Association: In association rule learning we want to discover the rules that describe the data. Example: *a priori* algorithm.

Reinforcement Learning This is an area of machine learning which consists of an agent learning. Reinforcement learning is mostly based on a trial and error method where the agent performs various actions and learns from those actions. The agent assigns a reward to the actions performed. These rewards act as an experience. Reinforcement learning is used in various fields such as game theory, control theory *etc.* Reinforcement learning is used to make a machine to learn to play games. It is also used in autonomous vehicles (*e.g.* self-driving cars) and in robotics. Various reinforcement learning algorithms are Q-learning, Deep adversarial networks *etc.*

4.2.5.2 Machine Learning, Deep Learning Techniques

In this section we will cover the most commonly used ML techniques for fog/edge computing. And explain how they are helpful.

Naïve Bayes (NB) Naïve Bayes is a classification model used to classify the data. NB is based on Bayes's theorem. It assumes that there is no dependency between the features. *i.e.* given a set of features, each feature is independent of the others. Bayes theorem is used find out the probability of an event occurring and it returns a probability value. Similarly, NB returns a probability value.

$$y = argmax_{\{y\}} P(y) \prod_{\{i=1\}}^{\{n\}} P(x_i|y)$$

Here, Y is the output and x_i is the input values $P(y)$ is the class probability and $P(x_i|y)$ is conditional probability. There other Naïve Bayes classifiers which are quite famous namely multinomial Naïve Bayes (use multinomial distribution), Gaussian Naïve Bayes (Gaussian distribution) and Bernoulli Naïve Bayes (features are binary variables).

In fog/edge computing, we will be having different IOT sensors, fog nodes, edge servers and edge devices. All these devices will generate huge amounts of data, sending this data to the cloud will cause latency and privacy issues. With the help of fog/edge computing we can overcome these issues. Naïve Bayes is used for on-device classification and sends only necessary information to the cloud device. This also preserves privacy by storing the data in the edge devices mostly in medical imagining and monitoring applications.

Kulkarni *et al.* (2014) proposed a method to preserve privacy in sensor–fog networks. The authors worked on Smart TV remote control devices, identifying the user based on the way they use the TV remote. Using Naïve Bayes, they achieved an accuracy of 77% for classifying the classes which means there are almost 200 features that are not needed to extract and transit by the device. They reduced the energy usage, time required and improved privacy by not sending data. They also added some noise to the data so that the data cannot be compromised

Support Vector Machines (SVM) Support Vectors Machines can be used for classification and regression. Mostly used for classification, SVM classifies the data by drawing and hyperplane. Here we need to find out the best hyperplane that separates the two different classes very well. We need to select the right hyperplane. We can do that by calculating the distance from the nearest data points. We select the hyperplane having the highest distance, if we select a lower distance hyperplane there is a chance of misclassification. In some of the cases we can see some outliers and SVM is robust to the outliers.

If the dataset is not linearly separable then we have to gain a linear separation by mapping the data to higher dimensions. SVM also contains various kernels namely RBF, Gaussian and polynomial kernel functions.

To deal with the remote areas we use sensors. For example, predicting a fire in the forest. The main challenge here is to predict quickly *i.e.* predicting a fire before it starts. In general scenarios the sensors will collect the data, this data is aggregated and sent to a cloud database which takes a longer time for prediction and can also have latency issues. A solution to these problems can be achieved using fog/edge computing. Aakash *et al.* (2019) proposed a data mining method based on fog computing. They used SVM to predict forest fires. High priority data was sent to the fog computing nodes/serve where prediction using SVM was done so that they can achieve low response time.

Al-Rakhami *et al.* (2018) proposed a cost efficient edge computing framework. They applied this in human activity recognition. They used SVM to recognize the human activity. Yhey used raspberry pi as the edge device. In the edge device the data preprocessing and recognition of human activity was performed, and the result data was sent to the central data repository.

Hou and Huang (2019) proposed a method to detect the network security in an edge computing system. They used Alibaba ECS for simulation of a smart home system. They classified the data into regular and mutation codes using SVM on the edge device. The approach gave an accuracy of 91% using RBF kernel.

K-nearest Neighbor KNN is a supervised ML algorithm which can used for regression and classification. *K* values are used to segregate the boundaries of the classes. In KNN we calculate the distance between the data points. KNN is most effective in pattern recognition, intrusion detection *etc.* Given a set of points/data we first define the *k* value and we perform classification using distance. Distance can be Euclidean or Manhattan distance.

In fog/edge computing the fog nodes or the edge devices have far fewer resources, *i.e.* they don't have high computational resources and storage. So, if any high resource intensive task is identified we need to offload the task. Transferring one task to another devices or server is known as offloading. Manukumar and Muthuswamy (2020) proposed a resource management framework using machine learning algorithms. A KNN algorithm was used to classify the fog nodes based on price, utilization of CPU, length of queue and instance. Based on these values KNN decided whether the service was available or not. If service was not available then the task was transferred to cloud server.

Yang *et al.* (2018) proposed a PMU fog for early anomaly detection using KNN. In wide area monitoring and control PMU provides high frequency high accuracy phasor measurements. They used KNN for anomaly detection.

Vijayakumar *et al.* (2019) proposed a healthcare system for the detection of diseases caused by mosquitos using fog computing. They used fuzzy based KNN for classification of the disease. Early detection of the disease is possible by reducing the communication overhead between cloud and user. So, fog computing is used, and classification is done in fog layers and results are sent to the cloud. This helps in reducing latency and communication time.

K-means K-means is an unsupervised learning algorithm. K-means is used to solve clustering problems. In k-means we first initialize the k clusters. We select centroids randomly by shuffling the dataset. Then we iterate through all the data points until there is no change in centroids, calculate the sum of squared distance between data points and centroid and assign to the closest cluster (centroid). K-means is a well-known and effective algorithm. It can be used in various applications namely document clustering, image segmentation, image compression *etc.*

In edge computing we will have edge servers and edge nodes (for example base stations and sensors). There will be a large number of edge nodes so, we should deploy an edge server at an optimal location such that there are no latency, bandwidth issues. Li *et al.* (2018) proposed an edge server deployment algorithm using k-means. Their study compares the different clustering algorithms and shows that k-means is accurate for deployment of edge servers. They have clustered the edge nodes using k-means considering some features and, based on the clustering edge server, placement is done.

Rodrigues *et al.* (2019) proposed a study of hyperparameters of k-means and a particle swarm optimization technique to solve edge server deployment. They studied how hyperparameters can impact time and efficiency. Their results show that there is significant change in time and efficiency. The aim of their study was to prove the importance of the hyperparameters. Proper tuning can reduce the service delay by 25%.

There are other various clustering algorithms namely fuzzy clustering, DBSCAN, DENCLUE, density-based clustering *etc.* These clustering techniques are used to handle the huge amounts of data. The optimized fuzzy clustering method is proposed for resource scheduling (G. Li *et al.*, 2019).

4.2.5.3 Deep Learning Techniques

Deep learning is an area of machine learning which mimics the human brain. But it differs from human brains in various ways. Deep learning uses multiple layers known as artificial neural layers. Groups of these layers form a network known as an artificial neural network (ANN). An ANN is not dynamic and analog like human brains. In deep learning various types of architectures are present.

Artificial Neural Networks (ANN) ANNs are combinations of different layers consisting of nodes. There are different types of layers namely the input layer, hidden layer, and output layer. There can be any number of hidden layers but just a single input and output layer. Each layer can have any number nodes. All of these layers have weighted connections between them. These weights serve as knowledge to the ANN. These ANNs are further classified into different types. Fully connected neural network (FCNN) is the simplest of all neural networks. In FCNN, each and every node is connected to all nodes in the next layers. FCCN is used for feature extraction. However, FCNN provides less accuracy and less convergence.

Auto-Encoders Auto-encoders are unsupervised learning neural networks. The main parts of auto-encoders are the encoder, decoder and reconstruction loss. An auto-encoder is a combination of two NN one for encoding and the other decoding. A simple auto-encoder contains three layers, in the first layer the input data is encoded, in the middle layer contains the encoded and compressed data and the third layer is used to decode the compressed data accurately. Auto-encoders are very helpful in dimensionality reduction of the data.

In fog/edge computing the edge devices are low computational devices *i.e.* very low storage and computing capability. So, they cannot handle large amounts of data so in this case we use auto-encoders. Auto-encoders reduce the dimensionality of the data.

DL on edge devices require high energy operations. Low quality data is used intentionally during data acquisition for higher battery life which reduces the quality of the model. To overcome this problem Na *et al.* (2019) proposed a model know as mixture of pre-processing experts. They also used auto-encoders for preprocessing noisy images.

Convolutional variational auto-encoders (CNN-VAE) are used for anomaly detection. But model size is large which means less inference. To overcome this problem Kim *et al.* (2018) proposed a squeezed version of CNN-VAE. Their proposed model is efficient for low resource devices and platforms like fog/edge computing.

Convolutional Neural Network CNN are also called shift invariant artificial neural networks. CNNs are useful in feature extraction from images. CNNs are widely used in object detection, face recognition, and classification. CNNs give the most promising results so they are widely used in many areas. CNNs contain filters which are applied on the input images to extract the features of the image. Given an input matrix of height (h), width (w), and dimension (d) *i.e.* ($h*w*d$) a filter of dimension ($fh*fw*d$) is applied on an image and obtains an output ($h-fh+1)*(w-fw+1)*1$ image.

Generally CNN contains higher computations which is not efficient for fog/edge computing devices. Nikouei *et al.* (2018) proposed a lightweight CNN for real-time human detection on an edge service. They narrowed the classifier searching space to focus on the humans in a surveillance video. L-CNN showed promising results for computing intensive devices in fog/edge computing.

Lai and Suda (2018) proposed a 7-layer CNN model that can be used in memory constrained devices. They reduced the memory and number of operations taken by the model with the help of quantization.

Recurrent Neural Network RNNs are a class of ANN. RNNs have memory to store the required previous information. RNNs can carry the temporal differences. RNN behave dynamically to a task. The memory is used to store the sequence of temporal data. RNN is used in many fields like speech recognition, stock prediction *etc.* But RNNs can carry up to only eight temporal differences *i.e.* they can memorize up to eight input sequences only. RNN has gradient explosion problem. LSTM overcomes that problem by introducing gates so that model can remember all input sequences.

Remote health monitoring is one of the applications in IOT. Pena Queralta *et al.* (2019) proposed a fall detection system with fog computing using an LSTM network. Although there are traditional technologies like low power wide area network and LoRa protocol, they have less bandwidth transmission. So, to overcome that, these authors introduced data processing and compression to model, and fall detection was done by the LSTM network. Their system achieved 90% precision and 95% recall rates.

Generative Adversarial Networks Ian Goodfellow introduced generative adversarial networks. These networks contain a generator and a discriminator. The generator is used to generate the synthetic data provided by a noise vector. Discrimination between the real and fake images is done by the discriminator. Even the slightly imperfect synthetic data is helpful in increasing the accuracy of a model. Given a set of images to GAN it will generate a new set of synthetic fake images. The generator learns about the input data while discriminator checks whether the data is coming from the input data or the generator.

GANs have many applications such as fake human face generation, text to image generation *etc.* In edge computing using GAN we can preserve the privacy of the data with help of image steganography. Image steganography is hiding one image in another image and this is possible with the help of GAN. Cui *et al.* (2019) proposed a method for image steganography using GAN in edge devices. The secret data is embedded in the foreground area of the image. The results of their study show much less degradation of the images and achieves real time processing.

4.2.5.4 Efficient Deep Learning Algorithms for Inference

Pruning Neural Networks Pruning is a technique to reduce the size and complexity of the model. In DL models pruning of connections and weights reduces the model complexity. A deep compression model is proposed by Han *et al.* (2016). Their study can reduce the storage requirement of NN by 35x to 45x. Their technique is a three stage pipeline *i.e.* pruning, trained quantization and Huffman coding.

Pruning was first introduced by LeCun *et al.* (1990) . Their idea was to remove the unwanted weights. A hessian matrix is used to calculate the contribution of each parameter. The process of finding the contribution of parameters and removing the unwanted parameters is performed until the model achieves a reasonable performance.

While most of the methods are trying to prune the weights of the network Srinivas and Babu (2015) proposed a data-free parameter pruning in which nodes are pruned in the densely connected networks. Their experiments showed that, 85% and 35% of parameters can be removed from the MNIST-trained network, AlexNet respectively without affecting the performance.

Data Quantization Quantization is a technique to reduce the size of the model and time for inference without loss of accuracy. In quantization we reduce the precision of the model *i.e.* a 32-bit model is converted into a 16-bit or 8-bit model. There are various quantization techniques.

CNNs have a high number of applications in computer vison *etc.* Qiu et al. (2016) proposed a design to accelerate the CNNs on embedded FPGA for ImageNet image classification. They identified that convolutional layers are computational-centric and FCNN are memory-centric. They proposed a quantization technique called dynamic precision data quantization. This helped in improving the bandwidth and resource utilization.

Zhu *et al.* (2019) proposed a trained ternary quantization method. They reduced the precision of the weights into ternary values (−1, 0, 1), their model showed very good results for deploying large models in limited power devices.

4.2.6 Pros and Cons of ML Algorithms for Fog and Edge Computing

In this section we will discuss the importance of machine learning algorithms in fog/edge/pervasive computing. With machine learning we can minimize the delay, improve privacy and conserve the bandwidth of the systems.

4.2.6.1 Advantages using ML Algorithms

Inferencing In traditional applications the sensor's data is sent to the cloud for the data to be analyzed. Using machine learning and deep learning algorithms we can make inference in the edge devices. Image classification, motion analysis and predictive analytics *etc.* can be done on the edge device of fog nodes, *i.e.* there is no need to send data for analytics in the cloud. With inference we do on-device computation in the sensors itself. Using TensorFlow Lite we can quantize the model to make inference on the devices.

Bandwidth and Latency IoT communication technologies like LoRa, NB-IoT has a lower payload size. Using ML and DL algorithms we can only require data (fewer data) which removes the latency and bandwidth problems. Deep learning algorithms can be partitioned, the starting layers can be trained on the fog/edge devices and can extract the high-level features. These high-level features are very much smaller compared to the original data. Instead of sending the original data we only need to send limited or required data which solves the latency and bandwidth problems.

Improve Security and Privacy Health monitoring systems and medical imagining contain private information that cannot be shared. Using ML and DL algorithms we perform on-device analysis and send the analysis report instead of the private data. So, we can perform computation on the device itself to prevent attacks when sending data to the cloud. There many clustering and other ML algorithms for anomaly detection which provide security.

Generative Models Generative models like GANs can be helpful in the case of low data. GANs can learn from fewer input data and generate new data. With fewer input data IoT sensors or edge devices can generate more complex data.

Transfer Learning In transfer learning we use the pre-trained networks *i.e.* the network is trained on some particular data and used for some other data. For example, a model trained on cat or dog images can be used to train on cancer images data. This idea of transfer learning helps to use the same edge device enabling it to work on different applications.

4.2.6.2 Disadvantages of using ML Algorithms
High Hardware Cost ML and DL algorithms contain a high number of computations and parameters which cannot run on the normal CPU devices. There is requirement of GPU devices which increase the cost of hardware.

High Energy Consumption Training of DL algorithms required more power. Fog/edge and IoT devices are low battery devices. Training and updating the DL models require high power. This can result in dead fog nodes or edge devices which in turn may have redundancy issues.

4.2.7 Hybrid ML Model for Smart IoT Applications

In this section we will cover the hybrid ML model for smart IoT applications. Hybrid models can reduce individual limitations of basic models. Hybrid machine learning methods are described in the following sections.

4.2.7.1 Multi-Task Learning
MTL is the most popular and trending topic in machine learning. MTL is a combination of one or more tasks which share some common features. In deep learning, one or more tasks can share a few layers. Tasks such as feature extraction, segmentation and classification can share some common CNN layers. Combining these

individual tasks helps to reduce computation, time and improve the robustness of the model.

Hard parameter sharing In hard parameter sharing, certain tasks may share a few common hidden layers at the start of the network and task specific layers are kept at the end of the model. This also reduces the risk of overfitting.

Soft parameter sharing Each model contains different weights and biases. In order to combine these models, we need to calculate the differences between these layers and make all parameters similar. Some constrained layers are used to define these distances and achieve similarity.

4.2.7.2 Ensemble Learning

Ensemble learning is a statistical and machine learning technique. In ensemble learning we combine the multiple learning algorithms to improve the predictions. Multiple models are trained and combined to get better results. Algorithms can be any ML algorithms such as logistic regression, Naïve Bayes, KNN *etc.* These multiple algorithms are combined and called ensembles. There various ensemble algorithms available namely AdaBoost, GBM, XGBM, Light GBM, random forest. There are various ensemble techniques available which are described in the following sections.

Simple Ensemble Techniques: These techniques are simple but effective ensemble techniques

Averaging. We calculate the average of predictions made by the multiple models. For example, if we want to rate a product. Using different algorithms, we will get different ratings, we calculate the average of the rating by each model and use it as the final result.

Voting. Voting is similar to the averaging technique but in these we do not calculate the average. Here we consider the max voting *i.e.* we want to find out the sentiment of a tweet and, for example, using logistic regression we predict it as positive (1), Naive Bayes as positive (1) and regression as negative (-1). Since the majority prediction is positive *i.e.* 2 out of 3 algorithms predicted the tweet as positive, we treat it as a positive tweet.

Weighted Average. Weighted average is the advancement of the average technique. In this we assign some weights based on the importance of the model and then calculate the average.

Advanced Ensemble Techniques

Stacking. In stacking we build a new model based on the predictions of the multiple models. We divide the data into a training set and a test set. For example, we are using random forest, KNN, SVM. These models are fitted on partitioned training sets. Predictions are made on both training and test set by each model. We stack these training predictions sets and test prediction sets together. Now these new datasets are used to build the new model *i.e.* we train the new model on the training prediction set and test in the test prediction set.

Bagging. In bagging we combine the predictions of multiple models. With the help of bootstrapping we make subsets of the original data. Each subset is trained by different models independently and this training is done in parallel. Then predictions of each model are combined to get the actual prediction.

Boosting. In boosting we make use of weights. Boosting is a sequential approach. Like bagging, we divide the data into sub-parts and weights are assigned to the data. First the model is trained on the subset and the error is calculated. Weights of wrongly predicted data are increased. Then a second model is trained in the same way. Each model reduces the error of previous model. Using the weighted mean of all model's a final result is calculated.

4.2.8   Possible Applications in Fog Era using Machine Learning

In this section we will cover the applications of fog computing using machine learning. We will describe the various applications in which machine learning is important.

4.2.8.1   Computer Vision

With the advancement of the computer, object detection, image classification, and intrusion detection is made easy. Combing computer vision with edge computing it is possible to bring object detection and classification on edge devices with low latency. This is very useful in video surveillance. For example, where we want to analyze customer queues in retail environments or detection of unauthorized persons in restricted areas.

4.2.8.2   ML- Assisted Healthcare Monitoring System

Automatic monitoring the health of the patients in hospitals or at home. Combining machine learning we can perform real time health care motoring. In hospitals the health of each patient based on their movements, heart rates *etc.* can be monitored in real time so that immediate help can be provided to the patient. Smart health monitoring devices can be used to track the health report of a patient and these reports are sent their personal or family doctors; in this way the doctors can monitor their patients.

4.2.8.3   Smart Homes

Adding intelligence to the homes. Using information from different sensors and devices we can trigger other actions. For example, smart doors, AI assistant, temperature controlling devices. Consider a scenario, a person leaves their home, with the help of ML-edge computing GPS data from their smartphone can be used to set the temperature of the house when they return home. Smart doors can be used to automatically open the doors when recognizing the home owner. By monitoring the daily actions of the home owner some actions can be automated, *e.g.* playing music at coffee time.

4.2.8.4 Behavior Analyses

Combining the fog/edge computing with ML can be used to analyse the behavior of customers in retail businesses. By analysing the behavior of the customer (such as their purchases, transactions, coupons applied) this can be used to improve the engagement of the customer. Based on the customer purchase promotions can be made using near edge devices. It is also helpful in redesigning store layouts.

4.2.8.5 Monitoring in Remote Areas and Industries

Remote areas are those areas where humans cannot reach or stay like dense forests. It is possible that accidents like fires can occur in remote areas and even in industries. Using edge computing we reduce the latency and improve response time of the devices which helps real time monitoring and prediction. We predict a fire early by analysing the sensor data and using machine learning in edge devices. In industries like gas and oil many accidents can occur. Early prediction of these accidents can help to prevent them. Deploying deep learning-based edge devices helps to monitor such situations in these industries.

4.2.8.6 Self-Driving Cars

Self-driving cars work based on various sensors. The decisions made by self-driving car is a combination of information provided by the different sensors. Gathering sensor information, sending it to a central repository, analyse it, and perform action is a time-consuming process. A slight delay in action can cause accidents could take the life of a person, therefore latency and response time are very important in self-driving cars. With the help of fog and edge computing we can reduce latency and bandwidth issues so that we can prevent accidents caused by self-driving cars.

Bibliography

Aakash, R. S., Nishanth, M., Rajageethan, R., Rao, R., & Ezhilarasie, R. (2019). Data Mining Approach to Predict Forest Fire Using Fog Computing. *Proceedings of the 2nd International Conference on Intelligent Computing and Control Systems, ICICCS 2018, (Iciccs)*, 1582–1587. https://doi.org/10.1109/ICCONS.2018.8663160

Al-Rakhami, M., Alsahli, M., Hassan, M. M., Alamri, A., Guerrieri, A., & Fortino, G. (2018). Cost efficient edge intelligence framework using docker containers. *Proceedings - IEEE 16th International Conference on Dependable, Autonomic and Secure Computing, IEEE 16th International Conference on Pervasive Intelligence and Computing, IEEE 4th International Conference on Big Data Intelligence and Computing and IEEE 3*, 792–799. https://doi.org/10.1109/DASC/PiCom/DataCom/CyberSciTec.2018.00138

Cui, Q., Zhou, Z., Fu, Z., Meng, R., Sun, X., & Jonathan Wu, Q. M. (2019). Image Steganography Based on Foreground Object Generation by Generative Adversarial Networks in Mobile Edge Computing with Internet of Things. *IEEE Access, 7*, 90815–90824. https://doi.org/10.1109/ACCESS.2019.2913895

Han, S., Mao, H., & Dally, W. J. (2016). Deep compression: Compressing deep neural networks with pruning, trained quantization and Huffman coding. *4th International Conference on Learning Representations, ICLR 2016 - Conference Track Proceedings*, 1–14.

Hou, S., & Huang, X. (2019). Use of Machine Learning in Detecting Network Security of Edge Computing System. *2019 4th IEEE International Conference on Big Data Analytics, ICBDA 2019*, 252–256. https://doi.org/10.1109/ICBDA.2019.8713237

Kim, D., Yang, H., Chung, M., Cho, S., Kim, H., Kim, M., … Kim, E. (2018). Squeezed Convolutional Variational AutoEncoder for unsupervised anomaly detection in edge device industrial Internet of Things. *2018 International Conference on Information and Computer Technologies, ICICT 2018*, 67–71. https://doi.org/10.1109/INFOCT.2018.8356842

Kulkarni, S., Saha, S., & Hockenbury, R. (2014). Preserving privacy in sensor-fog networks. *2014 9th International Conference for Internet Technology and Secured Transactions, ICITST 2014*, (1), 96–99. https://doi.org/10.1109/ICITST.2014.7038785

Lai, L., & Suda, N. (2018). Enabling deep learning at the IoT edge. *IEEE/ACM International Conference on Computer-Aided Design, Digest of Technical Papers, ICCAD*. https://doi.org/10.1145/3240765.3243473

LeCun, Y., Denker, J. S., & Solla, S. A. (1990). Optimal Brain Damage (Pruning). *Advances in Neural Information Processing Systems*, 598–605. Retrieved from http://papers.nips.cc/paper/250-optimal-brain-damage.pdf%0Ahttps://papers.nips.cc/paper/250-optimal-brain-damage

Li, B., Wang, K., Xue, D., & Pei, Y. (2018). K-Means based edge server deployment algorithm for edge computing environments. *Proceedings - 2018 IEEE SmartWorld, Ubiquitous Intelligence and Computing, Advanced and Trusted Computing, Scalable Computing and Communications, Cloud and Big Data Computing, Internet of People and Smart City Innovations, SmartWorld/UIC/ATC/ScalCom/CBDCo*, 1169–1174. https://doi.org/10.1109/SmartWorld.2018.00203

Li, G., Liu, Y., Wu, J., Lin, D., & Zhao, S. (2019). Methods of resource scheduling based on optimized fuzzy clustering in fog computing. *Sensors (Switzerland)*, 19(9). https://doi.org/10.3390/s19092122

Manukumar, S. T., & Muthuswamy, V. (2020). *A Novel Resource Management Framework for Fog Computing by Using Machine Learning Algorithm.* https://doi .org/10.4018/978-1-7998-0194-8.ch002

Na, T., Lee, M., Mudassar, B. A., Saha, P., Ko, J. H., & Mukhopadhyay, S. (2019). Mixture of Pre-processing Experts Model for Noise Robust Deep Learning on Resource Constrained Platforms. *Proceedings of the International Joint Conference on Neural Networks, 2019-July*, 1–10. https://doi.org/10.1109/IJCNN.2019.8851932

Nikouei, S. Y., Chen, Y., Song, S., Xu, R., Choi, B. Y., & Faughnan, T. R. (2018). Real-time human detection as an edge service enabled by a lightweight CNN. *Proceedings - 2018 IEEE International Conference on Edge Computing, EDGE 2018 - Part of the 2018 IEEE World Congress on Services*, 125–129. https://doi.org/10.1109/ EDGE.2018.00025

Pena Queralta, J., Gia, T. N., Tenhunen, H., & Westerlund, T. (2019). Edge-AI in LoRa-based health monitoring: Fall detection system with fog computing and LSTM recurrent neural networks. *2019 42nd International Conference on Telecommunications and Signal Processing, TSP 2019*, 601–604. https://doi.org/10 .1109/TSP.2019.8768883

Qiu, J., Wang, J., Yao, S., Guo, K., & Li, B. (2016). *Going Deeper with Embedded FPGA Platform for Convolutional Neural Network • Deep Learning and Convolutional Neural Network – V2 : Brief introduction.* 26–35.

Rodrigues, T. K., Suto, K., & Kato, N. (2019). Hyperparameter Study of Machine Learning Solutions for the Edge Server Deployment Problem. *2019 IEEE 90th Vehicular Technology Conference (VTC2019-Fall)*, (18), 1–5.

Srinivas, S., & Babu, R. V. (2015). *Data-free Parameter Pruning for Deep Neural Networks.* 31.1-31.12. https://doi.org/10.5244/c.29.31

Vijayakumar, V., Malathi, D., Subramaniyaswamy, V., Saravanan, P., & Logesh, R. (2019). Fog computing-based intelligent healthcare system for the detection and prevention of mosquito-borne diseases. *Computers in Human Behavior*, 100, 275–285. https://doi.org/10.1016/j.chb.2018.12.009

Yang, Z., Chen, N., Chen, Y., & Zhou, N. (2018). A Novel PMU Fog Based Early Anomaly Detection for an Efficient Wide Area PMU Network. *2018 IEEE 2nd International Conference on Fog and Edge Computing, ICFEC 2018 - In Conjunction with 18th IEEE/ACM International Symposium on Cluster, Cloud and Grid Computing, IEEE/ACM CCGrid 2018*, 1–10. https://doi.org/10.1109/CFEC.2018 .8358730

Zhu, C., Mao, H., Han, S., & Dally, W. J. (2019). Trained ternary quantization. *5th International Conference on Learning Representations, ICLR 2017 - Conference Track Proceedings*, 1–10.

5

Integrated Cloud Based Library Management in Intelligent IoT driven Applications

Md Robiul Alam Robel[1], Subrato Bharati[2,*], Prajoy Podder[3], and M. Rubaiyat Hossain Mondal[3]

[1] Department of CSE, Cumilla University, Cumilla, Bangladesh
[2] Department of EEE, Ranada Prasad Shaha University, -1400, Narayanganj, Bangladesh
[3] Institute of ICT, Bangladesh University of Engineering and Technology, Dhaka, Bangladesh

Abstract

With the increasing number of students in colleges and universities, maintaining a library using the traditional management systems is too complex and time consuming. In order to get rid of this situation in most of the middle level countries like Bangladesh, an automated digital library management system has been evolved. The main features of the library management software are user interactive graphical representation of the total number of books requested, books up-loaded, yearly books count, *etc.* The main contribution of this chapter is the description of the software which has three modules: student, librarian and admin. These modules have unique features of searching the library books with either the title, author's name, subject, ISBN/ISSN, *etc.* The interfaces of the software are shown as images in the chapter which is an abstraction that may be developed on the available mobile operating system like iOS, Android, *etc.* The interfaces are designed keeping in mind that they will be used on cross platform environments fulfilling minimum requirements using the Internet of Things available in the market. Furthermore, the overall information is preserved with the help of cloud storage while keeping the parallel options for physical storage on the destination master computer. The cloud-based system has given the library management a new dimension while giving a new feature referred as "management on the go" as a web or abstract GUI.

Keywords *IoT; GUI; Library management; Mobile based system*

*Corresponding Author:subratobharati1@gmail.com

Fog, Edge, and Pervasive Computing in Intelligent IoT Driven Applications, First Edition.
Edited by Deepak Gupta and Aditya Khamparia.

5.1 Introduction

Nowadays, the mobile phone has become the replacement for the wallet that once fit in the pocket of mankind. People cannot even imagine being, for a moment, without that tiny packed device evolved by technology [1]. In every way people are trying to introduce the use of mobile phones. In the education sector the use of mobile phone is also very useful [2, 3]. A big part of the education system lies fully or partially with the use of libraries. Libraries can make our life versatile. But going to a physical library to collect any information is difficult for everyone. So, they search for alternative ways. One way is the virtual library with or without augmented reality [4–7]. Every feature of the physical library will be available in that virtual library [8, 9]. A virtual library can be developed in real life applications regardless of the platform which will enable the features of the actual library. Hence the main focus of this chapter is the study of a platform independent virtual library. In addition, cloud-based computing is discussed in this chapter.

5.1.1 Execution Plan for the Mobile Application

There is a predefined plan for the execution of mobile based library management where cloud storage will be used. Firstly, there should be feature similarities with the desktop application having minimum complexities. Besides practical applications, options in manual analogue library management system should be introduced as main features in the mobile application. Ease of access should be another main feature of the mobile based library management. User report will be added as an option in the menu list of the applications. The facility of updating applications will also be available.

5.1.2 Main Contribution

This chapter will describe mobile based IoT driven library management with an abstraction of GUI flushed by virtual computing. This virtual computing can be regarded as cloud computing. This chapter reflects the common cross platform user interface involving IoT and cloud computing. The reader will get an in-depth idea about the capability of an application in managing a physical library with the help of mobile devices. Securities in order to avoid unauthorized use of the data are also ensured through using OTP (one-time password). Note that OTP is season based. Season based user authentication will also ensure that the relevant users are relaxed even if the device is accidentally in the wrong hands. IoT infrastructure is also briefly described in this chapter to understand the inner infrastructure of the proposed application.

5.2 Understanding Library Management

Library management is defined as the management of a physical library. Library management is found mainly in school, colleges, universities, offices, *etc.* But over time, the definition of library management has become somewhat different. Nowadays the concept of library management has changed depending on the evolution of the scientific changes. The mobility feature of anything attracts the people so they also want to get that feature in the library management system. Library management mainly involves three categories of people who are

  i) administrators
 ii) librarians
iii) users

These three categories combine the basic structure of the physical library too. So, people want these things to also be introduced into the mobile platform though it exists in the desktop-based platform. The following diagram gives a clear concept of the three important modules of the application. Figure 5.1 is a pictorial representation of the above-mentioned modules.

In Figure 5.1 the three main aspects of a physical library are closely connected to one another to get the best output in an organization. The goal of using these modules is to make the use of a physical library easier ensuring updating of virtual information. That makes the system dynamic in such a way that no user or

Figure 5.1 Sketch of the three modules of Library management

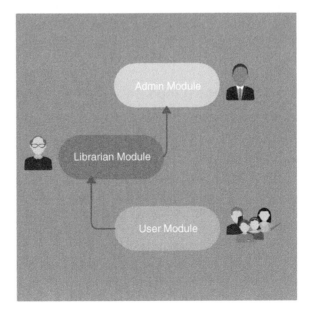

individual can feel unwilling to use the virtual version. The three modules can play different roles in both versions of the proposed concept. They are internally linked with one another to deliver the best services. These modules also work with the security and virtual computing for the betterment of the services provided by both versions of the library management. The main activities are described briefly in the latter part of this chapter.

5.3 Integration of Mobile Platform with the Physical Library- Brief Concept

Real life management of the library will be integrated on the mobile platform. It starts with the access to the physical library. This concept can be visualized by the following feature of the mobile application which will be discussed in the module part of the library.

Figure 5.2 can be referred to as the blueprint of the entrance to the application in the same was that the user uses the physical door of the physical library. When a user calls, it sends short message service (SMS) in mobile phones. The user is sending signals without a wired connection to nearby cell towers. These cell towers receive the user signals and send signals back down to the user. Mobile towers use EM waves in order to send information. Each mobile operator has different frequency bands. Carriers are pushing forward with new cellular technologies such as narrowband Internet of Things (NB-IoT) and M-type LTE. These technologies will provide low bandwidth and low power consumption that will enable a surplus of new IoT.

5.4 Database (Cloud Based) - A Must have Component for Library Automation

The database is the most important component in developing a library management automation whatever the developing environment is. In every module of automation, the database plays an important role. Figure 5.3 is the E–R diagram of modules associated with a database. In Figure 5.3, the activity plan for the whole IoT driven mobile based library management system is shown. Each module has individual activities with certain limited action through any database model. But the comfort of using the automation fully lies on the graphical user interface (GUI). It is facilitated by the use of JAVA and the modules are now described with real time screenshots taken while making the design using that object-oriented language.

Figure 5.2 Proposed mobile user interface that will appear after opening the application

5.5 IoT Driven Mobile Based Library Management - General Concept

Figure 5.4 is a general outline for any IoT driven mobile communication. It shows how a mobile communication is established. It also gives the general idea about the components needed to establish a mobile communication. Any digital communication operated device is run by a user or administrator who actually performs the next operation. The operation includes connecting to the network, downloading or installing the preloaded software, creating the environment for running the required application or checking whether the device is ready for use or not. Moreover, it is also required that the user has enough knowledge to check the networking devices. Figure 5.4 also shows the outline of the interaction of the general users with the communicating devices. The main components are base station, network towers, mobile devices, satellite and home networking devices

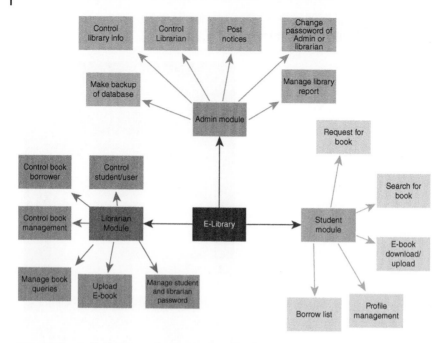

Figure 5.3 An E–R diagram for the database for the automation of E-Library/library management

like router, switch, modem, hub, *etc.* The connection can be made depending on the convention or the networking connection. Random setup of components is not allowed because that will increase ambiguity. As the best outcome is expected from any setup of network components, errors are not encouraged here. The main operational instructions are given by the people generally noted as "administrator and user". The administrator can be categorized in two ways. These are: (i) an admin which is noted as "admin module", and (ii) an administrator or the main hand behind the overall setup for the overall connection establishment. It is not necessary that the components should be arranged in the order shown on the figure rather they can be arranged in any way, but it should be an acceptable way to provide the best service. Any efficient methods can be followed for the sake of best use of the concept "Internet of Things" in this field.

Figure 5.5 is a general diagram of an IoT driven cloud computing for any communicating device such as mobile, tablet, *etc.* Fusion of IoT and cloud computing can play an important role in order to increase the efficiency of daily tasks. The IoT generates a massive amount of data and cloud computing provides a pathway for that data to travel to its destination. The IoT is the technological revolution in

Figure 5.4 IoT infrastructure for mobile based communication

this digital computing system. Cloud computing with the fusion of mobile (IoT enabled device) cloud computing provides on demand self-service, a wide range of network access and resource pooling.

Figure 5.6 represents a pictorial view where a cloud server acts as a channel or medium between a mobile app and the IoT. A cloud server is a powerful physical or virtual infrastructure that performs various operations and information-processing storage. Cloud servers are created using virtualization

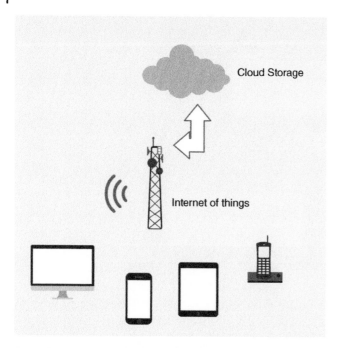

Figure 5.5 General design for IoT driven cloud computing for any communicating device

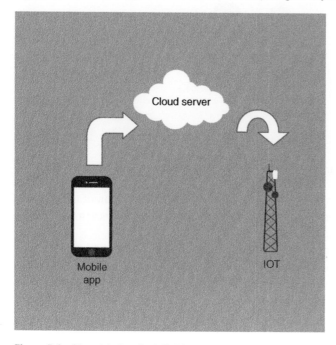

Figure 5.6 Pictorial view for IoT driven mobile application (any)

software to divide a physical (bare metal) server into multiple virtual servers. Cloud servers have three important features. They are:

(i) automated services are accessed on demand with the help of an API
(ii) cost effective system
(iii) enables users to process high workloads as well as store large volumes of information.

The general concept of the role of a cloud server as a middleware is now described. The physical representation need not be setup in the sketched way but the main focus is to make the service more usable and user friendly.

The following sequences of images are actually real time user interfaces (cross platform mobile application) at different levels which will be available to the user (though it is available to any module user).

5.6 IoT Involved Real Time GUI (Cross Platform) Available to User

The following stream of images is a general concept which can be modified at the application development level.

Figure 5.7 illustrates the log-in panel. There are three options:

1) admin log-in;
2) librarian log-in;
3) student /general user log-in.

OTP will be provided in the registered mobile number after tapping any of the sliders. A one-time password (OTP) is an automatically generated numeric or alphanumeric string of characters. It is used to authenticate and verify users before a transaction or a session in applications or website.

Figure 5.8 is the GUI interface of the admin module. The admin is the person who decides authentication and authorization for all the different users of the application. The module consists of the following categories:

(i) control library info;
(ii) manage library report;
(iii) post notices;
(iv) control librarian;
(v) update database to cloud, and
(vi) control login credentials;

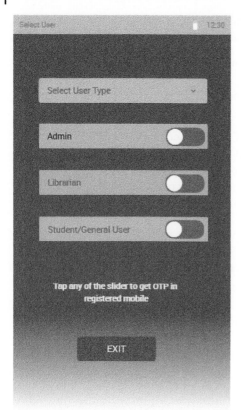

Figure 5.7 The very first GUI for a registered user

An admin can add or remove members from the website. They can update or delete member information. There are also some important features of the admin module:

(a) add or delete books, journals, magazines or thesis
(b) update the details of books, journals or magazines
(c) check in returned materials using a barcode scanner.

Figure 5.9 describes the features of "update database to cloud". There are three options: See current database, save as PDF/DOC/XLS, insert new credentials. In the option save as PDF/DOC/XLS, admin can save the book as PDF or DOC format. They also can add new credentials or remove any existing credentials.

Figure 5.10 illustrates that a new user can be created after inserting the necessary information in the control login credentials portal. The information of the existing user can be updated in the update existing user bar. Passwords can also be recovered after following some steps or queries this portal.

Figure 5.8 GUI for Admin module

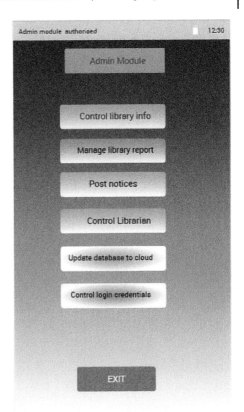

Figure 5.11 appears after a verified user logs in. IoT driven features are available here. The features are:

(i) search for a book

(ii) request a book

(iii) download a book

(iv) upload a book.

The user also can check their borrowed list if any book is lent by them from the physical library. Updating of user profiles can also be done through the proper mechanism. While searching for a book, the user interacts with the cloud-based storage that includes the Internet of Things. Every feature is a real time cloud-based example and it is impossible to implement these options without Internet communication or mobile communication. So, it is impossible without the Internet of Things.

Figure 5.9 GUI when Admin selects "update database to cloud"

Figure 5.12 describes the procedure for searching for a book. The module is linked directly from the home page interface of the system. The system must provide the facility of searching for books based on their ISBN, the name of the book, the author's name, some random value such as publishing year or publishing company or some key words of the book's title. There must be some filters available to search with keywords. A table view of the searches must be available. Information provided in this part is not static. They are dynamic values of the categories available here which directly interact with cloud-based storage. In this case, Internet is a must from any verified channel.

Figure 5.13 represents the download or upload book section. E-book can be downloaded in PDF format using this option. New books can also be uploaded using the admin id and password. The list of books which can be downloaded by a user is included in the database. PDF validator is needed when the digital certificate and digital signature associated with the PDF file have been verified on the system. Virtual storage or cloud storage is a must here for keeping the digital

Figure 5.10 GUI for Admin for selecting "control login credentials"

version of the physical books. There are lot of third-party cloud service providers in the market place. Free and paid options are also available, but verified cloud services are recommended. Figure 5.14 describes the borrow list in GUI.

Figure 5.15 describes the process of how a profile can be updated. The profile picture can be changed. Passwords can be changed based on the approval of user's request. Necessary information such as educational qualification, email id, phone number, *etc.* can be updated. All information is kept in a cloud storage and the Internet of Things is directly involved here.

Figure 5.16 mainly illustrates the overview of librarian module. This module has some important features:

(i) control book borrower which means that the librarian can easily see the necessary information about who can borrow book from the library

(ii) control book management which helps a librarian to efficiently manage books, journals, thesis paper, *etc.*

(iii) control user

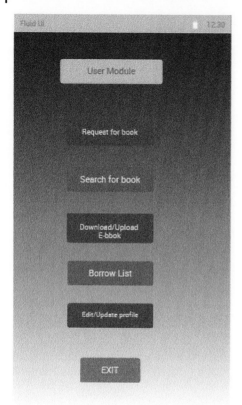

Figure 5.11 GUI for user module

(iv) manage book queries which helps the librarian to know the availability of a book in the library

(v) update e-book which helps to update the version of old books or insert new e-books considering the demand of users

(vi) manage user/self-password enables a librarian to update or change their password and reset the user's password if the user forgets.

5.7 IoT Challenges

Figure 5.17 summarizes IoT challenges for a mobile based library management system. An IoT based system is not so easy to maintain as there are many existing issues to overcome. These challenges can be classified into the three main categories descibed in the following sections.

Figure 5.12 GUI when user selects "search for a book" and "button" stands for "EXIT option"

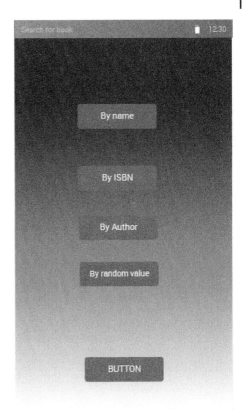

5.7.1 Infrastructure Challenges

The main IoT challenges actually start from the infrastructure as this includes the hardware setups on the physical application side. Interconnection of the communicating and other devices is one of the most ethical challenges on this side although there are also others. As data flows through every node, data flow control is also a big challenge. Besides, there may be individual device problems. Integrated virtual operation involving the physical devices may face problems to deliver uninterrupted services. So, these are the main notable things that will challenge the establishment of smooth and sound IoT based applications.

5.7.2 Security Challenges

Any IoT driven applications are concerned with security issues which can to be driven away by some available techniques. Security issues create challenges in different ways or dimensions as these are to be prevented in a manner where the

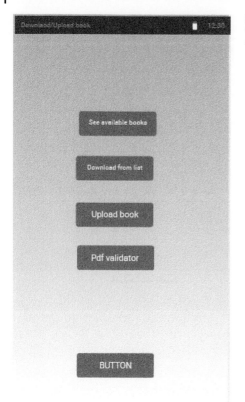

Figure 5.13 GUI when user selects "download/upload book" and "BUTTON" stands for exit options

running system cannot be interrupted. Specifically, privacy and authentication challenges are the two most notable challenges. Data misuse, payment gateways, *etc.* are also obstacles to overcome to establish an errorless system.

5.7.3 Societal Challenges

As it sounds, this is related with the challenges that deal mainly with cooperation challenges along with other issues like the matter of ownership, awareness among the users, ease of use challenges, *etc.* These factors are not so significant in terms of security and infrastructure issues, but the overcoming of societal challenges makes the IoT performance better. Ease of use is the most important key of the sub-category of societal challenges as it ensures the best use of the application. Application with complex use is of no use with respect to the general application with simple use. Besides awareness indicates the use of the application on a regular

Figure 5.14 GUI when user selects "borrow list"

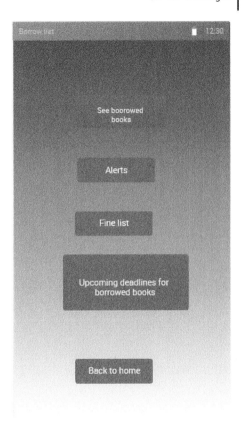

basis with neutral feedback, whereas ownership at the users end with subscription is also a challenge which is to be prevented by the developer or the available authority.

5.7.4 Commercial Challenges

Commercial challenges actually deal with financial things available to both the user and developer or authority. Individual revenue also causes challenges along with establishment of the devices as they are also costly. Revenue also causes problems if there are problems at the application level. Domain oriented earnings are those if the applications are served at different states with customizable features showing respect to the laws in that particular state. Chain of value also causes issues to the proper establishment of the application.

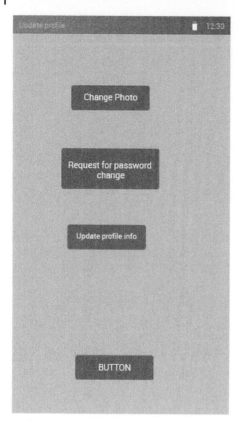

Figure 5.15 GUI when user selects "update profile" option and "BUTTON" is an abstraction for "EXIT" option

5.8 Conclusion

This chapter mainly discussed a mobile based IoT driven Library Management System. A mobile layout based (cross platform) operator collaborative database management scheme is effectively established, which is very applicable to any library in a school, college, university, office, *etc.* IoT driven mobile based layout enables users to be able to recognize the graphical intelligences of numerous transactions such as accessibility of books, number of copies, fines *etc.* Excellent security has also been preserved, to facilitate simply authorized users endorsed through the admin rights to use the services. This mobile based software can be used by mobile devices having minimum latest configuration. This chapter is a real-time implementation of cloud-based modules which is fully focused on the Internet of Things. In this chapter, all-important cross platform layout such as interface for generating OTP by librarian, borrower management by librarian and interface of

Figure 5.16 GUI for librarian module.

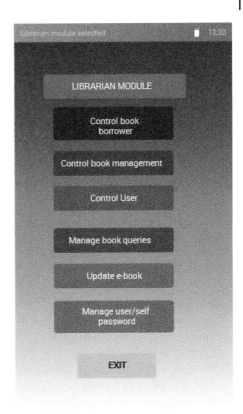

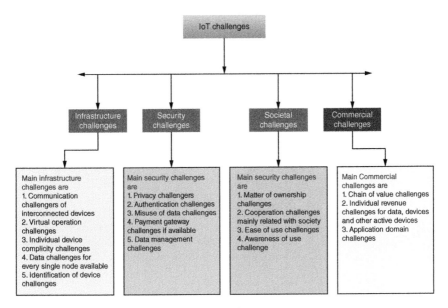

Figure 5.17 IoT challenges for mobile based library management system

admin section is developed using a cross platform (Android, ios) real-time GUI development method. In this chapter, a library management scheme is presented which is suitable for all educational institutions.

Bibliography

1 Jia F., Shi Y. (2013) Library Management System Based on Recommendation System. In: Yang Y., Ma M., Liu B. (eds) Information Computing and Applications. ICICA 2013. Communications in Computer and Information Science, vol 392. Springer, Berlin, Heidelberg.

2 Jing, M., Yu, Y.: Application of collaborative book recommendation model based on borrowing time score. Library and Information Service 3, 117–120 (2012)

3 Nie Y. (2011) Digital Library Billing Management System Design and Implementation. In: Shen G., Huang X. (eds) Advanced Research on Electronic Commerce, Web Application, and Communication. ECWAC 2011. Communications in Computer and Information Science, vol 143. Springer, Berlin, Heidelberg.

4 Asaduzzaman Noor and Md. Sharif Hossen. A Java based University Library Management System. International Journal of Computer Applications 180(29):37-45, March 2018.

5 Y. T. Yang, J. C. S. Kumamoto, "Data Warehouse Applications in Libraries - The Development of Library Management Reports," IEEE International Congress on Advanced Applied Informatics, Japan, 2016.

6 L. Mocean, V. P. Bresfelean, M. H. Macelaru, "A Proposal of an Academic Library Management System Based on an RDF Repository," International Conference on Business Information Systems, Springer, Cham, vol 263, 2017.

7 Frank M. Carrano: Data Structures and Abstractions with Java. Pearson/Prentice Hall (2007).

8 Sangsuree Vasupongayya, Kittisak Keawneam, Kittipong Sengloilaun, Patt Emmawat: Open Source Library Management System Software: A Review. International Journal of Computer, Electrical, Automation, Control and Information Engineering Vol:5, No:5, 2011.

9 Shasha Yu, Enhai Qiu and Mei Zhou: Research on Library Management System Based on Java. 7th International Conference on Social Network, Communication and Education (SNCE 2017). Advances in Computer Science Research, volume 82, pp 946-949 (2017).

6

A Systematic and Structured Review of Intelligent Systems for Diagnosis of Renal Cancer

Nikita[1], Harsh Sadawarti[1], Balwinder Kaur[2], and Jimmy Singla[2,*]

[1] School of Engineering & Technology, CT University, Ludhiana, India
[2] School of Computer Science and Engineering, Lovely Professional University, Phagwara, India

Abstract

Background and objective: The most prevalent type of kidney cancer is renal cell cancer which typically takes place at advanced or final stages of cancer. Renal cell cancer spreads rapidly and can be fatal if the disease is not spotted in the early stages. Hence, it is crucial to recognize kidney deformities before it is in its final phase. However, in this context, only a few review articles have been published almost 12 years ago. Hence, a systematic review was conducted to determine work done by various researchers on kidney cancer and to identify the research gaps between the studies so far.

Methods: To accomplish a structured review, all the relevant papers were categorized based on the authors' names, year of publication, main objective of the research, system input and output variables and finally remarks which gave erudite information apropos the research which has been reported in various publications. After that, the numerous conclusions regarding diagnosing kidney cancer have been scrutinized.

Result: The outcome of this study permitted the effective diagnosis of kidney cancer or renal cancer to be done using the adaptive neuro fuzzy method with a 94% accuracy. Although, many data mining techniques were applied by authors, the accuracy of these methods was lower than the adaptive neuro fuzzy method. This method is useful to identify the aspects of diagnosis of renal cancer more rigorously and ameliorate.

Conclusion: Overall, an appropriate platform is described by this chapter to detect the research gaps in the field to analyze kidney or renal cancer for further studies or researches.

Keywords *Renal Cancer; Medical expert system; Artificial Intelligence*

* Corresponding author: jimmy.21733@lpu.co.in

Fog, Edge, and Pervasive Computing in Intelligent IoT Driven Applications, First Edition.
Edited by Deepak Gupta and Aditya Khamparia.

6.1 Introduction

The kidney is one of the indispensable human body organs. The main responsibility of this organ is to filter the useless products from metabolic activities. The kidney plays an important role in the eradication of foreign elements like urea and creatinine from the human body. Thus, kidney failure can cause huge damage to the human body [1].

Renal cancer is the 13th most common cancer. It is more frequent in men and is the 9th most common cancer in the world. As of 2012, 214 000 new cases of this cancer were diagnosed in men whereas 124 000 new cases were diagnosed in women. These cases appeared in approximately 70% of countries with very high levels of socio-economic development. Consequently, renal cancer is more persistent in men compared to women and it is very rare in children. Kidney cancer was also the 16th most common cause of death due to cancer worldwide with 143 000 deaths documented in 2012 of which 91 000 were men and 52 000 were women [2].

Kidney cancer is a life-threatening disease and it is spreading very rapidly. The main contribution to the spread of kidney cancer is smoking. Therefore, men have a higher risk of renal cancer than women [3]. In the general practice of nephrology, the most regularly seen incidence is of the disease known as kidney or renal cell cancer. Regardless of these states, this interesting and terrible disease is also absent from the training curriculum of nephrologists [4].

Renal cancer can be hereditary. According to accounts, 3% to 5% of renal cancers are hereditary kidney cancers but this fraction is not understood yet. Ten inherited cancers are related to inherited danger of renal cancer and 12 genes have also been identified. The number of families which are suffering from hereditary kidney cancer is continuing to increase [5]. The re-occurrence of kidney cancer is possible. Although, even a number of years after the diagnosis of kidney cancer, there is the possibility of recurrence. Hence this must be considered both by the patient as well as the specialist. A special chart known as nomogram is used to predict the probability of recurrence [6].

Various techniques are used to procure a high-resolution image of the kidney and then image processing techniques are applied to pinpoint renal cancer by skin markers. In particular, techniques based on ultrasound imaging, magnetic resonance imaging (MRI) and computerised tomography (CT) are used to acquire images with high resolution that will be capable of differentiating density of tissue [7]. The histogram was also used for the detection of renal cancer. After segmentation of image, type of lesions by random sampling is classified by histogram of curvature related features [8]. The adapted neuro fuzzy inference system was modelled for the discernment of renal cancer and a comparative study indicated that this technique has an accuracy of 94% to give correct and exact result [9].

6.2 Related Works

This section consists of various work reviews which have already been done by researchers on kidney or renal cancer.

- Golodetz, Voiculescu, and Cameron explored the need for a decision-support system to make a tool that will detect renal cancer. This system was initially earmarked for renal cancer but it can also be used for other cancers in future. The plan was to exploit improvements in image segmentation, image registration, fusion and visualization and to use these techniques of computer science in a way that will help to make more informative and fruitful results for doctors or specialists to diagnose the growth of tumors in order to treat patients more effectively. Hence, this system will help clinicians to make decision about how to best treat best the patients in their care.
- Tang, Dillenseger, and Luo proposed a registration method by local mutual information maximization. The extraction of kidney volumes will be done from the abdomen and instead of performing registration on entire volumes it will be performed only on the selected kidneys. The accuracy of registration will be improved as illustrated by the experimental results.
- Linguraru *et al.* introduced a computer assisted clinical tool to appraise and distinguish renal tumors from acquired CT images and classified the renal tumor accordingly. The dimensions and shapes of lesions were refined step by step. From the segmented lesions, a histogram was constructed according to features which were utilized to categorize different kind of lesions using random sampling. The types of cancer are further classified into four categories. This clinical tool gives an exact and accurate classification of cancer from clinical cancer.
- Mdzinarishvili, Gleason, and Sherman proposed a new approach for the perception of the effect of time period and age on the incidence rate of pancreatic cancer (PC) and kidney cancer (KC). This rate was evaluated in white men and women using the tool SEER 9 and the data collected during the period 1975–2004. The yields of incidence rate classification at older ages were also found in KC and PC. Consideration of time observation, shows no systematic change in cases of PC. Whereas, for KC, there is systematic increase for time period while the coefficient of birth remains immutable.

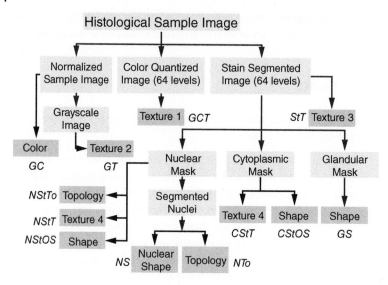

Figure 6.1 Flow diagram for extraction of image features [13].

- Kong developed and carried out a method used to track dimensional measurements from acquired medical images by using image morphing more efficiently. This method extricates the required features with edge detection and target image. After that, a nearest-neighbour algorithm was applied on it to map these required features. Hence, at the end these extracted features were used to measure the dimensions or to track dimensional measurement.
- Kothari *et al.* defined an extensive set of images in which 12 specific feature subsets were present. These features were evaluating heterogeneity among image properties. The author used a data-mining technique to ascertain feature subset and image properties. The main aim was to gauge the output or yield of the diagnostic system by using this extensive set of feature images sets in various applications. Figure 6.1 depicts the flow for extraction of image features.
- Fuchs *et al.* designed a detection system for kidney and bladder cancer using a Fresnel reflector. They developed three reflectors which were tested and simulated with the desired parameters. The main concentration during this simulation was on dependency of the main lobe gain on irradiation angle. They also focused on verification of position of focus point and the effect of number on Fresnel zone on gain.
- Fu *et al.* presented an investigation in which magneto-optical tweezers were used to detect biomechanical properties of human RCC. In this inspection, a

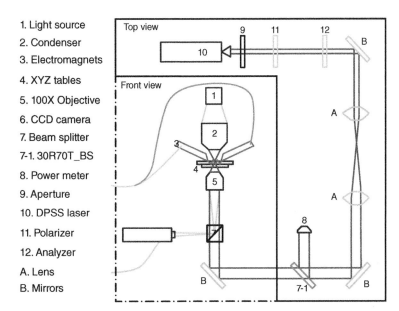

1. Light source
2. Condenser
3. Electromagnets
4. XYZ tables
5. 100X Objective
6. CCD camera
7. Beam splitter
7-1. 30R70T_BS
8. Power meter
9. Aperture
10. DPSS laser
11. Polarizer
12. Analyzer
A. Lens
B. Mirrors

Figure 6.2 Structure of magneto-optical tweezers [15].

couple of electromagnets were combined to construct a magneto-optical tweezers (Figure 6.2) with optical tweezers. These designed tweezers were employed on certain cancer cells to manipulate non-nano-sized magnetic particles. The magnetic fields were applied on RCC to compute biomechanical behaviour regularly. As a result, the author noted that the adaptability of cell member is thick but viscosity is less than cytoplasm.

- Abhilash and Chauhan presented a prediction technique to predict the movement of urological organ and the internal movement of the kidney that is targeted. The position is estimated by using an adaptive neuro-fuzzy inference system. By using this technique, the system achieved nonlinear mapping. This inspection approved that this technique gives more rigorous and accurate result with more than 94% accuracy.
- Haas and Nathanson identified the genes which are correlated with hereditary kidney cancer. This hereditary kidney cancer also led to an increase in pathological subtypes of kidney cancer. As this cancer occurs by virtue of mutation in various genes hence, it will lead to a new subtype of kidney cancer. In this, two new pathological subtypes of kidney cancer were examined and also kidney tumors and other syndromes associated with kidney cancer were understood.

- Schwartz *et al.* investigated the change in kidney cancer after five years by uni-model and multi model analysis. This analysis was done on data about RNA genes and exon. After this analysis, the accuracy was computed by area under the curve and by comparing two machine learning methods named the support vector machine (SVM) and the *k*-nearest neighbour method (KNN). The result of this study justified that the machine learning method KNN is more accurate than SVM to predict the difference in five year survival with wide range of gathered data.

- Tander, Ozmen, and Ozden introduced numerous post-operative re-occurrence estimation models called nomograms for the kidney cancer patients without any metastates and novel systems based on a multilayer perceptron neural network are designed to simplify and integrate the mentioned techniques which are believed to ease the physician's post-operative follow up procedures. A novel approach to nomograms, which are widely used by physicians to predict the freedom of recurrence after five years for the post-operative non-metastatic kidney cancer patients, is proposed. The system depends on a very popular neural network: MLP. Two MLPs are trained and tested with the Kattan's and Sorbellini's (Figure 6.3) nomogram data.

- Saribudak *et al.* introduced a method to calculate the growth of tumors mathematically using Voronoi features. These features of kidney cancer tissues were acquired by from H&E stained slides and this developed method is called personalized relevance parameterization. In this paper 13 different Voronoi diagrams are illustrated and the result shows that this method has an average error rate of about 12.8% and the standard deviation is less than 7%. This result

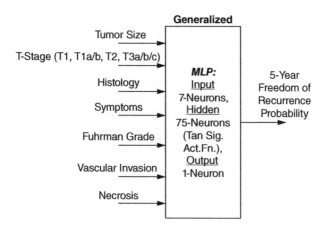

Figure 6.3 Multilayer perceptron is combined with Kattan's and Sorbellini's [6].

was calculated from various samples in which the error rate of three samples was above 20% while the error rate of five samples was below 10%.

- Suomi *et al.* investigated the effect on adequacy of high-intensity focused ultrasound (HIFU) therapy due to the effect of attenuation, reflection and refraction on various tissue types of kidney. The three-dimensional images were captured by computed tomography to perform HIFU therapy. As a result, it was found that the effect of attenuation due to refraction is the most significant which leads to decreases in the intensity of the ultrasound domain. Similarly, the effect of reflection by virtue of the interface of tissues was found to be negligible as compared to attenuation and refraction. Also, if there is a significant effect of reflection, this can be prevented by optimal positioning of the transducer.

- Vasilescu *et al.* presented some details about the affected paired organs due to the lack of detection of kidney cancer at the early stages. These details are about thermographic detection by which the organs near to the kidney can be prevented from the effects of cancer. The various thermal fields were applied on skin of nearby organs for detection purposes. Thermographic cameras and temperature sensors were used to get the value of temperature. It also analyzed the symmetry of the thermal field correlated with each and every organ to which it was applied for detection. The accuracy of results for this method are not concluded yet because many measurement errors are possible.

- Jones *et al.* represented the combination of various priorities of research for the sufferer, doctor and provider of health care in a sufficient way. This paper presents the top ten ambiguities and research priorities. For example, during the treatment of kidney cancer: what are the biomarkers for predicting the response; identify a tool which can be used by experts or specialist doctors to make the correct decision about the stage of cancer and help to guide the patient properly; identify the numerous causes and risk factor which lead to kidney cancer, and many more.

- Li *et al.* classified the correlation of metformin in the case of kidney cancer. The eight dataset was used to undertake the classification in which the total number of patients was 254 329. The dissimilarities between these patients has been analyzed and according to their results, the classification of association of metformin in the patients with kidney cancer was carried out. The overall survival (OS) and cancer-specific survival (CSS) of kidney cancer patients have improved with the use of metformin.

- Deng *et al.* generated a fused network to capture profitable information from various kinds of data to diagnose a disease more accurate and in more detail. This constructed fused network was established to determine the stage of kidney renal cell carcinoma (KIRC). The gene expression was also combined with DNA methylation for better detection of disease. Also inspect result of the network according to data from DNA methylation, gene expression and fused network. The classification of KIRC has been done in three categories respectively and the accuracy of this classification is higher than all the methods that were applied for prediction using fused network.

- Que *et al.* gave a precise and organized review of articles about kidney cancer that evaluated phosphatase and tensin homolog expression and outcome in clinics. In this, a total 35 articles were reviewed including 4532 patients suffering from kidney cancer. A meta-analysis was done and the result of this meta-analysis suggested that a decrease in overall survival was associated with low phosphatase and tensin homolog expression.

- Panuszewska, Minch, and Dzwinel created a model of kidney tumor established on a model of particle automata and also joined with an extra mechanism of viscosity. In this model, each object identifies the behaviour of kidney tissues. Each and every object of the model was affected by its environment and objects placed near to it and changes with time. These objects behave like finite state automation. The result of this model suggested that the mechanism of viscosity can also be used to mimic a model for identification of kidney tumors and kidney cancer.

- Tuncer and Alkan proposed a system that helps to make correct and exact decisions to detect renal cancer using images of abdominal and renal cell cancer tissues. The two main stages of this system are image segmentation and detection of kidney cancer. In the first step, an image is captured and segmentation of that image is done using image processing techniques to extract the kidney area from that image. In the second step, based on the extracted kidney area, the classification is made by the system to determine the stage of kidney or renal cancer. This method uses the support vector machine with an accuracy of 92%. Figure 6.4 illustrate the block diagram of decision support system.

- Atif *et al.* differentiate various kinds of biomolecular markers. This comparison was done between the patients who were suffering from kidney cancer and the healthy patients using fluorescence emission spectra and synchronous fluorescence excitation spectra. It also gave the rigorous classification of features of kidney cancer. The specificity and sensitivity of the method was 90%.

- Aljouie explored the different cases of kidney cancer for cross-validation and cross-study validation. It was found that with an accuracy of 71% and 72%, the support vector machine is the top ranked to obtain cross-validation in cases of in

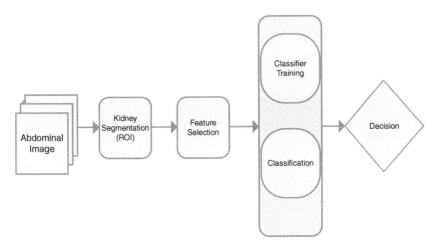

Figure 6.4 Block diagram of the decision support system.[25].

kidney renal papillary cell carcinoma (KIRP) and kidney chromophobe (KICH) respectively. After that, a model is trained for KIRP with an accuracy of 66%.

- Weiss furnished a familiar overview to determine kidney cancer by techniques of metabolomics and metabolic. However, the positive change in kidney tumor is correlated with metabolism in cases of kidney cancer. The illustration of metabolic disease has been labelled properly by a high degree of metabolic reprogramming in different biochemical ways.
- Leon, Kapur, and Pedrosa addressed heterogeneity for better tumor characterization. The RCC reveals the disease spectrum which is heterogeneous in nature. This spectrum may be a problem for patient management as well as the accurate detection of disease from image. This difficulty can be solved by analysis of the semiquantitative and quantitative data on the MR images. This analysis helps better tumor characterization.
- Technopole provided diagnosis and treatment of kidney cancer. There were minimal options for the treatment of kidney cancer in the early 2000s and in that period the victim could survive for 15 months approximately. The main domain for research in the case of kidney cancer is to find the biomarkers and the tools for predicting those biomarkers in the early stages so that specialists can save the life of victim. Better management for kidney cancer or renal cancer is also required. This would also provide causes of kidney cancer and proper drugs or medicines for it to overcome or reduce this disease.

Table 6.1 Comparison table on existing work done on diagnosis of kidney or renal cancer

#	Author	Year	Method	Objective	Input variable	Output	Remarks
1	Golodetz, Voiculescu, and Cameron [7]	2007	Decision-support system	To propose a prototype software system that will be useful for experts for the diagnose of renal cancer.	Ultrasound images, MRI and CT images	Track various homogeneous regions of kidney	Proposed prototype software system for renal cancer
2	Tang, Dillenseger, & Luo [10]	2007	Local mutual information maximization	To present a registration method by using local mutual information maximization	Synthetic volume of kidney	Kidney components are extracted	Effectiveness and accuracy of the registration method is improved
3	Linguraru et al. [8]	2009	Image processing	To introduce a tool and this will help to classify different renal or kidney tumors	Computerized tomography (CT) images	Analyzed five types of renal contusion	A method was presented for classification of renal and kidney cancer
4	Mdzinarishvili, Gleason, and Sherman [11]	2010	Not defined	To propose an approach which is used to evaluate the effect of time period and birth effect on age-specific incidence rates of cancer	Age and gender	The effect of time period and age is estimated in white men and women	The incidence rates of pancreatic and kidney cancer in white males and females are analyzed
5	Kong [12]	2010	Image-morphing based method	To determine the performance of a diagnostic tool that tracks the dimension of the tumor	CT scan images	Size of tumor	Track the change in dimensional size of tumor kidney

6	Kothari et al. [13]	2011	Data-mining approach	To develop a system that will undertake histological diagnosis using a set of images	Color, global texture, global color texture, stain texture, nuclear shape, etc.	Image features and its properties	A comprehensive system has been developed
7	Fuchs et al. [14]	2011	Not defined	To deal with design for Fresnel reflectors for detection of kidney and bladder cancer	Not defined	Not defined	Cancer detection can be done by three Fresnel reflectors.
8	Fu et al. [15]	2012	Not defined	To utilize the novel magneto-optical tweezers for renal cancer	Image of renal	Biomechanical properties of human RCC.	A novel magneto-optical tweezers designed
9	Abhilash and Chauhan [16]	2012	Adaptive neuro-fuzzy inference system	To develop an adaptive neuro-fuzzy inference system for diagnose of kidney cancer	Kidney images	The movement of the kidney was estimated from the skin markers	Prediction was done by the system with an accuracy of more than 94%
10	Haas and Nathanson [5]	2014	Genetic algorithm	To identify the genes related to hereditary kidney cancer	Various symptoms of kidney cancer which correlated with increase in risk	Hereditary kidney cancer recognized	Hereditary kidney cancer has been identified
11	Schwartz et al. [17]	2015	Support vector machine (SVM) and k-nearest neighbor (KNN) methods	To inspect multi-view learning methods	Genetic features	Comparison of results calculated by using SVM and KNN method	KNN method is more accurate than SVM method

(Continued)

Table 6.1 (Continued)

#	Author	Year	Method	Objective	Input variable	Output	Remarks
12	Tander, Ozmen, and Ozden [6]	2015	Multilayer perceptron neural network	To figure out the recurrence of kidney cancer after treatment	Tumor size. T-stage, histology, Fuhrman grade, vascular invasion and necrosis	Probability of recurrence	Two multilayer perceptrons are trained and tested
13	Saribudak et al. [19]	2015	Artificial intelligence	To introduce a method used to compute parameters of tumor growth mathematically	Voronoi features	Growth of tumor of kidney cancer	Gives growth of tumor with an average error rate of 12.8% and standard deviation less than 7%
14	Suomi, Jaros, Treeby, & Cleveland [18]	2016	High-intensity focused ultrasound (HIFU) therapy	To explore the effects of attenuation, reflection and refraction on different tissues	Computed tomography (CT) images	Effects of attenuation, reflection and refraction	The effects of attenuation, reflection and refraction on the efficacy of HIFU therapy in kidney were investigated
15	Vasilescu et al., [20]	2017	Thermographic	Thermographic detection of initial stage cancer that can affect paired organs	Measurement of temperature	Cancer stage has been detected	Thermographic assumption has been tested
16	Jones et al. [21]	2017	Not defined	To set up research priorities for kidney cancer	Not defined	Not defined	Gives top ten uncertainties and resulting research priorities

#	Author	Year	Technique	Objective	Dataset	Output	Findings
17	Y. Li, Hu, Xia, Yuan, & Mi [22]	2017	Not defined	To categorize the correlation of metformin with kidney cancer	Dataset from other publications about kidney cancer	Different effects of metformin on patients	The OS and CSS of kidney cancer patient has been improved with the use of metformin
18	Deng et al. [23]	2017	Classification	To construct a network by combining genes and DNA to determine KRCC stage	Clinical and pathological features, genomic alterations, DNA methylation profiles, and RNA and proteomic signatures	Predicts cancer stage	The KRCC stages are classified into three classes
19	Que et al. [24]	2018	Systematic review of literature	To spot the biomarkers used for diagnosis of kidney cancer	Not defined	Not defined	Phosphatase and tensin homologue (PTEN) gene is identified
20	Panuszewska, Minch, and Dzwinel [25]	2018	Particle automata model	To construct a model for renal cancer by using PAM model	Blood vessels, renal cells and cancer cells	Growth of tumor with and without viscosity	Mechanism of viscosity can be used to mimic various tumor behaviours
21	Tuncer and Alkan [26]	2018	Image processing and decision support systems	To propose a decision support system that detects renal cell cance	Input contains 130 different CT images of renal cell cancer patients	Detects existence or non-existence of RCC in CT images	Support vector machine was used for classification of RCC with an accuracy of 92%

(Continued)

Table 6.1 (Continued)

#	Author	Year	Method	Objective	Input variable	Output	Remarks
22	Atif et al. [3]	2018	Fluorescence emission spectra and synchronous fluorescence excitation spectra	To compare various kind of biomarkers in kidney cancer and healthy patients	Blood components, blood plasma, red blood cell, and urine	Diagnosis of kidney cancer	The accuracy of the method was 90%
23	Aljouie [27]	2018	Machine learning	To predict the accurate cancer risk	Dataset from Kidney Renal Papillary Cell Carcinoma (KIRP) and kidney chromphobe (KICH)	Two datasets are used for cross study validation	Kidney chromphobe carcinoma can be identified by using linear SVM with an accuracy of 66%
24	Weiss [4]	2018	Metabolic reprogramming	To provides a general overview of the technique of metabolomics used to identify kidney cancer	Not defined	Not defined	Metabolic and metabolomics are reprogrammed in renal cell carcinoma
25	Leon, Kapur, and Pedrosa [28]	2019	Radiomics	To give dissimilarities for better tumor categorization	MR images	Characteristics of tumor	Radiomic analysis method is used for characterizing the tumor
26	Technopole [2]	2019	Not defined	Gives diagnose and treatment of kidney cancer	Clear cell renal cell carcinomas (ccRCCs), papillary renal cell carcinoma (PRCC)	Identification of kidney cancer	Gives information about biomarkers and genes

6.3 Conclusion

This chapter has reviewed previous studies that were orchestrated to detect renal or kidney cancer. In the review, the main objective was to identify the research gaps in the domain of renal cancer diagnosis. Hence, many published scientific papers in the period of 2007 to 2019 were stipulated and reviewed to discover the research needs. To achieve the goal of this study, all organized papers were catalogued by author name, year of publication, the method used for recognition of renal cancer, the main objective of the research, input and output variables of the system and final remarks of the paper. The key point to note is that image processing data mining techniques have been used for the detection of renal cancer by skin markers so far.

Techniques such as support vector machine (SVM), *k*-nearest neighbour (KNN) and many more were used. The technique with accuracy 94% was modelled in 2012 by using adapted neuro fuzzy inference system.

Therefore, future studies on fuzzy expert system in terms of cancer diagnosis with extensive view and calculations of the major risk factors that influences the kidney which consequently leads renal cancer should be carried out. However, this study has successfully conducted a systematic review of renal cancer diagnose with certain impediment.

Bibliography

1 Batra, A., Batra, U., & Singh, V. (2016, March). A review to predictive methodology to diagnose chronic kidney disease. In *2016 3rd International Conference on Computing for Sustainable Global Development (INDIACom)* (pp. 2760–2763). IEEE.

2 K. Szymańska, (2018). Kidney Cancer: Diagnosis and Treatment. *Reference Module in Biomedical Sciences.* vol. 2, pp. 325–331, 2018.

3 Atif, M., AlSalhi, M. S., Devanesan, S., Masilamani, V., Farhat, K., & Rabah, D. (2018). A study for the detection of kidney cancer using fluorescence emission spectra and synchronous fluorescence excitation spectra of blood and urine. *Photodiagnosis and photodynamic therapy*, 23, 40–44.

4 Weiss, R. H. (2018, March). Metabolomics and metabolic reprogramming in kidney cancer. In *Seminars in nephrology* (Vol. 38, No. 2, pp. 175–182). WB Saunders.

5 Haas, N. B., & Nathanson, K. L. (2014). Hereditary kidney cancer syndromes. *Advances in chronic kidney disease*, 21(1), 81–90.

6 Tander, B., Özmen, A., & Özden, E. (2015, November). Neural network design for the recurrence prediction of post-operative non-metastatic kidney cancer

patients. In *2015 9th International Conference on Electrical and Electronics Engineering (ELECO)* (pp. 162–165). IEEE.

7 Golodetz, S., Voiculescu, I., & Cameron, S. (2007, October). A proposed decision-support system for (renal) cancer imaging. In *2007 Frontiers in the Convergence of Bioscience and Information Technologies* (pp. 361–366). IEEE.

8 Linguraru, M. G., Wang, S., Shah, F., Gautam, R., Peterson, J., Linehan, W. M., & Summers, R. M. (2009, September). Computer-aided renal cancer quantification and classification from contrast-enhanced CT via histograms of curvature-related features. In *2009 Annual International Conference of the IEEE Engineering in Medicine and Biology Society* (pp. 6679–6682). IEEE.

9 Ahmadi, H., Gholamzadeh, M., Shahmoradi, L., Nilashi, M., & Rashvand, P. (2018). Diseases diagnosis using fuzzy logic methods: A systematic and meta-analysis review. *Computer Methods and Programs in Biomedicine*, 161, 145–172.

10 Tang, H., Dillenseger, J. L., & Luo, L. M. (2007, August). Intra subject 3D/3D kidney registration using local mutual information maximization. In *2007 29th Annual International Conference of the IEEE Engineering in Medicine and Biology Society* (pp. 6379–6382). IEEE.

11 Mdzinarishvili, T., Gleason, M. X., & Sherman, S. (2010, January). Influence of Time Period and Birth Cohort Effects on Age-Specific Incidence Rates of Pancreatic and Kidney Cancer. In *2010 43rd Hawaii International Conference on System Sciences* (pp. 1–9). IEEE.

12 Strawn, N., & Yao, J. (2010, September). Tracking kidney tumor dimensional measurements via image morphing. In *2010 IEEE International Conference on Image Processing* (pp. 1721–1724). IEEE.

13 Kothari, S., Phan, J. H., Young, A. N., & Wang, M. D. (2011, November). Histological image feature mining reveals emergent diagnostic properties for renal cancer. In *2011 IEEE International Conference on Bioinformatics and Biomedicine* (pp. 422–425). IEEE.

14 Fuchs, M., Fernández, T., Tazón, A., & Vassal'lo, J. (2011, April). Design of Fresnel plate reflector for kidney cancer detection system. In *Proceedings of 21st International Conference Radioelektronika 2011* (pp. 1–4). IEEE.

15 Fu, C. M., Han, C. M., Cheng, C. W., & Chou, C. S. (2012, October). Bio-mechanical properties of human renal cancer cells probed by magneto-optical tweezers. In *SENSORS, 2012 IEEE* (pp. 1–4). IEEE.

16 Abhilash, R. H., & Chauhan, S. (2012). Respiration-induced movement correlation for synchronous noninvasive renal cancer surgery. *IEEE transactions on ultrasonics, ferroelectrics, and frequency control*, 59(7), 1478–1486.

17 Schwartz, M., Park, M., Phan, J. H., & Wang, M. D. (2015, November). Integration of multimodal RNA-seq data for prediction of kidney cancer survival. In

2015 IEEE International Conference on Bioinformatics and Biomedicine (BIBM) (pp. 1591–1595). IEEE.

18 Suomi, V., Jaros, J., Treeby, B., & Cleveland, R. (2016, August). Nonlinear 3-D simulation of high-intensity focused ultrasound therapy in the kidney. In *2016 38th Annual International Conference of the IEEE Engineering in Medicine and Biology Society (EMBC)* (pp. 5648–5651). IEEE.

19 Saribudak, A., Dong, Y., Gundry, S., Hsieh, J., & Uyar, M. Ü. (2015, August). Mathematical models of tumor growth using Voronoi tessellations in pathology slides of kidney cancer. In *2015 37th Annual International Conference of the IEEE Engineering in Medicine and Biology Society (EMBC)* (pp. 4454–4457). IEEE.

20 Vasilescu, G. M., Cauni, V., Kacso, G., Stănculescu, M., Marin, M. E., Maricaru, M., & Hănţilă, I. F. (2017, March). Issues on early cancer detection using thermographic methods. In *2017 10th International Symposium on Advanced Topics in Electrical Engineering (ATEE)* (pp. 283–286). IEEE.

21 Basappa, N. S., Basiuk, J., Canil, C., Al-Asaaed, S., Heng, D. Y., Wood, L., ... & Jewett, M. A. (2017). Setting research priorities for kidney cancer. *European urology*, 72, 861–864.

22 Li, Y., Hu, L., Xia, Q., Yuan, Y., & Mi, Y. (2017). The impact of metformin use on survival in kidney cancer patients with diabetes: a meta-analysis. *International urology and nephrology*, 49(6), 975–981.

23 Deng, S. P., Cao, S., Huang, D. S., & Wang, Y. P. (2016). Identifying stages of kidney renal cell carcinoma by combining gene expression and dna methylation data. *IEEE/ACM transactions on computational biology and bioinformatics*, 14(5), 1147–1153.

24 Que, W. C., Qiu, H. Q., Cheng, Y., Liu, M. B., & Wu, C. Y. (2018). PTEN in kidney cancer: A review and meta-analysis. *Clinica chimica acta*, 480, 92–98.

25 Panuszewska, M., Minch, B., & Dzwinel, W. (2018, May). Particle automata model of renal cancer progression. In *2018 International Interdisciplinary PhD Workshop (IIPhDW)* (pp. 254–256). IEEE.

26 Tuncer, S. A., & Alkan, A. (2018). A decision support system for detection of the renal cell cancer in the kidney. *Measurement*, 123, 298–303.

27 Aljouie, A., Patel, N., & Roshan, U. (2018, May). Cross-validation and cross-study validation of kidney cancer with machine learning and whole exome sequences from the National Cancer Institute. In *2018 IEEE Conference on Computational Intelligence in Bioinformatics and Computational Biology (CIBCB)* (pp. 1–6). IEEE.

28 de Leon, A. D., Kapur, P., & Pedrosa, I. (2019). Radiomics in kidney Cancer: MR imaging. *Magnetic Resonance Imaging Clinics*, 27(1), 1–13.

7

Location Driven Edge Assisted Device and Solutions for Intelligent Transportation

Saravjeet Singh * *and Jaiteg Singh*

Chitkara University Institute of Engineering and Technology, Chitkara University, Rajpura, Punjab 140401, India

Abstract

Abstract: With the development of the internet and mobile computing, software applications are migrating toward cloud computing and are using cloud services in different forms. Huge demand for centralized cloud computing poses severe challenges like degraded spectral efficiency, high latency, poor connection, and security issues. To handle these issues, fog computing and edge computing has come into existence. One application of cloud computing is Location Based Services (LBS). Intelligent transport systems being the important application of LBS are relying on the GPS, sensors, and spatial database for convenient transport facilities. These location-based applications are highly dependent on some external system like GPS device and map APIs (cloud support) for the spatial data and location information. These applications fetch spatial data using APIs from different proprietary service providers. The dependency on the APIs and GPS devices, create challenges for effective fleet management and routing process in the dead zones. Dead zones are the areas where no cellular coverage exists. Without the cellular data, devices are unable to fetch spatial data (from the cloud) for route finding and fleet management, similarly without the GPS signal devices cannot update the current location information. Using fog and edge computing we can counter this problem. Utilizing fog and edge computing for transportation applications can prevent issues related to dead zones and performance. In this chapter, we explained our approach to use edge computing in transportation and route-finding process so as to handle the performance issues.

Keywords   *GPS; Route finding; Dijkstra's algorithm; map matching; Spatial data; OpenStreetMap*

*Corresponding Author: Saravjeet Singh; saravjeet.singh@chitkara.edu.in

Fog, Edge, and Pervasive Computing in Intelligent IoT Driven Applications, First Edition.
Edited by Deepak Gupta and Aditya Khamparia.
© 2021 John Wiley & Sons, Inc. Published 2021 by John Wiley & Sons, Inc.

7.1 Introduction to Fog and Edge Computing

The popularity of internet-based applications highly affects the performance of services provided by cloud-based systems. These internet-based application are completely dependent on the cloud-based systems and generate high network traffic. Billions of devices use the internet and as a consequence a huge amount of data is being processed by the cloud data center. These centralized cloud data centers have capabilities to process the data but the communication bottleneck and transmission delay degrade the performance to the end user. To overcome the problem of high latency, a new framework named edge and fog computing has been introduced as shown in Figure 7.1. Edge and fog computing were introduced as a framework to help the original cloud platform. Fog computing provides a hierarchy of components between the end device and the cloud to improve performance and accuracy. Edge computing shifts the partial functioning of the centralized cloud to edge devices of the network. These edge devices include sensors, routers, switches, mobile devices and autonomous vehicles, *etc.* Nowadays fog computing is becoming the central part of cloud-based applications. Edge and fog computing have increased the responsibility and processing of the resources near the end user. The computing have introduced the proximity-based services and new protocols for load balancing. They also provide faster access and low latency (Bonomi *et al.* 2014, Shi and Dustdar, 2016).

7.1.1 Need for Fog and Edge Computing

In a traditional cloud-oriented approach, end devices are responsible for data transmission and displaying the output to the user. The actual spatial data storage for LBS is done on the cloud platform. High dependency on the cloud creates many performance and accuracy issues. These are explained as follows.

High Demand for Cloud Services Keeping all data and resources on the central system acted as an optimal way of resource utilization. Using the cloud as a central system for processing and resources has been heavily accepted by the user. With the passage of time, cloud services are being highly used by the user but the network bandwidth is remains steady. The steady state of network bandwidth created a bottleneck of data and congestion on the network. So huge traffic on the communication network, creates a challenge for the cloud services provider and acts as the key point for the invention of fog and edge computing.

Heterogeneous Numerous Cloud Interactions Due to heavily use of cloud services in the processing power, many applications are using the cloud environment. These applications include heavy processes to small device like IOT devices. The

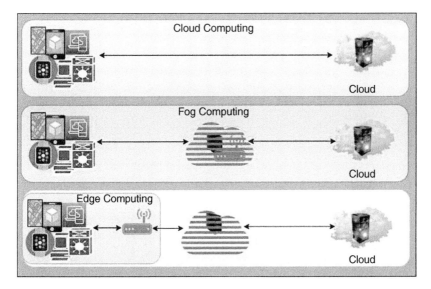

Figure 7.1 Computing technology

heterogeneity of the devices creates privacy and security issues for the cloud service providers. This heterogeneity issue can be resolve by using a homogeneous environment at the cloud interaction side and keep the heterogeneous access near to end user.

7.1.2 Fog Computing

Fog and edge computing are two interrelated concepts evolved to enhance the efficiency of cloud-based services. According to NIST fog computing can be defined as a horizontal, physical or virtual resource paradigm that resides between smart end-devices and traditional cloud or data centers also shown in Figure 7.2. This paradigm supports vertically-isolated, latency-sensitive applications by providing ubiquitous, scalable, layered, federated, and distributed computing, storage, and network connectivity (Iorga *et al.*, 2017).

7.1.2.1 Application Areas of Fog Computing

Augmented Reality Popular augmented reality (AR)-based projects like Google Glass, Smart Eye Glass and HoloLens require high processing power to process the real time videos and further data processing and transmission. Slightly higher transmission delay will mismatch the result of the augmented reality experience and ruin the actual user output. To handle such a scenario AR application-based on fog computing will definitely provide better results. Usage of Cloudlet (a

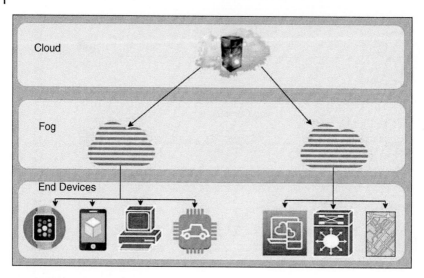

Figure 7.2   Fog computing

component of fog computing) improves the latency delay by offloading the computation at the centralized cloud.

Web Optimization   Fog computing improves the performance of websites that used content delivery networks (CDN) and data caching. Dependency of the client-side of a website on the server using CDN, creates performance issues while synchronizing the server side changes. Shifting of the fog server into the vicinity of the client makes synchronization easy. Similarly caching can be optimized by optimizing the bandwidth and latency using fog computing.

Data Analysis   Fog computing provides an easy way to handle big data and its analysis by using elastic resources near the client's machine. The transmission delay and bandwidth issues of the cloud can be easily handled by the fog.

7.1.3   Edge Computing

Mobile edge computing provides an IT service environment and cloud computing capabilities at the edge of the mobile network, within the radio access network (RAN) and in close proximity to mobile subscribers (Hu *et al.*, 2015). Edge computing provides the capabilities of cloud computing in the radio access layer. Edge computing minimizes direct access between the cloud and end-user. It limits the mobile traffic between end-user and the cloud by enabling local access near the end-user. By minimizing the direct access to the cloud, edge computing improves system performance and avoids bottlenecks (Jararweh *et al.*, 2016).

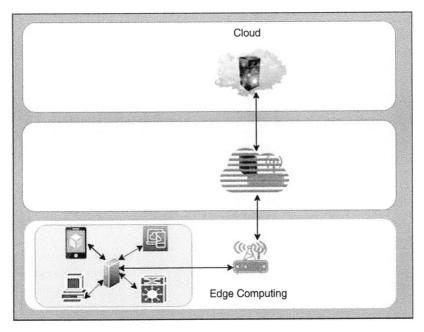

Figure 7.3   Edge computing

Edge computing acts as an enabling technology that makes it possible to perform computation near the end-user formally called at edge layers as shown in Figure 7.3. Edge computing works for both cloud and end user. For the cloud it reduces the overhead downstream of the cloud facility near the end-user and for end-user it reduces the latency doing creating optimized load balancing scenario. The devices working at the edge layer act as edge devices. These devices can be smart phones, mobile devices, cloudlets, routers or switches *etc.* The edge devices work in two modes termed as service provider and service taker. For end-users it acts as a service provider because it provides the necessary resources to end devices for the computation and end devices consume services provided by the edge device. For the cloud it acts as a service taker because it requests from the cloud the data whenever it is required. The edge device handles task like storage and cache handling, load balancing and proximity providing. Edge devices enable reliability, security, privacy, efficiency and availability (Shi and Dustdar, 2016, Svorobej *et al.*, 2019).

7.1.3.1   Advantages of Edge Computing
The key advantages of using edge computing are described in the following paragraphs (Hunt and Scott, 1998).

On-Site Edge computing segregates the processing into two parts that is near the end user and at the cloud. This segregation is possible by isolating the edge devices from the cloud. These edge devices can connect to each other using local network. On-site features make possible machine to machine connectivity.

Resource Togetherness Edge computing allows partial processing near the end device and this is only possible if all processing resources are available near the end user. Edge computing provides frequently used resources near the end user, so it provides benefits to the devices that require faster resource access. Big data analytics, virtual and augment reality devices and autonomous vehicles are the key areas which require resources within the proximity and edge computing enables all these things.

Improved Performance Edge computing architecture provides services near to end user and results in faster access to resources. Edge computing isolates the processing from the centralized system and this leads to low latency, better performance and decrease in traffic near the central system.

Location Dependency Edge computing devices primarily work in local area network. These devices create machine-to-machine scenarios by using the location information from local access network. In edge computing it is possible to track the user location, behavior, mobility condition *etc.*

Optimized User Experience Edge computing provides all possible resources near to the end user and satisfies user needs. One of the core features of edge computing is to provide quality of service to users by providing necessary services with lower latency and higher quality.

Energy Efficient The high demand of the network for day-to-day activities requires more energy to fulfill user needs. In such environments, there is a need to efficiently manage energy consumption. Lower network traffic and reduced central access can be one of the alternatives to optimally use the energy. So, this task can be achieved by off-loading the centralized cloud server and providing the services near the end user and the same can be achieved by edge computing (Kumar and Lu, 2010).

Load Balancing Edge computing supports load balancing near the end user. A first level of load balancing is introduced between centralized cloud and edge devices. A second level of load balancing is done between end devices as per the user requirement, quality of experience and quality of service parameters (Kosta *et al.*, 2012).

7.1.3.2   Application Areas of Fog Computing

Video Analytics   Use of motion pictures and videos in social media has become a major topic of analytics. Many companies use social media data to predict user trends. For video processing and analytics, edge computing act as an efficient way to overcome the problem of transmission delay and overloading arising in cloud computing (Chun *et al.*, 2011).

IoT-based Projects   Edge computing is an efficient way to handle the projects related to IoT. These include smart home appliances, smart cities project, camera-based analytics application and many other projects that uses sensor data. The data associated with the privacy of person is not appropriate to upload on the cloud environment e.g. images captured by the camera installed in the home. So, edge computing acts as more secure option for privacy related projects (Kosta *et al.*, 2012).

Navigation Applications   Edge computing is also used in navigation applications to enhance the system performance and to overcome the problem of latency (Shi *et al.*, 2016).

Collaborative Projects   In collaborative projects many different stakeholders are associated with the same data and process. To handle such projects in collaboration requires quick synchronization and processing. To avoid the problem of improper updates due to latency, edge computing provides a fast way for collaborative work (Rudenko *et al.*, 1998, Kumar and Lu, 2010).

7.2   Introduction to Transportation System

Transportation is the process to navigate from one position to another position. Transportation deals with the mobility of an entity from one place to another place. Transportation modes include water, land, sky, wire, enclosed vessel and space. For transportation, a system requires fixed infrastructure, which includes road and railway tracks, paths in the air for airways, air terminals, water paths, terminals for the water transport, seaports, ways for space transportation and satellite transportation. All the interrelated components of transportation system are shown in Figure 7.4.

Transportation systems are interrelated with navigation systems. The navigation system provides the study of motion and handling the motion of vehicle. It handles and controls the movement of objects in space. Navigation includes land navigation, water navigation, space navigation and aeronautic navigation. Navigation systems involve techniques like route finding, position finding, direction and orientation information, shortest path problems *etc.*

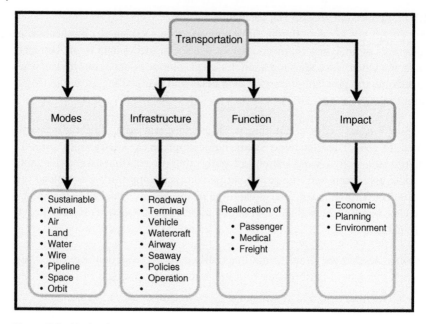

Figure 7.4 Navigation system and related components

Navigation systems involve following types:

- celestial navigation
- piloting
- marine chronometer
- inertial navigation
- electronic navigation
  - radio navigation
  - radar navigation
  - satellite navigation
- land navigation
- underwater navigation
- on water navigation

Transportation on land and land navigation are very much interrelated. Land transportation includes the procedures, policies and methods used for the operation of vehicles. Transportation is crucial for the globalization and economic growth of a country but also effects the environment across the world. In the current era global warning and pollution has become a challenging issue for everyone and it is also used by the vehicle transportation (Khoshmanesh and Nasr, 2016).

The pollution caused by land transportation can be reduced by using effective transportation and eco-friendly transport modes. For effective transportation, optimized route selection is one of the main components of navigation systems.

7.3 Route Finding Process

Routing or route finding is a process to find the path between two points or locations. The route-finding process is related to graph theory. Formally route finding can be defined in term of graph theory as a process to find the path between two vertexes of the graph that passes through n numbers of intermediate vertices.

If G(V,E) is a graph and V is set of vertices and E is set of edges then path P contains v_1 to v_n vertices. Consider Figure 7.5, this graph has 10 nodes (vertex) and 11 edges. An edge connects two nodes of the graph. If we want to find the path between the source node and the destination node then we need to find the connected nodes starting from the source node that are directed towards the destination node. For example, if we want to find the path between node 1 and 10 then there exist multiple paths. These paths are {1,2,3,4,8,10}, {1,2,3,4,8,7,9,10}, {1,23,7,9,10}, {1,2,6,7,9,10}, {1,5,6,7,89,10}. The path finding process selects a single path from all possible paths based upon selection criteria such as cost effectiveness, energy efficiency *etc.*

While navigating the location in real-time the path finding process is called routing or route-finding process. In route finding, the digital road network is taken as the graph. The road network contains ways and nodes. Combination of ways is called street or road. Nodes are the junction, end points or crossings in the road network. Figure 7.6 shows all the components of the digital map. The route between starting and ending nodes is highlighted in the figure 7.6.

Figure 7.5 Sample graph

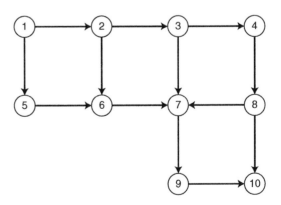

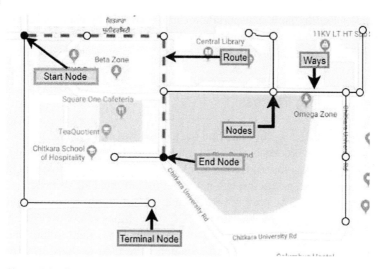

Figure 7.6   Route diagram

7.3.1   Challenges Associated with Land Navigation and Routing Process

Accuracy of the Digital Map   The digital map data is the crucial component of the navigation process. Initial mapping of the GPS signal onto the road network is highly affected by the accuracy of the digital map data. Any inaccuracy in the digital map leads to the wrong mapping result and an incorrect routing process

Unavailability of Internet Connectivity   As many routing systems and navigation activities depend on the online map data or cloud platform to fetch the data from the cloud requires access to the cloud using an internet connection. During the routing process if the internet connection is lost then the routing or navigation process is highly affected. Due to poor internet connectivity the navigation process can stop or provide low quality result.

Weak GPS Signal   The navigation process highly depends on the current location of the user. Automatic location identification can be implemented using the GPS signal output. To identify the current location of the user using GPS, a minimum of four deferent signals are required. In areas having no GPS or having weak GPS signal, location identification becomes very difficult.

Low Storage Capacity   To avoid dependency on the online spatial data, a few navigation applications use the concept of offline data. These applications store data on the device, but to store the complete data on the navigating devices is not a good

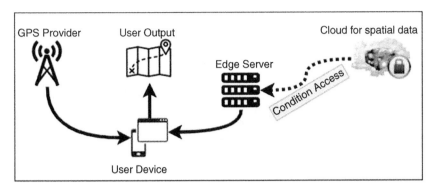

GPS Provider   User Output   Cloud for spatial data

                                    Edge Server

                                    Condition Access

                    User Device

Figure 7.7   Architecture used for the edge-based route-finding process

option and requires huge storage space. Secondly, it's very difficult to update and synchronize the online and local data.

Transmission Delay   In an online navigation system, transmission delay during the data fetches creates many performance issues. Sometimes due to poor bandwidth, data transmission process becomes very slow.

7.4   Edge Architecture for Route Finding

In our approach, we used a mobile device as an edge device for spatial data storage. When a user wants to find a route and other information, the device first checks the data at the edge device, if data is available, then the device performs processing at the edge device, otherwise it fetches the data from the cloud. However, it is not necessary to connect to the cloud for every change in the vehicle position and route information. Many operations on a vehicle can be easily handled by the edge device. Edge computing ensures the vehicle can make decisions as needed, in real-time and later, once free from the dead zone, sync with the cloud and update the global fleet manager's view.

In this research, we used the edge-based architecture in the route-finding domain. According to this architecture spatial data related to the route-finding process is transferred to the local edge device such as a mobile device, memory of an automated vehicle or memory of a device used for the routing process. In this architecture both outdoor and indoor navigation are covered.

In outdoor navigation, for most applications, spatial data for the routing process is stored on some cloud or external source. During the routing process, in the first step, it is required to fetch the data from the external source either using APIs, content network or physical download. This data fetching process occurs during

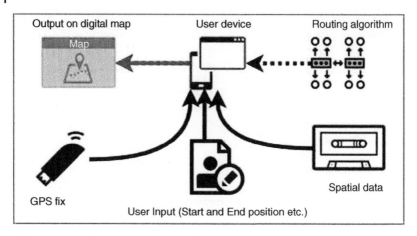

Figure 7.8 Input component for edge device

the routing process execution and induces some additional delay. This additional delay is due to the transfer of spatial data between cloud and device. To avoid this, spatial data is stored in the edge device. The edge device will be responsible for the downloading of data before stating the routing process. Before starting the routing process, the edge device downloads the spatial data as per the location of the device. So during the routing process, complete spatial data is available in the edge device. Figure 7.7 shows the architecture used for the route finding process using edge computing. In this architecture, the following components are used.

GPS Provider A GPS provider is used for the current location identification. The routing device has the capability to receive the GPS signal. The GPS receiver present in the routing device continuously accepts the GPS signals from the GPS satellite using associated devices and protocols. These GPS signals are used to find the current location of the device on the digital map. To find the current location a minimum of four signals from the satellite are required. More than four satellite signals enhance the performance of the location identification process and help to attain better accuracy.

Cloud for Spatial Data The cloud is used for the storage of spatial data. In this architecture the cloud platform stores the complete spatial data of the world. Spatial data contains complete information of the geography of the Earth. It includes details of the road network, land coverage, buildings, amenity and much more. This information can be fetched from the cloud service providers using different data accessing modes. This information is of two types: proprietary and non-proprietary.

Edge Server   The edge server is the main component of this architecture. According to this architecture, this component is used to add the functionality of edge computing in this system. This component conditionally accesses the cloud data and maintains a local copy of that data. Data fetched by this component from the cloud is small in size as per the short-term requirement. This server further provides the data to the routing device. The edge server acts as an intermediate between the cloud and user's device and maintains the high performance by reducing the access time.

User Device   The user's device is the actual device used to perform the routing functionality. This device can be any device like a mobile phone, tablet, computer system, *etc.* having the capability to process and receive the GPS signal. This device has a processor which can process the GPS receiver output, user input like starting and ending location, spatial data from the edge server. Complete flow of input and output data using the user's device is shown in Figure 7.8. As shown in Figure 7.7, the user's device accepts data from the GPS receiver, edge server and from the user input interface. The user's device processes the data using the routing algorithm and provides the output to the output interface.

User Output   The user output component provides the output to the end user in the form of a digital map. This output shows the route on the digital map. This component receives the output from the user's device component.

7.5   Technique Used

According to the proposed architecture, the edge server is used between the cloud and the user's device. This server enhances performance by reducing the data transmission time and providing easy availability of the data. According to this approach, the user's device sends the data request to the edge server and the edge server satisfies the user requirement by interacting with the cloud. The complete process for this technique is shown in Figure 7.9 using the interaction diagram.

This process has six major steps described below.

Step 1   The user sends a request for route guidance to the user service. In this step, the user sends the start and end location of the required route. Additional input that is the GPS receiver output is provided to the user's device in this step.

Step 2   In the second step, the user's device accepts the user input that starts and ends location, the GPS receiver output and locally processes the data. As per the processed user input, the device sends the request for the spatial data. In this step the device requests the spatial data according to the vicinity of the start and

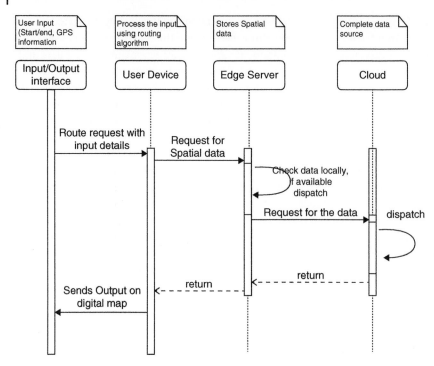

Figure 7.9 Interaction diagram for the complete process involved in edge-based route identification

end location. In this experiment, we used the road information on all the roads between the start and end point.

Step 3 In this step, the edge server accepts the spatial data request from the user's device and checks that data locally. If data is present local to the edge server then the edge server directly returns the data to the user's device. If the data is locally not present in the edge server then edge server connects to the cloud server for the data.

Step 4 In this step, the cloud server after accepting the request of the edge server, returns the required data to the edge server. Here again the edge server requests the data according to the requirements of the user's device not the complete data.

Step 5 After accepting the data from the cloud, the edge server sends the data to the user's device and remains active with the user's device.

Step 6 The user's device processes the spatial data using the routing algorithm and other inputs and provides the output to the output interface. After accepting the data, the user's device continuously sends the location updates to the edge server. The edge server keeps updating the local data as per the user location updates.

According to this technique, the edge server always tries to provide the data locally to the user's device. To maintain the maximum hit ratio (that is the availability of local data at the edge serer) the edge server works in two modes; on-demand mode and periodic mode. These modes are defined as below:

On-demand Mode In this mode, the edge server interacts with the cloud server only when the user's device requests the data and the data is not available at the edge server. In maximum cases the edge server works in this mode only when users initiate the routing process or some failure occurs. In this mode, the edge server demands the maximum possible data from the cloud. This mode may induce some additional waiting time for the user's device. This mode comes into existence only when misses occur while locally checking the required data.

Periodic Mode In this mode the edge server regularly updates the local data as per the current location of the user. This mode runs in the background. This mode always works independent to the working of the user's device. In periodic mode, the edge server always updates the local data. This mode helps to improve the performance of the edge server by reducing the miss ratio.

7.6 Algorithms Used for the Location Identification and Route Finding Process

In the route-finding process, the path between two positions is calculated using the digital road network data. Figure 7.6 shows the output generated by the route finding process. Figure 7.6 shows all possible routes between the starting and ending location. From all possible paths, the route-finding algorithm selects the best possible path based upon the shortest distance or economic factors. The routing process involves two techniques; one is location identification and the second is path generation. These two techniques are explained below.

7.6.1 Location Identification

The location identification technique finds the exact mapping of the GPS receiver output on the digital road network or special database. In this technique, location of the geographic entity is mapped to the digital map. In the routing process, movement of a moving entity is used to find the route so location identification plays an important role. For location identification many algorithms are proposed by the research community. These algorithms involve geometric-based, topology-based, probability theory-based or advanced algorithms. For accurate and effective location identification many factors are involve like GPS receiver output, vehicle movement information, speed of vehicle, geography information, *etc.* (Singh *et al.*, 2019). In this work, we used a topological location identification algorithm as shown in Figure 7.10. This algorithm uses the movement information of the vehicle, topological information of the road network and GPS receiver output to find the location identification. This algorithm works according to following steps:

1) Initial selection of link (road) on the digital map according vehicle position.
2) Identification of junctions, if a junction is near to GPS point then select that junction otherwise the nearest point on the identified road (already selected in step 1).
3) Nearest point is perpendicular point on link from GPS point as shown in Figure 7.11.
4) Before mapping the GPS point on to identified point on link check speed and direction of the vehicle.

The complete process of location identification using a topological algorithm is shown in Figure 7.10. To find the distance between GPS fix and road nodes, the perpendicular distance between the GPS fix and each node on the road were calculated. The point with minimum distance can be considered as target matched point as shown in Figure 7.11.

If GPS point P (p1,p2) and node N_i(ni1,ni2) are in digital map, then the distance between them is calculated using equation 1.1.

$$Di = \sqrt{\left(\left((g1 - ni1) * (g1 - ni1) + (g2 - ni2) * (g2 - ni2)\right)\right)} \qquad (7.1)$$

7.6.2 Path Generation Technique

The path generation technique finds all possible paths between the source and destination and selects the most suitable path. For the best path selection, these techniques used multiple criteria like shortest distance, fuel efficiency or eco-friendly paths. In this technique we used Dijkstra's shortest path algorithm. Dijkstra's algorithm is an algorithm that gives the shortest path between two nodes of the graph. Consider graph Gp(N,L) having N nodes and L links between the nodes, where n $\in$ N and lij is path between two nodes ni; nj Whereas ni; nj $\in$ N and lij $\in$ L and B, E are the beginning and ending nodes respectively. Each link lij has some associated cost, which is distance between two vertices ni and nj. Then Dijkstra's algorithm gives the shortest path SP between B and E by considering L, N and cost of each link (Zhang *et al.*, 2016, Jianya, 1999).

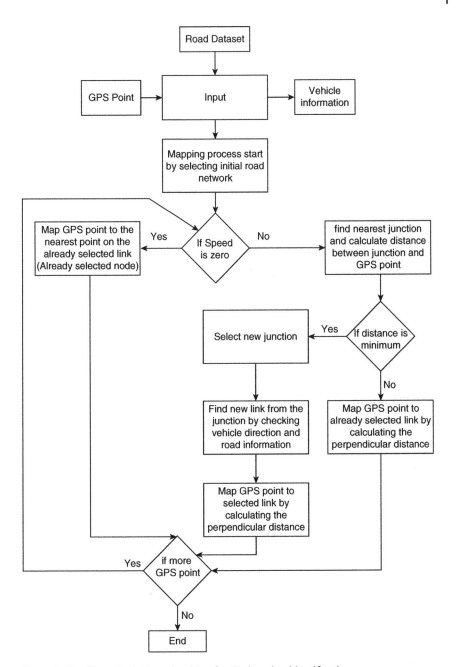

Figure 7.10 Flow chart of an algorithm for the location identification

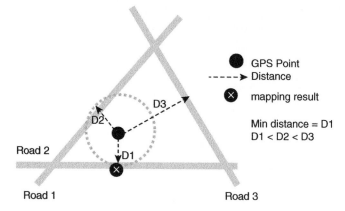

Figure 7.11 Sample scenario to find the distance

7.7 Results and Discussions

To empirically authenticate and analyze the accuracy and performance of the proposed architecture we performed a case study. Based on the available resources, the limited road network of Punjab (India) was used for the experiment purpose. For this experiment, a hand-held mobile device was used as the edge server. Required functioning, for the edge server was implemented using Python code and embedded in an android device. During this experiment the vehicle was travelling in urban, suburban and rural areas with a speed limit of 5–90 km h^{-1}. For the experiment OSM (OpenSteetMap) dataset was used. For comparison purposes both Google maps and OSM data were used. The experiment was performed on ten different routes with a minimum length of 5 km and a maximum length of 50 km. The area covered for this experiment was in the bounding box of (lat : 30.6355, lon : 76.7937) to (lat : 30.3699, lon : 76.4751) as shown in Figure 7.12

7.7.1 Output

The process was executed using the digital map data and GPS receiver output. Sample output of the routing process using offline OSM dataset and OSM API is shown in Figure 7.13 and Figure 7.14 respectively. To analyze the performance and accuracy of the proposed architecture, the two matrices of execution time and accuracy ratio were used. Execution time is defined as the total time to execute the process. Execution time excludes input and output delay and can be calculated using equation (7.2). Start time is when the process actually starts the execution (after accepting the data) and end time is the time when the process actual ends

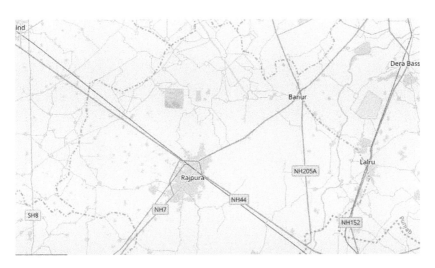

Figure 7.12   Considered area for the experiment

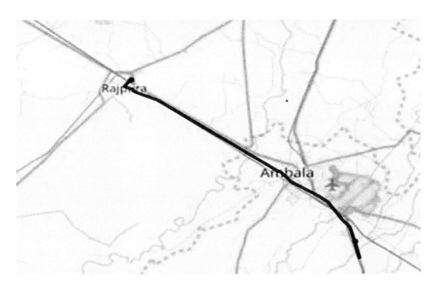

Figure 7.13   Sample output 1 generated by using the offline OSM data using routing process

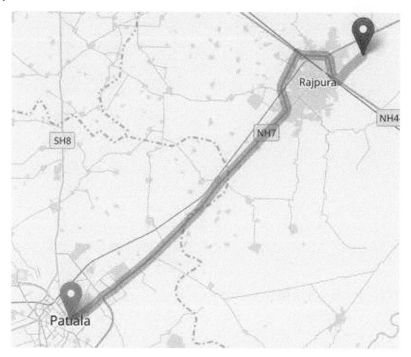

Figure 7.14   Output generated by the OSM API using routing process

(before displaying the output) (Puschner and Koza, 1989).

$$\text{Execution Time} = \text{End Time}{-}\text{Start Time} \tag{7.2}$$

Accuracy ratio is defined as the ratio of completely identified road nodes to the total number of road nodes. Road nodes are points or junctions in the road network. In graph theory, these road nodes are considered as vertices. Accuracy ratio can be premeditated by using the equation (7.3).

$$\text{Accuracy Ratio} = \text{completely identified road nodes}/ \\ \text{Total number of considered road node} \tag{7.3}$$

The effect of edge-based architecture on the execution time of routing process is shown in Figure 7.15. In this analysis average execution time is considered for the routing process. The routing process for both edge-based and without edge-based architecture was tested on ten routes, three times each route. The total routing process was executed 30 times and the average execution time was taken for the analysis. According to Figure 7.15 the execution time of the edge-based architecture is much better than without edge-based architecture. Performance of edge-based architecture is better due to less dependency on the cloud, which

Figure 7.15
Comparison-based on the execution time

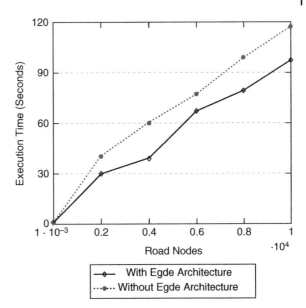

results in the less waiting time for fetching the spatial data. In this process during the routing process, the algorithm requires data for the routing process. According to accuracy-based analysis edge-based architecture has approximately the same accuracy in comparison to without edge-based architecture. Edge-based architecture lacks the accuracy at the large data set as shown in Figure 7.16. This can be due to less space to store the data at the edge server. The accuracy can be optimized by enhancing the capacity to store the data at the edge server. Comparison of with and without edge architecture-based routing processes in rural and urban-based on accuracy ratio is shown in Figure 7.17.

7.7.2 Benefits of Edge-based Routing

Improved Performance Edge-based architecture enhanced the performance of the routing process as shown in Figure 7.15. This is due to the easy availability of spatial data near to the end user and results in faster access to spatial data. Edge-based architecture isolates the fetching of spatial from the cloud from the processing device; and this leads to low latency and better performance.

Better Accuracy During Internet Disconnection Edge-based architecture improves system performance in areas having low internet connectivity. In the low internet connectivity areas, normal routing systems are unable to fetch the data from the cloud server but edge-based architecture made available the data near the actual process device. This leads to easy and maximum availability of data.

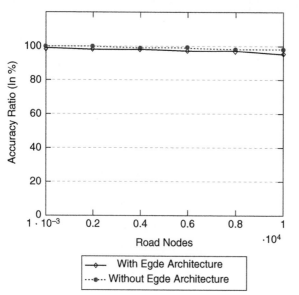

Figure 7.16 Comparison based on the accuracy ratio

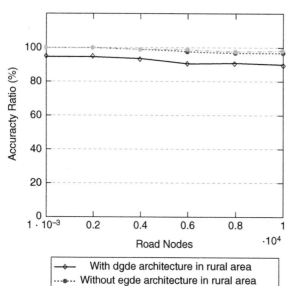

Figure 7.17 Comparison of with and without edge architecture-based routing process in rural and urban based on accuracy ratio.

Optimized User Experience Due the easy availability and direct edge server connectivity, the routing process gives better accuracy and performance, this leads to better user experience.

Resource Management Single edge server can provide the services to many computing devices using the local area network. With this, the same edge server can provide services to many users and provide better resource management.

Applications in Multiple Areas Edge-based routing can be used in many areas like routing during travel, by defense in remote areas, in rural areas, for location-based services during emergencies, during natural distastes, *etc.*

7.8 Conclusion

Route finding processes are widely used in navigation systems and other location-based services. Due to full dependency of routing processes on the central data storage through the internet many performance issues are created. In many standard location-based services, routing processes completely depend on the internet and cloud, so this creates issues during low internet connection. When there is no internet connection, the routing process stops and the system is unable to show the output. In daily life during routing in remote areas, routes on digital maps can suddenly disappear or the routing process can suddenly stop. This happens due to the unavailability of the required spatial data for the routing process. Edge-based architecture tries to improve the routing process and provides the required resources to the routing process for better performance. Using the edge server near the routing device, the server fetches the data from the cloud before starting the routing process. The data is fetched using the current location of the routing device, start and end location of the routing process. The data at the edge server can be customized and changed based on the requirements. The edge server removes the direct connection between the routing device and the cloud by avoiding direct steaming of the data between the cloud and the routing devices. The routing device requests the data from the edge server and the edge server fetches the data from the cloud. In this architecture, the edge server wisely works to provide the data to routing devices by periodically updating the local data from the cloud. In this chapter, we presented the edge-base architecture for the routing process.

The performance and accuracy of the presented architecture is further analyzed in this chapter. According to the presented analysis, this architecture provides better performance and same accuracy with non-edge-based architecture. In this experiment we provide the limited functionality to edge server-based on the available resources. Further this process can be improved by enhancing the capabilities of the edge server.

Bibliography

Flavio Bonomi, Rodolfo Milito, Preethi Natarajan, and Jiang Zhu. Fog computing: A platform for internet of things and analytics. In Big data and internet of things: A roadmap for smart environments, pages 169-186. Springer, 2014.

Byung-Gon Chun, Sunghwan Ihm, Petros Maniatis, Mayur Naik, and Ashwin Patti. Clonecloud: elastic execution between mobile device and cloud. In Proceedings of the sixth conference on Computer systems, pages 301-314.ACM, 2011.

Yun Chao Hu, Milan Patel, Dario Sabella, Nurit Sprecher, and Valerie Young. Mobile edge computinga key technology towards 5g. ETSI white paper, 11(11):1-16, 2015.

Galen C Hunt and Michael L Scott. A guided tour of the coign automatic distributed partitioning system. In Proceedings Second International Enterprise Distributed Object Computing (Cat. No. 98EX244), pages 252-262. IEEE, 1998.

Michaela Iorga, Larry Feldman, Robert Barton, Michael Martin, Nedim Goren, and Charif Mahmoudi. The nist definition of fog computing. Technical report, National Institute of Standards and Technology, 2017.

Yaser Jararweh, Ahmad Doulat, Omar AlQudah, Ejaz Ahmed, Mahmoud Al Ayyoub, and Elhadj Benkhelifa. The future of mobile cloud computing: Integrating cloudlets and mobile edge computing. In 2016 23rd International conference on telecommunications (ICT), pages 1-5. IEEE, 2016.

Yue Yang Gong Jianya. An efficient implementation of shortest path algorithm based on dijkstra algorithm [j]. Journal of Wuhan Technical University of Surveying and Mapping (Wtusm), 3(004), 1999.

B Khoshmanesh and S Nasr. The impact of urban transportation on air pollution and the role of subway inits control (tehran, beijing, barcelona). *Journal of Fundamental and Applied Sciences*, 8(2):1700-1708, 2016.

Sokol Kosta, Andrius Aucinas, Pan Hui, Richard Mortier, and Xinwen Zhang. Thinkair: Dynamic resource allocation and parallel execution in the cloud for mobile code offloading. In 2012 Proceedings IEEE Infocom, pages 945-953. IEEE, 2012.

Karthik Kumar and Yung-Hsiang Lu. Cloud computing for mobile users: Can o_oading computation save energy? Computer, (4):51-56, 2010.

Peter Puschner and Ch Koza. Calculating the maximum execution time of real-time programs. Real-time systems, 1(2):159-176, 1989.

Alexey Rudenko, Peter Reiher, Gerald J Popek, and Geoffrey H Kuenning. Saving portable computer battery power through remote process execution. ACM SIGMOBILE Mobile Computing and Communications Review, 2(1): 19-26, 1998.

Weisong Shi and Schahram Dustdar. The promise of edge computing. Computer, 49(5):78-81, 2016.

Weisong Shi, Jie Cao, Quan Zhang, Youhuizi Li, and Lanyu Xu. Edge computing: Vision and challenges. IEEE Internet of Things Journal, 3(5):637-646, 2016.

Jaiteg Singh, Saravjeet Singh, Sukhjit Singh, and Hardeep Singh. Evaluating the performance of map matching algorithms for navigation systems: an empirical study. Spatial Information Research, 27(1):63-74, 2019a.

Saravjeet Singh, Jaiteg Singh, and Sukhjit Singh Sehra. Genetic-inspired map matching algorithm for real-time gps trajectories. Arabian Journal for Science and Engineering, pages 1-17, 2019b.

Sergej Svorobej, Patricia Takako Endo, Malika Bendechache, Christos Filelis Papadopoulos, Konstantinos M Giannoutakis, George A Gravvanis, Dimitrios Tzovaras, James Byrne, and Theo Lynn. Simulating fog and edge computing scenarios: An overview and research challenges. Future Internet, 11(3):55, 2019.

Jin-dong Zhang, Yu-jie Feng, Fei-fei Shi, Gang Wang, Bin Ma, Rui-sheng Li, and Xiao-yan Jia. Vehicle routing in urban areas based on the oil consumption weight-dijkstra algorithm. IET Intelligent Transport Systems, 10(7): 495-502, 2016.

8

Design and Simulation of MEMS for Automobile Condition Monitoring Using COMSOL Multiphysics Simulator

Natasha Tiwari[1], Anil Kumar[2], Pallavi Asthana[2], Sumita Mishra[2], and Bramah Hazela[2]

[1]*University of Oxford, UK*
[2]*Amity University, Uttar Pradesh, India*

Abstract

In this chapter, simulation and designing of an optimized low-cost comb drive based acoustic MEMS sensor has been explained. These sensors would be useful for the condition monitoring of the automobile on the basis of changes in sound waves emerging from malfunctioned or defected parts of automobiles. These sensors can be developed from silicon substrates. Simulation is done using COMSOL Multiphysics simulation software based on finite element analysis. This optimized sensor is sensitive for the frequency range between 30 Hz to 300 Hz. This frequency range was obtained after the FFT analysis of various signals received from engines using MATLAB software. Simulation result for optimized four comb MEM sensor showed better sensitivity in terms of displacement by acoustic waves than a two pad comb based structure and a four pad comb based structure.

Keywords *Acoustic Waves; piezoresistive MEMS; comb drive; COMSOL Multiphysics*

8.1 Introduction

Automotive engines consist of various parts like reciprocating pistons, rotating cranks and camshafts. These are crucial parts of an engine as they are responsible for converting the reciprocating movement of piston into circular motion. The camshaft opens and closes the inlet and exhaust valves of the engine with the exact stroke which is driven by the crankshaft through the gearwheels. The crankshaft

Corresponding author: bramhhazela77@gmail.com

Fog, Edge, and Pervasive Computing in Intelligent IoT Driven Applications, First Edition.
Edited by Deepak Gupta and Aditya Khamparia.

rotates through the ball bearings and the crankpins rotate through the rod bearings. Rotating motion is imparted to the crank (arm attached at a right angle) through a rotating shaft. For each rotational cycle of the automobile, all these parts undergo coordinated continuous movement. Excessive vibrations may occur in these parts as a result of coupling misalignment, failure of bearing or failure in gearwheels causing catastrophes like breakdown in any part of engine that may even require a huge part replacement [6].

To keep all these parts in good condition, it is essential that they are monitored regularly. Regular monitoring will help in timely identification and detection of faults. When identified, these faults can be corrected at the initial stages of damage, thus saving huge cost of replacements in terms of money and time. This makes vehicle condition monitoring an important factor for drivers, dispatchers and support personnel by making them aware of defects or the defective piece of equipment. With this awareness, proactive steps can be taken to repair the known defect or substitute the defective parts of engine [7].

But, this requires a real time monitoring system to monitor the vibration of the mechanical system. It is important to mention that these faults occur at characteristic frequencies, for example, ball bearings may fail due to bearing race defect, inner bearing race defect or ball defect occurring at a characteristic frequency. During the vibrations, amplitude becomes large at these characteristic frequencies, hence, if the amplitude of the relevant frequencies can be monitored, it would be helpful in determining the state of the mechanical system. Vibrations in coupling misalignments occur at speed of shaft rotation or its harmonics. Vibration velocity is the main cause of fatigue in automobiles that leads to failure. Thus, it is possible to identify faults in automobiles by real time monitoring of vibrations of its mechanical systems [8, 9].

Vibrations that occur due to misalignments or deformation releases energy from the source within the material that generates the transient elastic waves known as acoustic waves. These acoustic emissions occur over a varying frequency range that varies from a few KHz to some MHz. Hence with the help of suitable sensors, AE can be monitored to find the fault in the automobile system. The AE method can detect frequencies in the ultrasonic range. It detects the dynamic energy emitted by the flaws in materials [10, 11]. The AE method shows the following advantages:

(i) Progress of plastic deformation and microscopic collapse in real time can be observed.

(ii) To locate a flaw by using several AE sensors.

(iii) To diagnose facilities while they are in operation

8.2   Related Work

Navid *et al.* in their paper titled "Micromachined Inertial Sensors" presented a detailed review of silicon based micromachined accelerometers and gyroscopes with a brief introduction of their operating principles and specifications, various device structures, fabrication technologies, device designs and packaging [1].

They stated that resonance frequency of a micro-machined accelerometer structure can be increased by increasing the spring constant and decreasing the proof mass and the quality factor of the device can be increased by reducing the damping and by increasing the proof mass and the spring constant

William Fleming in his paper "Overview of Automotive Sensors" has given a comprehensive review of current production and emerging state-of-the-art automotive sensor technologies [2].

Rebello Joel in his paper "Design and analysis of a mems vibration sensor for automotive mechanical systems" has presented a theoretical analysis and experiment results of MEMS sensors designed for low frequency vibration sensing application. Every sensor consists of a proof mass connected to a folded beam micro-flexure, with an attached capacitive comb drive for displacement sensing [3].

Nii O *et al.* have developed a framework for the application of MEMS technology in pavement condition monitoring and evaluation in the paper "MEMS Application in Pavement Condition Monitoring – Challenges". Microsensors are embedded unobtrusively and inconspicuously in structures to monitor parameters critical to the safe operation and performance [4].

Amita Gupta and Amir Ahmad in their paper "Microsensors Based on MEMS Technology" have presented an overview of the currently available MEMS sensors, materials for sensors and their processing technologies, together with integralism of sensors and electronics. In smart microsystems, many types of sensors and actuators are used. Typical sensors consist of strain gauges, silicon cantilever-based accelerometers, optical fibers, piezoelectric films, and piezoceramics [5].

8.3   Vehicle Condition Monitoring through Acoustic Emission

When any part or component in an engine undergoes a permanent or plastic change in its internal structure, it emits acoustic waves. These waves are surface waves that travel over the surfaces and reach the sensor. The acoustic emission technique is an important non-destructive method of fault detection from its source (damaged or defected part). This technique has an advantage of real time monitoring, dynamic and high detection sensitivity. But, during propagation, elastic waves couple with frequency response characteristics of the sensor and they are also impacted by environmental noise, hence making the signal complicated and difficult to analyze [12].

Hence, a functional relationship is developed between the acoustic emission signal and material condition for a given structure and material. Acoustic emissions are time domain waveforms that range over a wide frequency between 20 Hz to 1 MHz. Now, it is important to know that each fault occurs at a characteristic frequency so the theoretical relationship between the parameters of acoustic emission and their characteristics with machine faults needs to be formulated through time–frequency analysis methods. The state of the mechanical system is determined by monitoring the amplitudes of relevant frequencies. Sensors are attuned to characteristic frequencies to monitor the displacement occurring due stress waves [3]. If AE occurs, these piezo-resistive sensors respond to the stress by changing their charge profile, hence, detecting the fault.

In this paper, FFT analysis of an engine is carried out to find out the characteristic frequency of the acoustic emission. We have designed three sensors with different specifications for their Eigen frequencies. Eigen frequencies are those frequencies at which structure deforms into corresponding shape. Comb drive actuators are electrostatic sensors based on the principle of parallel plate capacitor in which one plate is fixed and other is moveable in the form of interdigitated fingers. Any deformation occurring due to AE is registered as displacement in these sensors, hence, varying the capacitance of the complete arrangement [1].

8.4 Piezo-resistive Micro Electromechanical Sensors for Monitoring the Faults Through AE

Acoustic emissions can be sensed with the help of piezo-resistive accelerometers. These accelerometers are composed on a semiconductor-based silicon chip that are created by micro machining. They show good performance for local responses that are even characterized by high frequency content [13].

Piezo-resistive based MEMS are mounted due to their accuracy, precision, robustness and high SNR. These sensors are capable of working in extreme conditions prevailing inside the engine compartment. The temperature of a combustion engine varies between 195 degrees Fahrenheit – 220 degrees Fahrenheit and high frequency range.

In this chapter, we have explained the designing and simulation of a sensor that is low cost and mass producible. This piezo-resistivity based sensor is sensitive to automotive frequency range of 30–300 Hz. Design rules have been validated through COMSOL multi physics simulator and frequency range has been identified by fast Fourier transform analysis of a sample signal in MATLAB.

The sensors designed in this work are piezo-resistive micro accelerometers. These accelerometers incorporate piezo-resistors made up of silicon substrate in their suspension beam.

The cross-section structure of the sensor is connected as a cantilever array, which is designed to resonate at a specific frequency through an integrated proof mass. Each cantilever is a rectangular shaped bar of Si having a thickness of less than 1 micrometer. Whenever these cantilevers encounter a vibration, acceleration is applied to the beam structure, and effective mass converts it into force. Suspension beams are connected to the support frame that moves relative to proof mass and based on this force, either suspension beams elongate or shorten. These movements cause tension or compression inducing a charge shift and charge accumulation and change in resistivity of embedded piezo-resistive sensors due to a change in stress profile. These sensors are connected in a bridge arrangement, so, any change in charge is collected by electrodes generating a low output impedance voltage which is proportional to the stress that caused the relative displacement. Stress variations are found to be maximum at the edge of the support rim and proof mass, hence, sensors are placed in these two positions [14, 15].

8.5 Designing of MEM Sensor

Flow chart of designing and simulation of the sensor has been shown in Figure 8.1, it consists of the fast Fourier analysis of engine sound recording though MATLAB, designing of a comb structure of capacitive sensor using COMSOL simulator and optimizing it after comparing its four pad structure and optimizing the design for resonant frequency.

FLOWCHART OF THE DESIGNING PROCESS OF SENSOR:

8.6 Experimental Setup:

This project was completed in two stages:

(i) Finding the frequency range of AE of Diesel engine with time frequency analysis using FFT.
(ii) Designing a suitable sensor which would be more sensitive at the given frequency range.

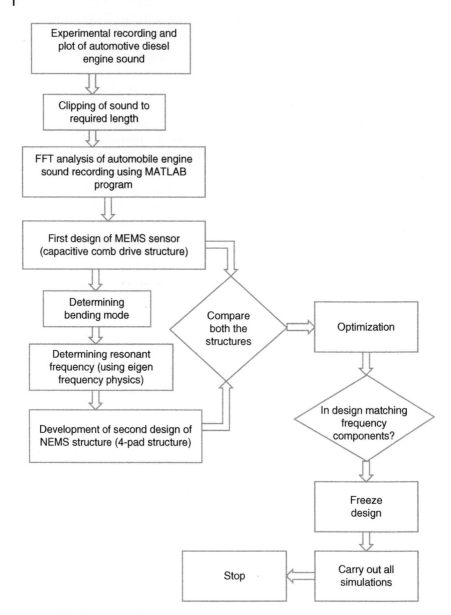

Figure 8.1   Flow chart of the design of MEMS.

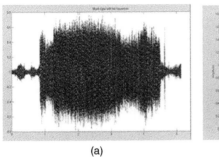
(a)
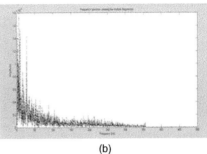
(b)

Figure 8.2 (a) Tata Manza engine sound frequency plot (b) Amplitude verses frequency plot of signal.

8.6.1 FFT Analysis of Automotive Diesel Engine Sound Recording using MATLAB

Rolling parts of the automobile have the element bearings and in many cases it has been observed that failure occurs due the degradation of these bearings. Some characteristic frequencies are associated with the change geometries and construction of bearings. To attain the knowledge of these frequencies, it is mandatory to convert the time domain signal into a frequency domain signal as shown in Figure 8.2 and Figure 8.3.

In most vibration analysis instruments; discrete Fourier transform is utilized for the conversion of time domain signals to equivalent frequency domain signal.

In this work, we have considered the diesel engine. An experimental recording of Tata Manza engine was done. Recording was clipped to a desired length to avoid unnecessary random frequency components to avoid noise Excess noise can add random frequencies increasing overall frequency response of the system. This mp4 was converted to a .wave file for easy computation through FFT algorithm in MATLAB.

MATLAB code was developed for analyzing sound recording to a frequency versus amplitude plot. Using this MATLAB code FFT analysis of an experimental recording of Tata Manza diesel engine sound to identify various frequency components were carried out.

Characteristic Frequency Range:
FFT analysis of the signals confirmed that in diesel engine, frequency of 30–300 Hz is generated during malfunctioning; hence, the designed MEM sensor should be sensitive to this frequency range.

8.6.2 Design of MEMS Sensor using COMSOL Multiphysics

COMSOL Multiphysics is a finite element method technique based on finite element analysis. It is a dominant discretization technique in structural mechanics.

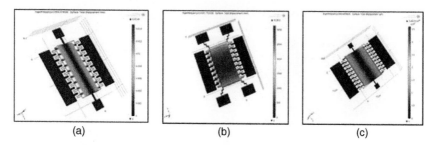

(a)	(b)	(c)

Figure 8.3 (a) Comb drive sensor. (b) Four pad structure. (c) Optimized tri-plate comb structure as designed on COMSOL Multiphysics simulator.

In this technique, mathematical is converted into discrete model by sub-dividing it into non-overlapping components of simple geometry known as finite elements. This method is useful for structural analysis of cantilevers employed in the MEM sensors for vehicle condition monitoring [16]. The material chosen for designing these sensors is silicon because it has about the same Young's modulus as steel, and it is as light as aluminum with a mass density of about 2.3 g cm^{-3}. Such materials can maintain a linear relationship between applied load and the induced deformations. The melting point of Si is 1400 ^{0}C which makes it suitable for working in elevated temperatures inside combustion engines [17].

The piezo-resistive sensor designed in this work is based on the principle of parallel plate capacitor. It consists of two conducting plates where one plate is fixed and other is moveable and the space between the plates is filled with insulating material known as dielectric 0.5 V was applied to these plates. We have designed comb drive sensors that consist of interdigitated comb fingers and this design increases the stray electric field known as the fringe field existing between closely spaced conductors.

8.6.3 Electrostatic Study Steps for the Optimized Tri-plate Comb Structure

To carry out electrostatic study simulation of capacitive comb drive the following methodology was adopted.

- LOAD APPLICATION: A load of 1 N meter^{-1} cube *i.e.* force per volume was applied on the comb structure under structural mechanics physics.
- VOLTAGE APPLICATION: A voltage is applied to one plate and the other is connected to ground. An electric field is created between the two plates. Under electrostatic physics voltage was applied to the combs. The middle plate was grounded, and the fixed pates were given a voltage of 5 volts.

Table 8.1 Description of three sensors designed on COMSOL Multiphysics simulator.

Type of MEMS Structure	Dimensions of the designed structure	Displacement (micrometer)	Eigen frequency (Hz)
Comb drive structure	2 mm x 1 mm, fixed plates is 2 mm x 0.5 mm and for comb fingers it is 0.2 mm x 0.05 mm. Thickness of structure on the whole is 0.03 mm.	1.4638	14 954
Four pad structure	Two fixed plates have four pads of 0.6 mm x 0.4 mm. Two fixed plates of 1.8 mm x 0.5 mm dimension consisting of eight combs and for movable plate dimensions are 1.8 mm x 1 mm with nine combs on either side.	3.128	14 301
Optimized tri-plate comb structure	The structure shape was same as the two-pad comb drive structure with the device dimensions of comb, fixed plate optimized for middle plate dimensions of 8 mm × 7 mm.	4.15	169

8.7 Result and Discussions

It is evident from Table −8.1, that three comb drive sensors were designed and simulated using COMSOL Multiphysics simulator. When these sensors were simulated for the Eigen frequency, it was found that an optimized tri-plate comb drive sensor was the most sensitive for the frequency range of 30–300 Hz. Stress waves within this frequency range created maximum deflection. Hence, this optimized structure was found most suitable to detect the defects that produces the AE within this range. This range was earlier established by the FFT analysis of the sound sample of a diesel engine.

Study Results

1. Stationary study: describes the stress distribution in comb drive structure. After applying a body load of 1 N, the stress developed was 27.729 N m^{-2}.
2. Eigenfrequency study: in this study displacements for six different frequencies were realized. For the eigenfrequency of 169 Hz, a displacement of 4.15 micrometers was observed.
3. Electrostatics study: in the electrostatic study the middle plate was grounded, and the fixed plates were given a voltage of 5 volts. The capacitance value obtained was 1.39086 e-17

8.8 Conclusion

FFT analysis of the sound of a diesel engine was conducted and it was found that most of the frequency change occurring due to faults were in the range 30–300 Hz. Hence, a tri-plate comb structure based on piezoresistive effect design was designed that was highly sensitive to this frequency range. Performance of this sensor in terms of the displacement was compared with four pad structure. The design has been simulated in COMSOL to ascertain its frequency response and bending mode analysis. This structure was analyzed for six eigenfrequencies on the basis of stress and eigenfrequency versus displacement value.

Bibliography

1 N. Yazdi, F. Ayazi, K. Najafi. 1998 "Micromachined inertial sensors", Proceedings of the IEEE, Vol 86, Issue 8, pp 1640–1659.

2 W.J Fleming, 2001. Overview of automotive sensors, IEEE Sensors Journal, vol.1, pp. 296–308.

3 Joel R., 2010. "Design and analysis of a low frequency MEMS vibration sensor for automotive fault detection", International Journal of Vehicle Design, Vol. 54, Issue -2, pp 93–110.

4 Nii O. A. O., and Stephen M., 2002. MEMS Application in Pavement Condition Monitoring-Challenges, *Proceedings of* 19th International Symposium on Automation and Robotics in Construction, Sepetember, Washington, DC, USA

5 Gupta A. and Ahmad A., 2007. Microsensors based on MEMS technology, Defence Science Journal, Vol 57, Issue 3, pp 225–229.

6 Abhirrao, N.S., Bhosle, S.P. and Nehete D.V., 2018. Dynamics and Vibration Measurements in Engines, *Procedia* Manufacturing 20, 2nd International Conference on Material Manufacturing and Design Engineering, pp 434–439.

7 Haoxiong F., Weijian Y., 2017. Propagation characteristics of acoustic emission wave in reinforced, Results in Physics 7, pp 3815–3819.

8 Sathish K. M., Ravindra H. V., 2008. "Estimation of AE Parameters for Monitoring Bearing in a Lathe Using Multiple Regression and GMDH", Volume 11: Mechanical Systems and Control, *Proceedings of* ASME 2008 International Mechanical Engineering Congress and Exposition, October 31–November 6, Boston, USA, pp 81–88.

9 Al-Obaidi, S., 2012 A Review of Acoustic Emission Technique for Machinery Condition Monitoring: Defects Detection & Diagnostic, Applied Mechanics and Materials, Vols. 229–231, pp 1476–1480.

10 Tan A., 2016. How can Acoustic Emission Signals be Used in Condition Monitoring and Diagnosis of Diesel Engine Condition?, Advances in automobile Engineering, ISSN:2167–7670, pp1–4.

11 Zadeh K., Kourosh, Trinchi A., Wlodarski W., Holland A., Galatsis K., Derek A., Alex H., and Vijay K. V., 2001. "Mass sensitivity of layered shear-horizontal surface acoustic wave devices for sensing applications", Electronics and Structures for MEMS II, *Proceedings of* International Symposium on Microelectronics and MEMS, vol 4591, Adelaide, Australia.

12 Bedon C., Bergamo E.ID, Izzi M.ID and Noè S., 2018. Prototyping and Validation of MEMS Accelerometers for Structural Health Monitoring—The Case Study of the Pietratagliata Cable-Stayed Bridge, Journal of Sensor and Actuator Networks, vol-7, Issue-30, pp 1–18.

13 Hua Y., Jielin Z., Licheng D.,and Zhiyu W., 2014. A Vibration-Based MEMS Piezoelectric Energy Harvester and Power Conditioning Circuit, Sensors, vol-14 Issue-2, pp 3323–41.

14 Bhatt G., Manoharan K., Chauhan P. S. and Bhattacharya S., 2019. MEMS Sensors for Automotive Applications: A Review, Sensors for Automotive and Aerospace, Springer Nature Singapore Pte Limited, pp 223–239.

15 T. D. Tan, L. M. Ha, and N. T. Anh, 2010. "A Real-time Vibration Monitoring for Vehicle Based on 3-DOF MEMS Accelerometer", Proceedings of International Conference on Computational Intelligence and Vehicular System (CIVS2010), Cau Giay, Ha, Noi, VietNam.

16 Choudhary R. and Singh P., 2014. MEMS Comb Drive Actuator: A Comparison of Power Dissipation Using Different Structural Design and Materials Using COMSOL 3.5b, Advance in Electronic and Electric Engineering. ISSN 2231–1297, Vol. 4, Issue 5, pp. 507–512.

17 Saad, M. Musa. Strontium Titanate Perovskite SrTiO3 as a Promising Ideal Material for MEMS 9 Applications : First - Principles FP - LMTO + LSDA Study, Journal of Natural Sciences and Mathematics. - 2014, Vol. 7, Issue 1, pp. 1–12.

9

IoT Driven Healthcare Monitoring System

Md Robiul Alam Robel[1], Subrato Bharati[2,], Prajoy Podder[3], and M. Rubaiyat Hossain Mondal[3]*

[1]*Department of CSE, Cumilla University, Cumilla, Bangladesh*
[2]*Department of EEE, Ranada Prasad Shaha University, Narayanganj-1400, Bangladesh*
[3]*Institute of ICT, Bangladesh University of Engineering and Technology, Dhaka, Bangladesh*

Abstract

Technological evolution in the area of healthcare enables different ways of managing patients' medical records thus improving the quality of healthcare. In this regard, this chapter offers an outline of developing the Internet of Things (IoT) technology in the area of healthcare as a flourishing research and experimental trend at the present time. The main advantages and benefits have been considered in this chapter. Moreover, it has been established that most of the hospitals in a number of countries are still covering many issues concerning their exchange of health information. In recent times, several studies in the healthcare information system proposed that the shortcomings of health information is one of the most significant challenges with the management of patient medical records. As a result, in this chapter, we provide a detailed design and overview of the IoT healthcare system and its architecture. Furthermore, a full explanation of merits and demerits has been focused on using IoT in-patients' healthcare.

Keywords *IoT; Health care; e-Health; Health information exchange; Patient level architecture*

9.1 Introduction

Healthcare services are continuously causing problems since various diseases are increasing day by day [1]. The Internet of Things (IoT) has been extensively used

*Corresponding Author: subratobharati1@gmail.com

Fog, Edge, and Pervasive Computing in Intelligent IoT Driven Applications, First Edition.
Edited by Deepak Gupta and Aditya Khamparia.

to interrelate obtainable medical properties as well as to offer reliable, operational, and smart healthcare facilities to patients. Several improvements have been made in monitoring healthcare [2]. These realizations have confirmed the usefulness and prospects of IoT in healthcare schemes [3]. Despite the present achievement, there are issues on how to promptly and methodically establish intellectual IoT-based healthcare schemes [4]. Pointing at exploiting the abilities of IoT in healthcare schemes, organizations and scientists have been dedicated to the improvement of IoT-based expertise. Moreover, developing IoT equipment has delivered numerous chances for emerging smart hospitals [5]. Also, challenges still occur in realizing security as well as operative healthcare presentations. Some of these issues are for instance self-improvement, self-learning, implantable and wearable sensors, privacy, standardization, and safety [6]. Consequently, the combination of the IoT using additional technologies, facilities, and solutions of communication is a precarious topic in the effectiveness and performance of IoT equipment [7, 8].

The development of the Internet offers the prospects for emerging environmental software structures in various sections of science. A novel paradigm known as the cloud IoT is offered combining IoT and cloud computing. Conversely, the cloud IoT in healthcare applications is demonstrating the fundamentals of this research. Cloud IoT in healthcare is an amalgamation of interconnected apps, communication technologies, things (sensors, devices, and so on), cloud, people that will track, monitor, and stock patients' data for continuing care and exploration.

Due to the lack of previous work [9–11] in cloud IoT in the health sector, there is a crucial requirement for a real-time health monitoring setup for evaluating patients' healthcare records to avoid unnecessary deaths. Cloud healthcare, healthcare industrial IoT, and cloud IoT are important for the understanding of such monitoring, for example, delivering encouraging solutions to track the facilities of the healthcare schemes.

9.1.1 Complementary Aspects of Cloud IoT in Healthcare Applications

IoT can achieve the resources of cloud computing to overcome its several challenges, such as storage, communications, as well as processing [12, 13]. In contrast, cloud computing can achieve the features of IoT to prolong its opportunity to protect the real-world objects in a dispersed way. In several cases, cloud computing offers a transitional layer among the objects and applications [14, 15]. In the future information collecting, processing, as well as transmission of the applications will carry novel experiments as a multi-cloud environment [16]. Furthermore, the IoT is considered by protocols and extraordinary heterogeneity levels of strategies. It needs various essential properties, for example, interoperability,

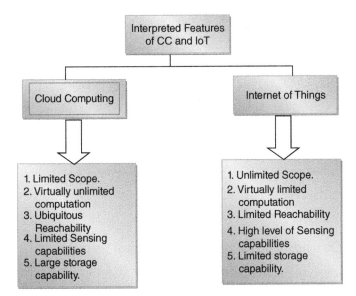

Figure 9.1 Interpreted features of cloud computing and IoT.

availability, scalability, reliability, flexibility, security, and efficiency. Subsequently, the cloud has been shown to offer these things, and we recognize them as some of the keys of cloud in IoT cores. Two additional cores are reduced cost and that both providers and users of services and applications find them easy to use. Certainly, the cloud simplifies the movement of IoT data processing and data collection and permits integration and rapid setup of novel equipment, while conserving low costs for complex data processing and deployment. Most of the IoT drivers of incorporation break down into three groups: storage, communication, and computation. The features of cloud computing and IoT is shown in Figure 9.1.

The cloud can be used to distribute computing among the edge and the remote devices in the context of IoT, whereas automation of both data distribution and collection can be managed at little cost. Cloud computing offers an operational method as well as tracking and connecting from everywhere.

IoT storage includes an explanation of several data sources (such as, sensors), which develop a massive amount of semi-structured or unstructured data. This massive amount of data has the general features of big data, velocity, and volume. Long-lived storage and large-scale, low-cost, effectively unlimited, and on-demand storage capability which is delivered by the cloud, characterizes an essential driver of cloud IoT. Cloud is the best cost-effective and expedient explanation to contend with data created by IoT. In addition, it produces new prospects for data integration, accumulation, and allocation by way of third parties.

9.1.2 Main Contribution

This chapter focuses on the issue of IoT based health monitoring systems. The contributions of this chapter can be summarized as follows.

(i) To explain the three tiers for the overall IoT healthcare system.
(ii) To design an IoT healthcare system where this chapter explains the patient level architecture, tier 2 for network communication architecture.
(iii) To model an IoT for an e-healthcare system and to outline a topology for IoT driven systems.
(iv) To describe a whole model in a single scenario and a process flow for the overall model.

9.2 General Concept for IoT Based Healthcare System

Figure 9.2 is a dissection of how the new innovation works with the overall concept for the upcoming new era where every single part of our daily activities will be improved with the help of modern technology and in particular by the IoT. In Figure 9.2, three blocks are used to show the internal relationship of the overall scenario where the design of innovation meets with both sources of innovation and the integrated application of the currently available IoT. Source of innovation indicates the actual available resources to implement the digital prescription method or in a word, e-healthcare. Current healthcare information, education, and employment information are the main parameters for the sources of innovation along with features like financial information and regional or whole nutritional information. These parameters are then combined with the IoT. Technological aspects are mainly integrated here to get the output for the e-healthcare system. As well, social integration is one of the most important aspects of the implementation of e-healthcare as the whole proposal is designed for the benefit of all users. On the other hand, the expectation of ease of use of the system varies from person to person where data from different scenarios should be collected and studied . This parameter thus results in satisfied users whose high expectations are fulfilled (although it is not possible to fulfill their expectations completely). The main target of this innovation is to serve remote areas where it is impossible to get the proper treatment in time. At times of natural disasters, IoT driven healthcare can play an important role. Not only in remote areas but also in modern towns, people are so busy that they find it difficult to make time to consult with doctors as the doctors also have a fixed schedule to adhere to. Hence, easy access to the best features of healthcare system is important.

Figure 9.3 shows how the system of an e-healthcare system works with the help of the IoT. The IoT generally works as a middleware that can be used in many ways to implement projects with ease. The user/patient, doctor or medical service

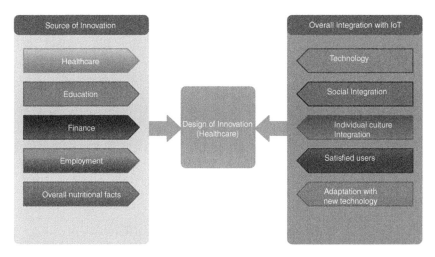

Figure 9.2 IoT based general concept for innovation at the healthcare system

provider are the key aspects of the proposed system. Here patients cannot communicate with the medical professionals directly, a network communication is a requirement. The patient will have mobile, tablet computer or some other communication device to connect with the network to see e available what services are available nearby. When the patient interacts with the medical service providers, the medical personnel will be notified and thus service will be provided after fulfilling certain pre-defined conditions (as shown in Figure 9.2). The information about the people involved with this service will be kept on a cloud server as well as in physical storage.

9.3   View of the Overall IoT Healthcare System- Tiers Explained

The general concept discussed above can be categorized into a three-tier architecture which is presented in the following. The three tiers can be classified as:

(1) Tier 1: user/ patient level architecture
(2) Tier 2: network-level architecture
(3) Tier 3: medical professional and hospital end architecture

Tier 1 deals with the patient level architecture where the patients' current situation is the most important aspect. Here the patient interacts with the doctor or other medical personnel indirectly because all communication there is established through the Internet amd therefore involves the IoT. Patients first log in to the app or any interface which is cross-platform and then create a profile so that the medical personnel on the other side can see the biodata of the patient. All updated information is kept on the cloud server as well as in physical storage.

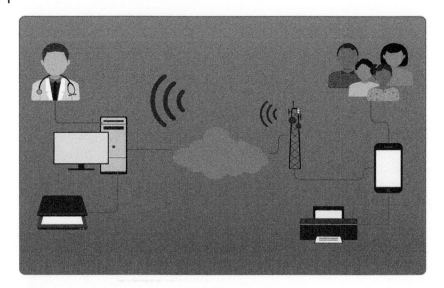

Figure 9.3   General concept for IoT based health care system

Tier 2 is about the network architecture, which actually works as the middle-ware of the whole system. It ensures that the communicating devices are active to establish a relationship between the medical service and the patients. General checks are done to ensure errorless network communication using certain algorithms. These algorithms vary from system to system depending on the architecture. Security is the main concern here. As an important issue, security credentials are looked after using some predefined parameters. The concept for IoT based health system is shown in Figure 9.3.

On the other hand, tier 3 covers the medical services that are provided depending on the demands of the patient. In short, the full system helps removes stresses from society to ensure a healthy generation.

9.4   A Brief Design of the IoT Healthcare Architecture-individual Block Explanation

In this section, three-block figures are illustrated. The first block sketched in Figure 9.5 shows the interaction of patients with an e-healthcare system. The whole system is designed keeping in mind that the system should be cross-platform. Ease of use at the user end is important irrespective of whether the system is using the latest technology or not. Patients can communicate using the network with the help of communicating devices. Various types of such communication devices are available nowadays. The network communication is

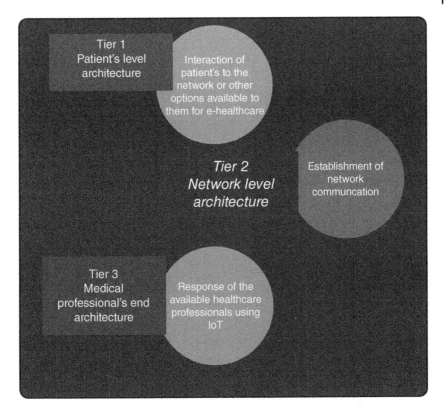

Figure 9.4 Three tiers for the e-healthcare architecture

established wired or wirelessly. Most connections occur using wireless technology, as this is the most common current technology. Routers play an important role in establishing wireless communication.

Figure 9.6 shows the action center for the automation required to implement an e-healthcare system. While complex actions are performed here to establish errorless network communication it is important to bear in mind that other tiers are also important and every part of the automation should be treated with equal importance. It is important that there is no possibility of error present here. In remote areas the availability of electricity is a big problem, base stations need to be in working order to ensure flawless communication in order to perform the required tasks. To overcome this problem there are some costly alternatives available in the market. Internet service providers, known as ISPs, provide the same service from their respective position for wired or wireless network services. Cloud storage is used here so security is a big issue bearing in mind that it is equally important to store the data provided by both user and the service provider. In order

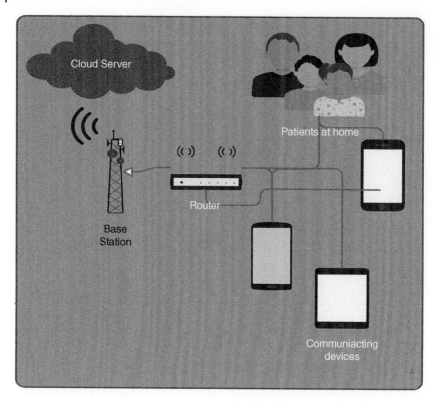

Figure 9.5 Patient level architecture of an IoT based e-health care system

to establish good security, a predefined firewall is needed to accomplish the best service in the e-healthcare system thus fulfilling the requirements of the proposed project. Figure 9.6 depicts an idea of the network interacting with a cloud server or storage and firewall. The main concept is sketched here to provide a practical understanding rather than just a theoretical idea. Moreover, base stations play the most important roles in establishing the whole scenario as an active problem. Sometimes there are natural hazards that cause network or communication failure. Service providers use several techniques to overcome these issues but it is important to ensure the best output from this scenario.

9.5 Models/Frameworks for IoT use in Healthcare

There are basically three types of component available with their framework as they are pictured

- infrastructure framework components

Figure 9.6 Tier 2 for a network communication architecture for e-healthcare

- topological framework components
- platform framework components

Infrastructure framework components include hardware configuration, connection topology and oriented maintenance personnel. In this case, hardware configuration is all about the available framework for hardware components or some predefined framework for hardware components. Moreover, these frameworks ensure the proper use of the hardware available in the whole automation. These generally give the proper guidelines for the use of the devices and the components related to them. Maintenance personnel also play an important in the automation, the main infrastructure is run by experts and they are responsible for every command at the hardware level so that the whole automation is a success. They actually define the way the connections between the devices work and also regulate the connections. Good maintenance ensures the best output so this component is critical.

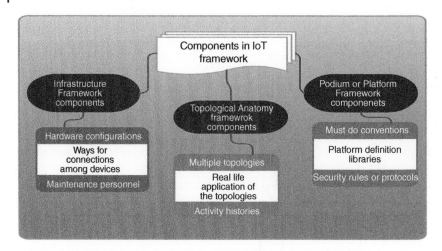

Figure 9.7   Basic IoT components and their frameworks in e-healthcare

Topology guarantees the way actions are performed. This component framework generally combines multiple topologies along with activity history and real-life topologies. Real-life application topologies indicate the best use of the topologies that are available to enable the control personnel to manage the automation. But these are the things that need to be predefined. Reviewing the activity history shows the previous performance and allows aspects of performance to be improved.

Platform framework components define the boundaries on which the whole automation should be performed in order for the automation to be successful. Bear in mind that no misuse of data will occur whether the control personnel are watching or not and these constraints are also predefined. All of the security rules and protocols are defined by this component framework. As the user is mainly concerned with security issues, proper security credentials and their definitions are as important as other frameworks. To communicate between android/iOS based applications and devices, wireless communication is an important method. As a result, data or information should be encrypted at the time of communication; otherwise local traffic might be divulged. The android/iOS based application should use a secure socket layer (SSL) protocol or a TLS protocol and validate the TLS certificate of the device. This validation process will protect the communication against a MITM attack. Interaction between web/ mobile applications and cloud services can be secured with TLS/SSL. This is performed by allowing its use from the cloud service; otherwise the attacker will capture the data passively. A strong password is mandatory.

Hardware configurations are required in the infrastructure framework. Standard communication technology may be required for an IoT system. In order to communicate properly between IoT devices and backend ISP, a communication technology is preferred. IoT devices are enabled with wireless connectivity. There are some preferred wireless connections for IoT, they are: IEEE 802.15.4 standard, Wi-Fi, and bluetooth, *etc.* Bluetooth has very low power consumption, in contrast, the data rate of Wi-Fi connections are very good compared to the IEEE 802.15.4 standard and Bluetooth [17]. In order to support IoT based smart applications different specialized protocols have been developed due to their efficient working performance with limited capacity of resource.

Overall activity topology is sketched in Figure 9.8. Figure 9.8 shows that all the components are working sequentially. It confirms that the patient cannot interact with the medical personnel without the help of network and cloud storage and vice versa. This is a one-to-one real life topology that actually works in all mobile communications where cloud computing is involved. The action time between the patient and the doctor is so small that a single error on any node or device can cause failure so the devices and rules for running the devices are predefined. In the whole topology, the user first interacts with the available mobile devices to connect to the network then the networking components provide them with the required privileges to connect to the server and then with the doctor or medical services. This is rather a complex scenario, but nowadays proper predefined rules are applied to show that the connection between doctor and patient is medical centre/hospital to home.

9.6 IoT e-Health System Model

Figure 9.9 is a pictorial view of the whole e-healthcare system. The actual work process is shown in this figure to clarify the whole automation. Components are symbolized here and communicating devices are also shown on an abstraction basis. Lines are used to show the connections or relationships between the components of the whole e-healthcare system. Successful communication between a patient and a doctor happens if the abstract components shown in the figure work as they are supposed to do. The shown components are a requirement for a successful e-healthcare system. The IoT is so important in this arena that if there is no Internet then there is no communication, therefore most e-healthcare system require the use of the IoT. A distributed database is hugely involved to ensure the cloud server or storage gives the best service. Individual healthycare records (HCR) in the distributed database through wireless/local area networks (W/LAN) are used to send user or patient information to the medical service provider as there is no direct communication between patients and the relevant doctors. These HCRS can be accessed from anywhere in the world as they are uploaded on virtual

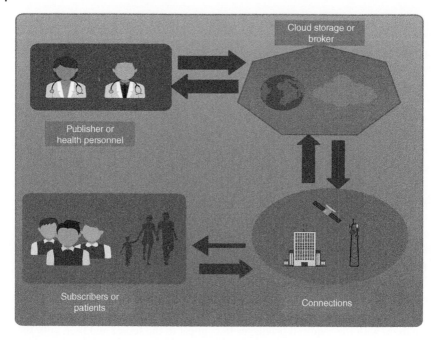

Figure 9.8 Topology for IoT driven healthcare framework

storage known as cloud storage. Advanced equipped is used to get the greatest benefit from the scenario that reduces stress from both remote medical patients as well as very busy people in modern society. Hence, in every sector of society, e-healthcare can ensure a better life after giving the best service possible. Since the treatment history is preserved with the help of cloud storage, data analysis can be performed later to ensure better treatment for patients.

9.7 Process Flow for the Overall Model

Figure 9.10 shows the model of the IoT e-health process for a patient. Figure 9.10 offers the IoT communication model as the key enabler of widely distributed healthcare applications. The crucial actors in this process consist of the observed patients, doctors and distributed information records. This finding contributes to the real application of a widespread healthcare process within IoT. It also highlights the significance of using various networks, devices, and procedures in analyzing the patient's progression. On the other hand, this model is not still completely obtainable, the modules present in the use case are at various stages of realization and the offered model does not extend runtime identifying information into healthcare databases. This overall model is abridged in Figure 9.10.

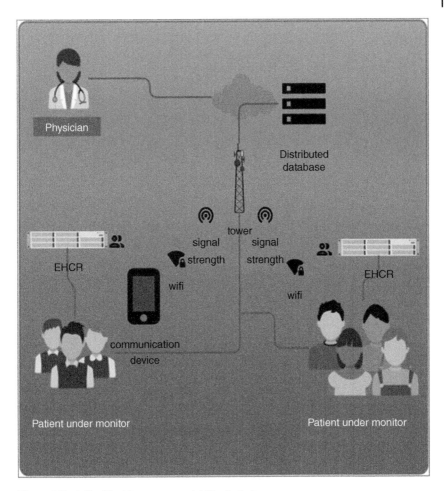

Figure 9.9 IoT e-Health system model illustrated

9.8 Conclusion

These of the IoT has developed an emergency need for public hospitals, patients, the technical section of a hospital and their management activities. A fruitful IoT use is affected by how well this technology improves the prospects of its users. The inventors of this technology must recognize the application requirements from the hospital management perspective as well and must align the application with the aims of hospitals so that it confirms a fruitful application and utilization. Precisely, some frameworks and models have been considered only for specific circumstances, contexts, and environments. For the moment, processes have only

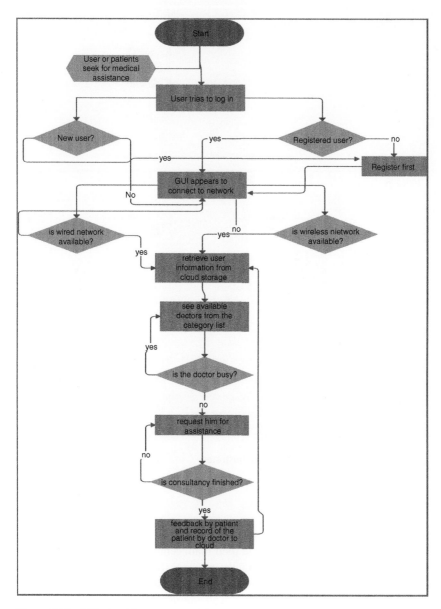

Figure 9.10 IoT e-health process model

offered frameworks/models without any post-evaluation or examination, making these frameworks/models inappropriate for investigating IoT use in health information exchange (HIE) for various reasons. This chapter not only emphasizes the HIE context, but also describes tiers for the e-healthcare architecture with the IoT e-healthcare system. For clarity, the IoT e-health process model with the flowchart is provided in this chapter.

Bibliography

1 Mosadeghrad, Ali Mohammad. "Factors influencing healthcare service quality." International journal of health policy and management vol. 3,2 77–89. 26 Jul. 2014, doi:10.15171/ijhpm.2014.65.

2 Mohammed Dauwed and Ahmed Meri, "IOT Service Utilisation in Healthcare." Internet of Things (IoT) for Automated and Smart Applications, 2019. DOI: 10.5772/intechopen.86014.

3 Evans, R S. "Electronic Health Records: Then, Now, and in the Future." Yearbook of medical informatics vol. Suppl 1,Suppl 1 S48–61. 20 May. 2016, doi:10.15265/IYS-2016-s006.

4 S. Tyagi, A. Agarwal and P. Maheshwari, "A conceptual framework for IoT–based healthcare system using cloud computing," 2016 6th International Conference – Cloud System and Big Data Engineering (Confluence), Noida, 2016, pp. 503–507. doi: 10.1109/CONFLUENCE.2016.7508172.

5 K. Yeh, "A Secure IoT–Based Healthcare System With Body Sensor Networks," in IEEE Access, vol. 4, pp. 10288–10299, 2016. doi: 10.1109/ACCESS.2016.2638038.

6 YuehongYIN, Yan Zeng, Xing Chen, Yuanjie Fan, "The internet of things in healthcare: An overview," Journal of Industrial Information Integration, Volume 1, pp. 3–13, March 2016.

7 G. Neagu, Ş. Preda, A. Stanciu and V. Florian, "A Cloud–IoT based sensing service for health monitoring," 2017 E–Health and Bioengineering Conference (EHB), Sinaia, 2017, pp. 53–56. doi: 10.1109/EHB.2017.7995359.

8 C. Bradley, S. El–Tawab and M. H. Heydari, "Security analysis of an IoT system used for indoor localization in healthcare facilities," 2018 Systems and Information Engineering Design Symposium (SIEDS), Charlottesville, VA, 2018, pp. 147–152. doi: 10.1109/SIEDS.2018.8374726.

9 Ghanavati, S., Abawajy, J.H., Izadi, D. et al. Cluster Comput 20: 1843, 2017. https://doi.org/10.1007/s10586-017-0847-y.

10 G. Muhammad, S. M. M. Rahman, A. Alelaiwi and A. Alamri, "Smart Health Solution Integrating IoT and Cloud: A Case Study of Voice Pathology Monitoring," in IEEE Communications Magazine, vol. 55, no. 1, pp. 69–73, January 2017. doi: 10.1109/MCOM.2017.1600425CM.

11 M. Hassanalieragh et al., "Health Monitoring and Management Using Internet–of–Things (IoT) Sensing with Cloud–Based Processing: Opportunities and Challenges," 2015 IEEE International Conference on Services Computing, New York, NY, 2015, pp. 285–292. doi: 10.1109/SCC.2015.47.

12 Darwish, A., Hassanien, A.E., Elhoseny, M. et al. J Ambient Intell Human Comput (2019) 10: 4151. https://doi.org/10.1007/s12652-017-0659-1.

13 Bandyopadhyay, D. & Sen, J. Wireless Pers Commun (2011) 58: 49. https://doi.org/10.1007/s11277-011-0288-5.

14 F. Fernandez and G. C. Pallis, "Opportunities and challenges of the Internet of Things for healthcare: Systems engineering perspective," 2014 4th International Conference on Wireless Mobile Communication and Healthcare – Transforming Healthcare Through Innovations in Mobile and Wireless Technologies (MOBIHEALTH), Athens, 2014, pp. 263–266. doi: 10.1109/MOBI-HEALTH.2014.7015961.

15 C. Doukas and I. Maglogiannis, "Bringing IoT and Cloud Computing towards Pervasive Healthcare," 2012 Sixth International Conference on Innovative Mobile and Internet Services in Ubiquitous Computing, Palermo, 2012, pp. 922–926. doi: 10.1109/IMIS.2012.26.

16 Shahariar Parvez A.H.M., Robiul Alam Robel M., Rouf M.A., Podder P., Bharati S., "Effect of Fault Tolerance in the Field of Cloud Computing," Inventive Computation Technologies. ICICIT 2019. Lecture Notes in Networks and Systems, vol 98. Springer, Cham, 2020. https://doi.org/10.1007/978-3-030-33846-6_34.

17 Maruf Pasha and Syed Muhammad Waqas Shah, "Framework for E–Health Systems in IoT–Based Environments," Wireless Communications and Mobile Computing, vol. 2018, Article ID 6183732, 11 pages, 2018. https://doi.org/10.1155/2018/6183732.

10

Fog Computing as Future Perspective in Vehicular Ad hoc Networks

Harjit Singh[1,], Dr. Vijay Laxmi[2], Dr. Arun Malik[3], and Dr. Isha[4]*

[1] Research Scholar, Guru Kashi University, Punjab, India
[2] Professor, Guru Kashi University, Punjab, India
[3] Associate Professor, Lovely Professional University, Punjab, India
[4] Associate Professor, Lovely Professional University, Punjab, India

Abstract

Vehicular *Ad Hoc* Networks (VANETs) have been intensively examined due to the broad spectrum of apps and facilities, including passenger safety, road effectiveness improvements, and infotainment. VANETs consist of an on-board unit (OBU), roadside unit (RSU) and mobile vehicles. As technology progresses and the number of intelligent cars grows, traditional VANETs face a number of technical difficulties due to less scalability, flexibility, insufficient intelligence, and connectivity intelligence when it is deployed and managed. As computer and communication techniques evolve quickly, VANETs grow into system revolutions either peripherally or progressively. Cloud computing has been used as an alternative in VANETs which need resources (*e.g.* computation, transport and networking). Special requirements for autonomous vehicles in the next generation VANETs cannot be solved by conventional cloud computing, such as strong flexibility, low latency, real time appliances and communication. Fog computing plays an important role to acquire optimal solutions of the existing problems in VANETs.. Recently the study of the position of fog computing in VANET has become more concerned with unique connectivity demands, place knowledge, and low latency. The combination of VANET and fog computing offers a range of options for cloud computing applications and facilities. Fog computing deals with high-virtualized VANET software and communication systems, where dynamic-speed vehicles travel. Mobile cars may also require low-latency fog computing in VANET and local connections within short distances. The modern state of the work and upcoming viewpoints of VANET fog computing are explored in this chapter. In addition, this chapter outlines the

Fog, Edge, and Pervasive Computing in Intelligent IoT Driven Applications, First Edition.
Edited by Deepak Gupta and Aditya Khamparia.

features of fog computing and fog-based services for VANETs. In addition to this, fog and cloud computing-based technology applications in VANET are discussed. Some possibilities for challenges and issues associated in connection with fog computing are also covered in this chapter.

Keywords *resource; collision; latency; cloud; access networks*

10.1 Introduction

The automotive market has grown considerably in the last five years. The development and deployment of different network forms, particularly for Vehicular *Ad Hoc* Networks (VANETs), are being updated with the enhancement of technology, hardware, and communications technologies. VANETs are special Mobile Ad Hoc Networks (MANETs) made up of on-board units (OBUs), roadside units (RSUs) that tend to hold automobiles as remote nodes (Zaidi, K., and Rajarajan, M., 2015). Traffic probelsm are persistent difficulties in both developed and developing countries and lead to massive mortality and property damage. VANETs have been developed by Intelligent Transportation Systems to create a safer road transport network, to address these obstacles, and to make travel easier, secure, effortless and enjoyable (Yang *et al.* 2014). VANETs concentrate on road safety and effective public road traffic management, thus making journeys convenient and enjoyable for drivers and passengers. Communications from Vehicle to Vehicle (V2V) and Vehicle to Infrastructure (V2I) can be carried out within VANETs by vehicles. OBUs that form MANETs connect with other vehicles, which enable fully distributed wireless communication while communicating with RSUs in the infrastructures (Whaiduzzaman *et al.* 2014). It provides drivers with security and safety when exchanging data between VANET entities, and only when vehicles are interconnected, is the true potential of connected vehicles realized.

Primarily, safety and non-safety are the two applications that are used in VANETs. VANET security systems are used to relay safety alerts, such as numerous warning messages assisting road vehicles, so that collisions and dangerous situations can be prevented. Health alerts include road mishap records, traffic congestion updates, road creation intelligence, and vehicle emergency alerts. These systems require high reliability and low latency. For instance, the fog scenario for the communication of safety messages in VANETs is shown in figure 10.1. In contrast, non-safety systems provide an effective and relaxed driving experience. Infotainment systems provide passengers with Internet access, for example storage of data, video streaming and video calling, for information and entertainment purposes. Unlike security systems, high performance and low latency are not needed for such applications. VANETs need to disseminate such data about critical events (*e.g.* incident reports) immediately and accurately. Although VANETs

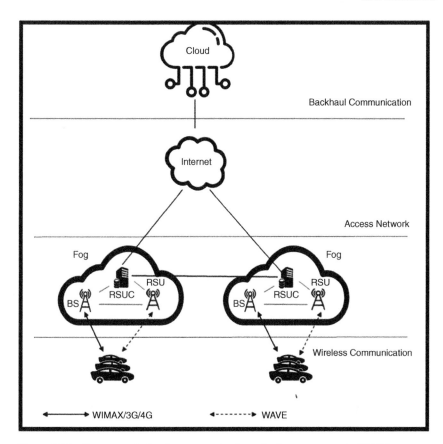

Figure 10.1 Fog scenario for the communication of safety messages in VANETs.

are able to distribute event information, it is always a challenge that critical messages are transmitted in a complex vehicle setting. This is because of the restricted DSRC communication distance and the controversial multi-collision control (CSMA/CA) method in the IEEE 802.11p specification. The critical messages face media access control (MAC) containment delays that are not appropriate for VANETs. If such critical time information is not disseminated in a timely and correct way, neighboring vehicles may be damaged in collateral ways.

The standard VANET hierarchical architecture cannot handle large volumes of traffic information produced by intelligent automobiles such as audiovisual and sensor records. Additional servers in distributed areas are required to gather and practice large numbers of instant traffic data. Cloud-based VANETs can be a suitable solution in these situations. A number of mobile devices have recently adopted cloud computing to handle complex local calculations. Much research in cloud-based VANET's has been carried out to further expand the capabilities of VANETs. In recent years, cloud computing has been the latest innovation. Vehicles in various scenarios are increasingly needed for resources like storage, networking and computing with the development of VANETs, with several cloud-based solutions emerging. Cloud computing systems offer thoroughly managed and flexible cloud databases, digital networks, computational services, processing assets and network resources, according to the user's requirements and what they are prepared to pay. In fact, information can be processed easily without the complexity of running massive space or computer systems. A wide range of content can be accessed or interacted with conveniently through cloud computing. It offers a clarification for contents that are distributed. A new concept of vehicle cloud computing is introduced, that benefits cloud computing to evaluate all of the infomration with regard to VANET drivers. However, all quality of service (QoS) requirements of VANETs cannot be met, so that new technologies and architectures are necessary.

10.2 Future VANET: Primary Issues and Specifications

VANETs are a crucial technology in secure, strong and insightful transportation systems. However, people spend a lot of time in automobiles. Intelligent VANET-based cars have been designed to ensure they are secure, effective and friendly. In this case, the sharing of critical events can provide greater security efficiently. Improving productivity is accomplished by decreasing traffic congestion then converting emissions and rendering travel time more reliable. In order to improve the journey through upload and access to social networks using software, VANET can also be connected to the Internet. Two forms of signals are used by VANETs: warning messages and security messages. Beacons are used by vehicles to transmit regularly and to publish information on the condition of nearby vehicles at 100 meter intervals. The sender tells the neighboring cars through rapid messages about the pace, position and pseudo-ID. On the other hand, safety messages provide vehicles on the highway with emergency information, to prevent accidents and to help save people from having their lives put in danger.

For data storage, replication, control, and complex computing, VANETs utilize standard clouds. The cloud infrastructure may be categorized into three specific distribution models, layers: infrastructure software, system platforms, and product infrastructure. Thus, users can utilize various codes and network services for better results. In this context, customers have almost unlimited resources based only on their budget. While VANETs use the cloud to resolve problems, there are still some major challenges for smart cars. Based on the very fast developments and growth patterns in VANETs, some primary issues and specifications are defined for future VANETs. A few of the main issues confronting potential VANETs are:

- **Irregular connectivity:** network connectivity control and maintenance between automobiles and networks is a key issue. Due to the high uncertainty and large losses in the packets, intermittent communications in vehicle networks should be halted.
- **Sensitivity to location and high mobility:** for future VANETs, the vehicles involved in communication must have awareness of location and high mobility. In order to deal with an emergency situation, the accurate location of every vehicle should be known to new vehicles on the network.
- **Safety:** the confidentiality of software contents and location of the client is always at risk. The vehicles operating within the network must enable users to choose which details to share and which data to keep private.
- **Intelligence aid for the network:** one of the challenges facing future VANETs is to support network intelligence. For future VANETs, most sensors will be installed in cars, and before connecting with other parts of the network, for instance, standard database servers, the edge cloud gathers and processes information.

Taking these difficulties into account, the main specifications of future VANETs can be defined as follwos.

Low latency is the key requirement for real-time applications in future VANETs. Future VANETs must support real-time applications including latency protection notifications.

For future VANETs, high-quality video streaming, infotainment and convenience applications will be in demand. Furthermore, regular automatic updates are needed for traffic applications such as 3D maps and navigation systems.

Future VANETs require a seamless link between connected vehicles to meet the high demands of interaction. Driverless or connected cars must retain contact between automobiles and fog sensors constant and that contact must be highly assured.

10.3 Fog Computing

Fog computing is a concept that applies to network edge cloud computing and services. Fog computing is targeted at networks with widely available implementations rather than the more hierarchical Internet, one of its strengths is that contextual, low latency, and improved situational awareness can be conveyed in fog. Like the Internet, fog provides end-users with information, measurement, inventory and software services (Bonomi *et al.* 2012). New architecture fog computing

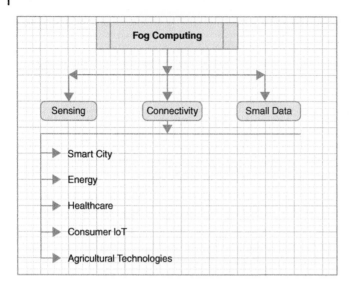

Figure 10.2 Application scenario of fog computing

is a strong VANET nominee to meet requirements, such as quick response to the underlying system, server cutbacks, cloud data flow real-time analysis, *etc.* Each computer with processing, memory and network access can be a fog node, containing road infrastructure and movable vehicles. VANETs collect fog nodes nearest to the edge of a network, and system designers determine how to use data from different styles. Fog computing can bring numerous advantages for different Internet of Things (IoT) applications as shown in Figure 10.2.

Fog devices are closely related to the IoT. The IoT generates a volume and diversity of data without precedence. When the information reaches the cloud, then security of information decreases. Cisco's White Paper for IT practitioners and operating engineering professionals clarifies a new model for interpreting and working on IoT information. Fog computing is:

- The most time-sensitive data processed close to where it is being produced rather than transferring large amounts of IoT information to the cloud.
- Providing the best performance on IoT information by considering appropriate policies.

The following sections will introduce the idea of fog computing, review the fog computing state of the art and present the fog computing characterization.

10.3.1 Fog Computing Concept

Cisco engineers introduced a fog computer system to spread cloud computing prototypes to the borders of IoT device wireless networks. Fog technology is powered by the Open Fog Project. The aim is to influence standard organizations to establish specifications so that edge equipment can safely communicate with other edge products, such as IoT or cloud services. Fog systems combine bait information with systems that generate warnings when a decoy is used incorrectly. After being adopted by Cisco and other firms, it has changed its sense that fog provides end-users with cloud-like data, computing, processing and database resources. Fog is a fully virtualized framework for measuring, processing and networking assets between end devices and data centres. Similar to cloud, fog provides consumers with information, measurements, processing and mobile services. Fog computing is a scenario in which large numbers of ubiquitous and decentralized heterogeneous devices communicate between themselves and potentially cooperate to perform storage and treatment without the involvement of third parties. Such activities may include support for basic network operations and modern sandboxing technologies and applications. Users who rent their computers to host such products are allowed to do so.

Implementation of IoT technologies in the double-stage cloud infrastructure does not meet the mobility, low latency or smart device localization criteria. It explores the development of the multi-faceted fog computing system. Users usually have to access their information from the cloud (multimedia files, reports, *etc.*). Data is stored in fog near the users with reduced latency and increased output on the fog servers. The ITS application in vehicles is used in the first stage. The second stage is the fog platform, which includes fog devices like RSU and wireless networks. The third stage is a large-scale traditional cloud datacenter, Fog computing for broadcasting and real time applications in cars, provides high throughput, low latency, position visibility and a quality of service. Cloud computing and fog computing are included in the simple IoT processing, storage and networking system. Figure 10.3 shows the cloud–fog relationship.

10.3.2 Fog Technology Characterization

Fog technology is also used to expand conventional cloud computing to its outskirts. It is fully virtualized and can deliver measurement, processing and networking between the end nodes and conventional clouds. Fog computing hosts are the network edge services or even end devices. It reduces fog network latency and increases QoS to provide better user experience. Some fog capabilities include: consumer location, geographical distribution and flexibility aids. A variety of fog processing functions run counter to your instincts. The foregoing are the various characteristics of fog computing:

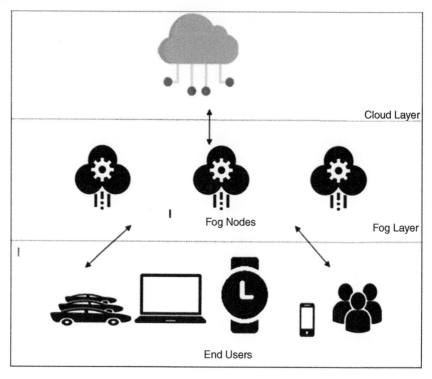

Figure 10.3   Cloud–fog relationship

Fog nodes can be used for ingesting data from different forms of terminal machines in the local area network. Fog nodes can use data from various terminal device types in the local area network.

Fog computing contains very numerous disseminated nodes, as demonstrated in general sensor networks and smart grids in particular, as a result of wider geo-distribution. For example, fog can play an important role in providing high quality streaming for connected vehicles by proxies and access points, the resources and the intent of the fog.

The locator/identification separation protocol is a Cisco systems-developed routing and addressing architecture. This offers accessibility tools such as host identification decoupled from place identity. Applications of fog can directly communicate with vehicles through this protocol. Fog allows administrators to check where and how users are receiving information, enables location-based mobility and improves performance and service quality.

In multi-tiered organizations, fog is highly dynamic and heterogeneous with low latency and scalability criteria in various levels of network hierarchy. Fog nodes in the shape of the physical and/or digital fog nodes can be deployed in a broad range of environments.

Fog modules have to be able to interact (such as streaming) with different service providers. Fog is creating new ways of rivalry and collaboration between suppliers.

10.4 Related Works in Cloud and Fog Computing

The researchers propose incorporating connected technologies into a car and using computer technology from the car cloud. The authors in this study suggested that Vehicle Clouds (VCs) are technologically viable and have an important social impact. Two VC-enabled cloud computing resources are mentioned: network as a service (NaaS) and data storage (SaaS). NaaS is relevant to situations in which a number of cars have an Internet connection on a road that shares Internet access with those cars that might need such a network. SaaS includes automobiles with substantial on-board storage and other vehicles that might require additional space to submit (Olariu, S., Khalil, I., & Abuelela, M. 2011).

In Alam *et al.* (2015) the authors suggest a cyber-physical IoV (CIoV) design. CIoV is a Social IoT (SIoT) vehicle example, where the main social entities are vehicles. This consists of a three-layer architecture: physical body, digital entity and social map interaction information on the physical and digital scales were enhanced in the social graph cloud, where each node is an individual, and connections are the interactions between data.

In Hussain *et al.* (2012) the researchers describe a VANET Cloud's architecture and splits the VANET Cloud into three structural design: Vehicle Clouds Network (VCN), Vehicular Utilizing Clouds (VUC) and Vehicular Fusion clouds. The VCN comprises VANET, gateways and agents. VCN is about building clouds with every VANET and cloud participants bring their resources together to create a rich essential climate. VUC provides fast cloud services. RSUs that act as gates to vehicles for cloud services deliver virtualization. Vehicular Hybrid Cloud is a mix of VCNs and VUCs.

In Speed, C., & Shingleton, D. (2012) the use of a single vehicle ID (vehicle number plates and bar code for quick responses) is suggested. The aim of this work is to detect automobiles as IoT objects with the potential to connect places, services, people and devices.

In Leng, Y., & Zhao, L. (2011) cloud computing-based IoT-based system called Vehicle Management Smart Internet (VMSI) system is defined. VMSI uses road traffic management techniques for mobile device, cloud computing, IoT and IT. The system aims to monitorg and manage the transport of vehicles in real-time.

In Peter, N. (2015) the author discusses fog computing and its applications in real time. This article shows that fog computers are able to handle the information that large numbers of IoT devices generate. Fog computation also demonstrates that noise and latency issues can be overcome. The paper also shows that fog is able to support an intelligent framework to handle modern IoT infrastructures, decentralized and real-time characteristics, and develop new network services that result in new market models and network operator opportunities.

The overview of fog integration with IoT was provided by M. Chiang and T. Zhang in Talib *et al.* (2017). They discussed new problems in the development of IoT systems and the difficulty in solving current IT and networking models. The paper then addressed the need for a modern computer, processing and networking infrastructure and how this technology could be used to build new business possibilities. The authors therefore investigated the characteristics and strengths of the fog model and proposed solutions to certain IoT issues.

In Puliafito *et al.* (2017) the authors discussed the significance for traffic control management of vehicle cloud computing in VANET. Traffic congestion produces additional negative effects. This influences daily work, health and the quality of life, drawing several different scientific concerns. In any urban planning, road traffic protection must be the key issue.

The overview of vehicles as communication and calculation infrastructure, known as vehicular fog computing (VFC), was presented in Yi *et al.* (2015). The authors proposed that this architecture be used to make communication and computation based on vehicle resources between end users easier. Four situations, such as interaction and network networks, involved moving and storing vehicles and a quantitative analysis of the skill of VFC were also explored.

Sookhak *et al.* (2017) proposed VFCs for the inefficient use of fog computing outlets by automobiles. Working out how decisions can be made and how different types of resource are shared across vehicles as fog knots, VFC suggested a cross layer architecture.

10.5 Fog and Cloud Computing-based Technology Applications in VANET

Because of the benefits of edge placement, fog computing will support low latency applications. For example, fog server-based gaming, increased real-time video stream processing and the fog-node. Like Vehicular Cloud Computing (VCC), fog computing appears to be in the first phase in its growth. Fog computation features for VANET are not as much like VCC software. VANET fog-based application scenarios are as follows:

1) Intelligent lights help to synchronize fog systems and give warning signals to approaching vehicles. In addition to WiFi, 3G, roadside units and intelligent

road lights, the interactions between vehicles and access points are improved. A large-scale traffic data collection with expanded spatial details has taken place. It is important for executing successful traffic policies to maintain a reasonable degree of coherence between the different aggregator points. Vehicles will affect the traffic lights as they drive by. A video device, for example, will instantly detect emergency flashing lights. Smart lights will help to combine fog systems for generating green traffic waves and sending warning signals to vehicles.

2) When the number of vehicles increases rapidly, traffic is in chaos. As a result, it is very difficult and expensive to find a parking space. The research by Kim was concerned with solving the problem of parking, reducing traffic congestion, minimizing air emissions and effective driving for IoT purposes. Fog computing and roadside clouds are therefore used to locate an empty parking spot. A parking space in many cases can be rented using these infrastructures. Throughout he analyzes the underlying principle to resolve the problem of parking. As well as helping drivers find an optimal available space, the initiative also supports the operators. It also presents drivers with an incentive. Simulations demonstrate that the method presented is a reliable solution for finding available parking lots.

3) Luan *et al.* (2015) applied fog computing software as an optimized broad-based network for the delivery of localized data. Figure 10.4 reflects the delivery of information through the use of fog wireless communications. A shop mounts a fog database in its parking lot to disperse a flyer; the store owner installs and manages the fog server to distribute the flyers. A bus station's fog server distributes bus information. Fog databases are dispersed to meet their own needs, and are built and run by different entities.

4) In contrast to the advancement of security products, the main focus for VANET systems is on the development of non-safety resources such as online gaming, interactive apps, video conferencing, and web sharing. Three layers rely on the VCC architecture: the vehicle interior, connectivity and the Internet. Vehicles on the road have been able to use excess computing power based on a cloud computing paradigm. It introduced the VCC concept. The VCC concept is a community of predominantly self-contained vehicles with the capacity to organize and delegate licensed users' business processing, sensing, interaction and physical resources. Examples of VCC application scenarios are:

- A datacenter airport.
- Cloud records space portion.
- Records center shopping mall.
- Management of active traffic signals.
- Traffic signals optimization.
- High-occupancy (HOV) self-organized streets.

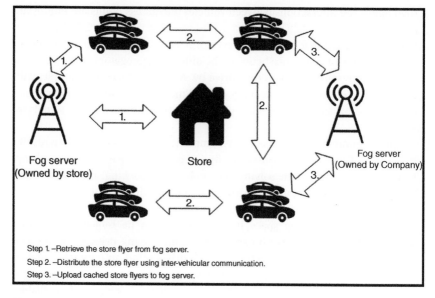

Figure 10.4 Information delivery through fog computing in VANET

- Evacuation management.
- Message for road safety.
- Eases congestion frequently.
- Management of parking.

10.6 Challenges of Fog Computing in VANET

This section describes and addresses major challenges in the sense of VANET fog computing. While the fog computing model compromises numerous advantages for various IoT applications, In terms of its successful implementation it faces numerous challenges.

1) **Calibration**: The number of cars is immense, producing a great deal of information, needing a great number of resources, including energy processing and storage. Fog servers should therefore be able to provide adequate backing for all these tools. The ability to respond to the speedy progress of automobiles and skills will be the real problem.

2) **Complexity**: Since many vehicles and sensors are designed by various manufacturers, the selection of the optimal components, including various configurations and specific criteria for software and hardware, is becoming

very complex. Furthermore, in some situations high safety applications require other equipment and protocols to work, which raises functional complexity.

3) **Dynamism**: The capacity to dynamically adjust speed and direction is one of the important features of VANET. This challenge will change VANET's internal characteristics and efficiency. Handheld devices often suffer from the ageing of software and hardware which leads to a shift in the actions and properties of operating flows. Fog nodes need to then reorganize the topology architecture and tools dynamically and smartly.

4) **Heterogeneity:** Most IoT products and sensors were produced by various manufacturers. Such systems have different communication capacities in radios, cameras, computer power, processing, *etc.* It will be important to manage and organize the networks of heterogeneous IoT devices and to choose the appropriate resources.

5) **Latency:** The primary reasons for cloud replacement with fog computing, especially in time sensitive applications, is to provide low latency. Nonetheless, there are many reasons that show high user latency or network quality on fog platforms. The high latency fog contributes to the dissatisfaction of the client.

6) **Management of Resources**: Fog terminals are often network appliances with extra power and storage. However, the capabilities of conventional databases and even the Internet are challenging for these tools to suit. For the effective action of the fog computing environment, responsive management of fog resources is thus important.

7) **Consumption of Power**: A large number of fog end devices are in the fog setting, the measurement is dispersed and can be less energy efficient than the hierarchical cloud computing method.

10.7 Issues of Fog Computing in VANET

Obviously, fog computing is a new technology that needs further research to meet all of these challenges. A summary of the open questions and direction of research related to fog computing and VANET integration is provided in this section.

1) **Fog and cloud transportations**: Fog processing is a cloud component, which is a main fog database device, which is mounted at diverse localities. The computer and system data are handled by the cloud. Cloud synchronized localized applications are provided only in fog. The cloud manages the whole device's code and information. In the Fog computing, the dual functionality of the cloud makes it difficult to contact information supply and change from cloud to fog during fog node generation. Choosing the right communication between fog and cloud, ensuring high performance and low fog-node latency, is a major challenge.

2) **Fog Network Communications:** Each fog database administers a pool of resources at various sites. In order to preserve facility delivery and content distribution among these providers, interaction and coordination between fog servers is essential. Improved interaction capacity would improve the performance of the whole network. For instance, in order to adapt to the various policies that are specified by holders, service policies are required where fog servers with various entities are deployed at several sites. In addition, connection functions must be taken into account when transmitting data between fog servers. Fog servers, in other words, must be able to connect *via* the wired or wireless Internet connection.

3) **Deployment of Fog Computing:** Fog computing builds extra computing and memory power on the edges of the network to effectively handle local service requests. When fog servers are based at various locations they have to change their control and maintenance process. The network operator of a fog network often needs to satisfy both IoT application requirements and fog service cooperation.

4) **Information Security:** Cloud computing has been researched and requires significant safety measures to protect the data. Nonetheless, because of different features such as flexibility, diversity and large-scale geo-distributor, these steps are not appropriate for fog computing. However, fog is an enticing tool for cyber criminals as the fog includes large volumes of sensitive data, both from the Internet and from IoT nodes. Further work is therefore needed to improve the protection of fog.

5) **Privacy for End User:** The security of the confidentiality of the end user is an important problem, because fog computing becomes closer to end users, allowing them to capture more sensitive data counting monetary proceedings, identities, use of services and locations. In addition, the maintenance of consolidated control is very difficult, because fog nodes are distributed across wide areas. Unprotected fog nodes may be an intruder entry point for accessing network information and stealing user data between fog entities. The security of the confidentiality of fog nodes needs further investigation.

10.8 Conclusion

This chapter presents an analysis of fog computing as future perspectives in VANET and represent fog computing as a border concept for VANETs. Moreover, this chapter describes and explores the structure, some fascinating software scenarios and problems in VANETs relevant to fog computing and cloud computing. This chapter also presents the concept and key features of fog computing and fog based technology applications in VANET. There are also difficulties in combining VANET with fog and transparent problems. The purpose of this chapter is to review the future perspective of fog computing in VANET and its implementations worldwide, as well as to explain potential avenues for study and open issues related to the integration of fog computing in VANET.

Bibliography

Zaidi, K., & Rajarajan, M. (2015). Vehicular internet: security & privacy challenges and opportunities. *Future Internet*, 7(3), 257–275.

Yang, F., Wang, S., Li, J., Liu, Z., & Sun, Q. (2014). An overview of internet of vehicles. *China communications*, 11(10), 1–15.

Whaiduzzaman, M., Sookhak, M., Gani, A., & Buyya, R. (2014). A survey on vehicular cloud computing. *Journal of Network and Computer applications*, 40, 325–344.

Bonomi, F., Milito, R., Zhu, J., & Addepalli, S. (2012, August). Fog computing and its role in the internet of things. In *Proceedings of the first edition of the MCC workshop on Mobile cloud computing* (pp. 13–16). ACM.

Olariu, S., Khalil, I., & Abuelela, M. (2011). Taking VANET to the clouds. *International Journal of Pervasive Computing and Communications*, 7(1), 7–21.

Alam, K. M., Saini, M., & El Saddik, A. (2015). Toward social internet of vehicles: Concept, architecture, and applications. *IEEE access*, 3, 343–357.

Hussain, R., Son, J., Eun, H., Kim, S., & Oh, H. (2012, December). Rethinking vehicular communications: Merging VANET with cloud computing. In *4th IEEE International Conference on Cloud Computing Technology and Science Proceedings* (pp. 606–609). IEEE.

Speed, C., & Shingleton, D. (2012, June). An internet of cars: connecting the flow of things to people, artefacts, environments and businesses. In *Proceedings of the 6th ACM workshop on Next generation mobile computing for dynamic personalised travel planning* (pp. 11–12). ACM.

Leng, Y., & Zhao, L. (2011, August). Novel design of intelligent internet-of-vehicles management system based on cloud-computing and internet-of-things. In *Proceedings of 2011 International Conference on Electronic & Mechanical Engineering and Information Technology* (Vol. 6, pp. 3190–3193). IEEE.

Peter, N. (2015). Fog computing and its real time applications. *International Journal of Emerging Technology and Advanced Engineering*, 5(6), 266–269.

Talib, M. S., Hussin, B., & Hassan, A. (2017). Converging VANET with vehicular cloud networks to reduce the traffic congestions: A review. *Int. J. Appl. Eng. Res*, 12(21), 10646–10654.

Puliafito, C., Mingozzi, E., & Anastasi, G. (2017, May). Fog computing for the internet of mobile things: issues and challenges. In *2017 IEEE International Conference on Smart Computing (SMARTCOMP)* (pp. 1–6). IEEE.

Yi, S., Hao, Z., Qin, Z., & Li, Q. (2015, November). Fog computing: Platform and applications. In *2015 Third IEEE Workshop on Hot Topics in Web Systems and Technologies (HotWeb)* (pp. 73–78). IEEE.

Sookhak, M., Yu, F. R., He, Y., Talebian, H., Safa, N. S., Zhao, N., … & Kumar, N. (2017). Fog vehicular computing: Augmentation of fog computing using vehicular cloud computing. *IEEE Vehicular Technology Magazine*, 12(3), 55–64.

Luan, T. H., Gao, L., Li, Z., Xiang, Y., Wei, G., & Sun, L. (2015). Fog computing: Focusing on mobile users at the edge. *arXiv preprint arXiv:1502.01815*.

11

An Overview to Design an Efficient and Secure Fog-assisted Data Collection Method in the Internet of Things

Sofia[1,], Arun Malik[2], Isha[3], and Aditya Khamparia[4]*

[1] *Research Scholar, Lovely Professional University, Punjab, India*
[2] *Associate Professor, Lovely Professional University, Punjab, India*
[3] *Associate Professor, Lovely Professional University, Punjab, India*
[4] *School of Computer Science and Engineering, Lovely Professional University, Punjab, India*

Abstract

This chapter outlines an idea to design an efficient data collection method in the IoT network. IoT technology deals with smart devices used to collect the data as well as to provide useful and accurate data to users. Many data collection methods already exist, but they still have some drawbacks and require further enhancements. This chapter outlines detailed information about designing a novel data collection method using fog computing in the IoT network. The main idea of using fog computing over cloud computing is to provide greater security to the data which is completely lacking in cloud computing. Today, security of data is one of the most necessary requirements. So, different security aspects are discussed in this chapter. Moreover, different aspects related to fog computing are also reviewed.

Keywords *Security; IoT; Fog computing; Data collection*

11.1 Introduction

The Internet of things (IoT), refers to the interconnection of various physical devices embedded with sensors. IoT technology deals with smart devices used to collect the data and also provide useful and accurate data to users. Due to the massive amount of data being generated by each device, a novel and secure data collection method is required in the IoT. There are various methods which

*Corresponding Author: sofiasingla97@gmail.com

Fog, Edge, and Pervasive Computing in Intelligent IoT Driven Applications, First Edition.
Edited by Deepak Gupta and Aditya Khamparia.

already exist, but more enhancements are needed to fetch data accurately and efficiently from devices and sensors. After the data has been collected, the next phase is storing the data in an efficient manner. Cloud and fog storage can be used for storing this, but cloud computing mainly uses a centralized mechanism which is completely lacking in security. Anyone can access the stored data in the cloud and no authentication is required to access cloud storage. However, in fog computing high security is present because it does not use the centralized mechanism and a large number of nodes or systems take part in the processing of data, making the system more complex and secure. At the time of collecting data, the biggest problem is latency. The data collection method should also be efficient in terms of latency. For improvements in latency, fog computing plays a vital role. In fog computing, fog or fog networks are generally very close to users geographically which helps to provide instant responses.

11.2 Related Works

In today's era, IoT smart devices are used to perform all tasks in an intelligent manner. The IoT is expanding more and more every day. Previously, more time was required to complete a task. But following the birth of the IoT, hard work was replaced with smart work. In the IoT network, a number of sensors are used which helps to sense the information and provides the output in a more intelligent way. Sensing of information and providing output comprises two techniques – data collection and data mining. Data collection is defined as the collection of data from various devices and data mining is providing accurate results to users.

Many data collection methods already exist and they provide various results to users in terms of resource consumption, security *etc.* To allow collection of data more securely and proficiently, two separate algorithms (message receiving and message extraction) have been proposed (Ullah, Said, Sher, & Ning, 2019). The message extraction algorithm fetches the message sensed from the devices whereas message receiving algorithm helps to receive the data at the destination. After the whole process is complete, aggregated data is stored at local repositories and later in global ones.

As network lifetime is one of the major problems in the IoT network, a novel multi-representative refusion method had been proposed to eliminate this problem (A. Liu *et al.*, 2017). Moreover it helps to collect the heterogeneous data from various devices. Results have been evaluated in terms energy consumption and network lifetime with the help extensive simulators. Later, a theory called compression sensing theory was proposed which is comprised of a construction of node clusters based on their characteristics and reconstruction of the data at the destination of the network. ADMM algorithm has been used for reconstruction of the data to maintain accuracy (G. Li *et al.*, 2019). This method provides a result which allows increase of the network lifetime.

As in an intelligent network, data nodes can be both stationary and mobile. For mobile devices, location tracking is the one of the most difficult tasks. That is why, for tracking mobile and wearable devices, a novel algorithm had been proposed (Alduais, Abdullah, & Jamil, 2019). This proposed algorithm considered three phases *i.e.* the initial phase, the IoT edge phase and the cloud phase. This method evaluates the result in terms of energy and is considered as an energy efficient method for tracking mobile and wearable devices (Alduais *et al.*, 2019).

Reliability is a basic parameter which should be in every phase of data collection. A hybrid scheme has been proposed for efficient and reliable transmission. This hybrid scheme is comprised of packet reproduction and hop-by-hop automatic repeat request (Zhang, Hu, & Long, 2019). and provides results using simulations.

While collecting data from various devices, data is usually collected in a heterogeneous way. In a network, various nodes are present which vary due to their characteristics and regions. These nodes provide results based on the needs of the user and characteristics which are heterogeneous in nature. So, to aggregate the heterogeneous data is a major problem in the IoT network. To resolve this problem, a model which based on the hidden Markov process had been proposed (S. Cheng, Li, Tian, Cheng, & Cheng, 2019). In this, researchers provided results based on power consumption. They used extensive simulators to achieve best level of accuracy.

When data is collected from the various devices, there can be delays due to pre-processing or processing mechanisms. To resolve this problem, communication mechanisms from many to one using probabilistic models has been proposed (Z. Li, Liu, Liu, Wang, & Liu, 2018). It takes care of the network lifetime, delays or intermediate delays, and energy consumption. In this mechanism, the TDMA (Time Division Multiple Access) approach was also involved.

Intelligent systems are being used in every application, but its major use is in healthcare applications. The biggest problem in healthcare is to maintain the confidentiality of the patient's sensitive data. Today, blockchain technology provides the best results in healthcare applications to secure the patient's data. In IoT, the HealthIIoT mechanism had been proposed (Hossain & Muhammad, 2016). This mechanism is a monitoring mechanism. It monitors the data collected and secures the data from various attacks by using various analytical techniques. Later, the APPA method was proposed by researchers to maintain the confidentiality of data and to secure the data from various attacks (Guan *et al.*, 2019). In this mechanism, fog assisted devices are used to take the real time data services for device updation and device registration.

Various data collection methods have been proposed, but some further enhancements are still required.

11.3   Overview of the Chapter

This chapter gives detailed information for designing a novel data collection method using fog computing. We have provided the best information by analyzing all aspects of the creation of data collection methods. The IoT follows a unique lifestyle comprising the following steps: collect, communicate, analyze , and act as shown in Figure 11.1.

This chapter is about data collection. Data collection is the very first step in the whole IoT lifestyle. It is one of the most challenging phases, because if faulty data is collected, then only faulty results will be obtained. So, there is a great need to design data collection methods with greater enhancements compared to existing methods. In this chapter, we explain the design of data collection method using fog computing.

Section 1.4 will explain what data collection is and why, as various data collection methods already exist, there is need to create more methods. It provides a tabular view of some existing data collection methods listing their advantages and disadvantages. Section 1.5 describes the basics of fog computing. It also explains why fog computing should be used to design an efficient and secure data collection method. It describes fog computing on the basis of various parameters like storage, latency, energy *etc.* Section 1.5.2 describes the complete design of fog computing and explains how data transmission takes place with the help of corresponding tiers in the architecture. Section 1.5.2 lists the benefits of using the fog computing mechanism in the whole IoT network. It explains various features such as security, latency, geographical distribution, decentralization, cost *etc.* If there are benefits, then there will obviously be some threats to steal those benefits. In the same manner, fog computing also has some threats *i.e.* security threats and privacy threats.

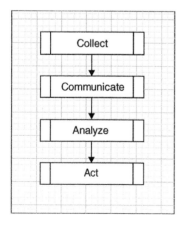

Figure 11.1   IoT lifestyle.

Section 1.5.5 explains the various areas where fog computing can be used in IoT networks. Section 1.6 describes the requirements for designing the data collection method after considering parameters such a security, resource efficiency, *etc.*

11.4   Data Collection in the IoT

The IoT network refers to the interconnection of the various physical devices or nodes having embedded sensors. These sensors make a device a smart device. These sensors play a vital role in collecting data from various devices or nodes. Data collection is one of the most challenging task in the IoT. As data will be collected from various devices or nodes, more time, energy and storage will be required. Various performance factors will be analysed for evaluation of the performance of the data collection. As this data collection phase is the most critical phase, optimal decisions need be taken. Numerous data collection methods already exist to provide the best results to users in various applications.

Example: the privacy protector framework and SW-SSS method were proposed to provide security on the data in the healthcare application. It provides the safety data with regard to various attacks (Luo *et al.*, 2018), the proxy-encryption method wad proposed to maintain the confidentiality of data while fetching or collecting, and storing the data from various devices. It can provide security for data both from insiders and outsiders (Wang, Xu, & Yang, 2018). Privacy preserving raw data collection was proposed in the preservation of raw data and removes the possibility of privacy leakage of data (Y. N. Liu, Wang, Wang, Xia, & Xu, 2019). These are some of the existing methods which have many pros as well as cons.

11.5   Fog Computing

The term fog computing defines the enhancement over cloud computing in which various heterogeneous, distributed and decentralized nodes communicate with one another and provides various services and efficient results to users without any need for a third party. Fog computing is always applied at the edge of a network. It doesn't disturb the whole network and does all of its processing at the edge of the network. It doesn't depend on centralized mechanisms, and it creates its own local repositories rather than depending on the cloud repositories. Fog computing methods and research direction is shown in Table 11.1

11.5.1   Why fog Computing for Data Collection in IoT?

One of the biggest problems that arises at the time of data collection is 'why fog computing?' As cloud computing is also available for data collection from various

Table 11.1 Some other current methods showing the advantages, disadvantages, and research directions.

Sr. No.	Method	Advantages	Disadvantages
1.	Context-awareness routing method (Shi, Adeel, Theodoridis, Haghighi, & Mccann, 2016)	Efficient energy and time consumption	Storage overflow problem is present
2.	LDPC (low density parity check) method (Jang, Jung, & Park, 2018)	Efficient energy and storage consumption is present. Acceptable delay while collecting the data is present	Security on data collection can be improved by using high-level encryption algorithms for securing the data from various attacks
3.	Concurrent tree method for data collection (C. T. Cheng, Ganganath, & Fok, 2017)	Negligible delay is present	Maximum storage and energy consumption is present
4.	SecureData data collection method (Hayajneh et al., 2018)	Securing the patients' sensitive data and providing efficient performance in terms of energy consumption, frequency and cost	Security of the data can be improved further from various advanced attacks/threats which would be applicable for various applications.
5.	Smart data collection using MQTT protocol (Cherradi, 2016)	Scalable and rapid IoT platform for the collection and transportation of the data	Easy target for high-end attacks

devices, what is the need to using fog computing over the cloud computing? Cloud computing works as a centralized mechanism and allows various users to gain access to different remote servers hosted on the internet for various functionalities such as storing and processing data. Fog computing is a platform which extends cloud computing functionalities with more efficiency and accuracy. Fog computing simply introduces a new layer between the user's nodes and cloud computing. This layer is known as the fog layer as shown in Figure 11.2. The fog layer contains all of the network devices such as routers, gateways, switches, access-points *etc.* It is a decentralized mechanism which has no centralized server. In this, all the fog nodes have self-organizing capabilities rather than having dependent mechanisms. Likewise, fog computing consumes less resources because it provides fog layer which is much closer to the IoT users. That is why fog computing saves storage space (by saving the most frequently used data in fog servers), uses less processing time (fog severs are very close to the user's node), provides better security (because no unauthorized user can access the data unlike in cloud computing), and provides mobility, unlike in cloud computing. So, this comprises the idea of using the fog computing over the cloud computing.

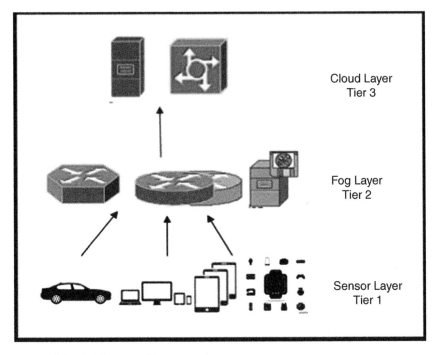

Figure 11.2　Architecture of fog computing.

11.5.2 Architecture of Fog Computing

Fog computing architecture is a three tier architecture, the three tiers used are the sensor layer, the fog layer and the cloud layer (Thareja & Singh, n.d.)

- **Sensor Layer:**
  The sensor layer is the tier 1 layer which is the lowest layer in the architecture of fog computing. It is also known as the user layer. Here all the devices of all users are present. In this layer, all devices are heterogeneous in nature and send heterogeneous data to the upper layer *i.e.* tier 2. In this layer, the devices have embedded sensors which helps to sense data from various sources. These devices collect the required data and after doing all pre-processing sends to the upper layer (Desai & Thakkar, n.d.).
- **Fog Layer:**
  The fog layer is the tier 2 layer which is the middle layer. This layer acts as the intermediate layer. It helps to keep the data in a homogeneous manner by creating clusters or using different fog nodes. This layer contains different network devices such as routers, gateways, access-points *etc.*
- **Cloud Layer:**
  Cloud layer is tier 3 and the top layer in the whole architecture of fog computing. It is one of the most important and sensitive layers in this architecture. It contains various data centres, large servers *etc.* It uses as the backend of the whole network but it is one of the most sensitive layers because it provides the centralized mechanism which can be an easy target for numerous attacks. It also provides the recovery of data when fog nodes experience failures.

11.5.3 Features of Fog Computing

The key features of fog computing over cloud computing are described below. It completely indicates the necessity to use fog computing over cloud computing. Figure 11.3 shows the various features of fog computing. By considering these features, an efficient and secure data collection method can be implemented.

- **Security:**
  Fog computing provides more security than cloud computing because cloud uses a centralized mechanism which allows unauthorized users to gain access to data, which violates the security principle, *i.e.* confidentiality. Fog computing also helps to filter malicious data at the edge network (Lu *et al.*, 2017).
- **Latency:**
  In fog computing, network latency is very low because the fog layer is very much closer to the sensor layer. As the cloud computing layer is far away from the user's layer then user data needs to travel more in the network and takes more time which is considered as delay (Rahman & Wen, 2018).

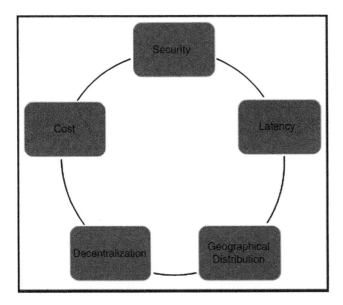

Figure 11.3 Features of fog computing.

- **Geographical distribution:**
  Fog computing has distributed geographical distribution. It offers the distributed deployment of nodes in order to provide the desired level of efficiency in the networks of both mobile and stationary devices to achieve better results (Dabhi, Raval, & Chaudhary, 2017).
- **Decentralization:**
  Fog computing is a decentralized mechanism because each node works independently rather than relying on a centralized infrastructure. In fog computing, no central server is present. All nodes have self-organizing capability by which they collaborate with one another and provide results to end users (Rahman & Wen, 2018).
- **Cost:**
  The cost in terms of resources is very small because fog computing consumes far fewer resources such as energy, time, and storage. As fog computing is one the distributed decentralized mechanisms there is no need for centralised infrastructures which reduces costs. In addition, fog computing has less computational cost (Rahman & Wen, 2018). For these reasons fog computing is much less expensive than cloud computing.

11.5.4 Threats of Fog Computing

There can be various threats in fog computing but the two broad categories in fog computing are threats to security and privacy.

- **Security threats:**
  As we know fog computing has fewer security threats than cloud computing. This is for two primary reasons. One is that in fog computing there is less need for Internet connection because the data is transferred locally from users to the fog devices or nodes which makes it harder for hackers or intruders to interpret the users' data. The second reason is that the data is not being transferred between the fog and cloud layers in real time which makes it difficult for intruders to keep track of users' sensitive data.
  However, experienced hackers can still breach the security of the data. Some security threats are given below:
  - **Jamming**: Jamming occurs when a large amount of fake or illegitimate messages occur and create congestion in the communication channel, blocking authorized user from various services of fog. As fog has fewer resources it is very common in the fog environment. It is also known as denial-of-service attack.
  - **Impersonation**: Impersonation is an attack in which an illegitimate or unauthorized user acts as an authorized or legal user. In this attack, either a fake user can utilize all of the resources on behalf of the legal user or a fake node can be inserted in the network which provides services to the legal user and then steal all of their information.
  - **Spam**: Spam is the addition or sending of fake messages, emails, sms to users or fog nodes which can destroy the normal behaviour of fog nodes and change it into abnormal behaviour.
  - **Tampering**: Tampering is an attack which produces a failure in the successful transmission of data to the destination. It either drops the data or modifies the data and creates a failure, the reason for which, is not understood by the user.
  - **Man in the middle attack**: This attack occurs when the intruder sits between the sender and the receiver. In this attack, the attacker concatenates its malicious code in the source's data to get the sensitive data of the whole communication. It can occur at the time of transmission of data from users to fog nodes or fog layer to destination nodes *i.e.* cloud layer.
- **Privacy Threats:**
  Today, a huge amount of data is being collected, stored and processed which needs a high level of confidentiality. As no user wants their data being viewed by a third party there should be no privacy leak. But still some intruders or hackers are able to access private data. Some privacy threats are listed below as shown in Figure 11.4:

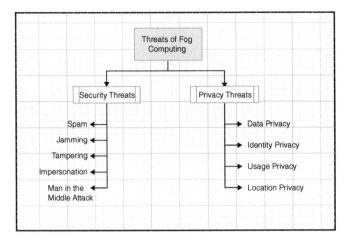

Figure 11.4 Threats of fog computing.

- **Location Privacy**: Due to social media networking, users sacrifice their location privacy. A user's location can easily be accessed while utilizing the services of the fog node. The fog node allows the user's location where they are connected to be easily trackable.
- **Usage Privacy**: Usage privacy is one of the most critical parameters because by intercepting the usage pattern an attacker can acquire information about the user's livelihood. Usage patterns can show when the user is usually transmitting data to the fog nodes, which type of data is mostly transmitted by the user, and how the data is being transmitted by the user.
- **Identity Privacy**: Breach of identity privacy leads to privacy leakage related to a user's sensitive data such as unique ID number, bank account details, login credentials *etc*. These data can be easily compromised while utilizing fog node services.
- **Data Privacy**: When a user transmits data to fog nodes, there can privacy leakage of that data. To avoid data privacy leakage, high-end data encrypting techniques should be used.

11.5.5 Applications of Fog Computing with the IoT

The IoT is expanding everywhere in the world. Fog computing is emerging with the IoT to increase the efficiency in the whole system. There exist various areas where fog computing is a vital requirement in the IoT network. So, a few basic applications of fog computing with IoT are given below as shown in Figure 11.5:

- **Smart Traffic:** Every day numerous accidents happen due to poor traffic management. To decrease or stop these accidents, intelligent systems need to be created. Fog computing has been emerging to create these intelligent systems.

Fog nodes have been used to send alerts to various users like pedestrians, cyclists *etc.* Fog nodes also provide the measure of distance and speed of vehicles which are close to traffic lights and those traffic lights share that information with other vehicles to manage traffic efficiently and to decrease the number of accidents.

- **Smart Home:** A smart home contains smart devices like smart appliances, smart security system, smart networks *etc.* For all these devices, fog computing is used. In smart security systems, fog nodes interpret abnormal activities which are very different from routine ones and generate alerts. In this way a home network becomes more secure.

- **Wireless Networks:** In wireless networks, wireless devices are used. These devices possess a self-organizing capability. They also include fog nodes. These fog nodes have used in designing various IoT based wireless sensor networks. Various parameters have been evaluated to get efficient results. Fog computing in IoT based wireless sensor networks makes the network energy efficient, storage efficient and delay efficient.

- **Healthcare:** In healthcare applications, fog computing is widely used. In this, one of the most important tasks is to provide security for patients' data. Fog computing emerges with the IoT and provides security in healthcare more smartly and by consuming fewer resources. So, fog computing is much more important in healthcare applications.

- **Cyber World Applications:** In the cyber world, numerous attacks happening every day. These attacks create problems to various organizations, other physical systems *etc.* So, fog nodes can develop intrusion detection and prevention capabilities to stop these attacks in the IoT network. Moreover, amalgamation of the IoT and cyber devices can create smart buildings, smart vehicle management, robotic innovations, agricultural innovations *etc.*

11.6 Requirements for Designing a Data Collection Method

Due to collection of data from various IoT devices, there will be some requirements to allow for a particular data collection method to work in an efficient way and be able to provide accurate results. Requirements vary according to the network, infrastructure, hardware, implementation environment *etc.* Some of these requirements are given below and their different perspectives are explained:

- One of the most important requirements is security of data. There are many encryption/decryption techniques. Data can be encrypted using various algorithms such as transposition cipher algorithms, AES, DES algorithms, caesar ciphers, hill cipher algorithms *etc.* Encryption can be done at the source

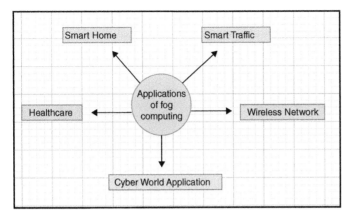

Figure 11.5 Applications of fog computing.

node and decryption at the destination. It increases the security of the data and there is much less chance of data hacking.

- As many networks form clusters according to their spatial correlation, regions *etc.* There is a very high possibility that an intruder will attempt to steal the data and mislead the user. That intruder can also start sending malicious code within the network which can create problems to other nodes. So, there needs to be an appropriate authentication mechanism for a particular node before entering to a cluster. Authentication can be a two-way authentication which will lead to higher security. A key management mechanism among different nodes can be used to provide the more security and higher authentication.
- Data transmission from source to destination should contain digital signatures with the data. Digital signatures are one of the most efficient techniques used to ensure the integrity of the data in the whole transmission process. Digital signatures contain hash values of the data which is sent with the data to the destination. At the destination, the hash of the same data is again calculated. If both the hashes (sent hash and calculated hash) match, then the data is accepted at the destination otherwise it is discarded.
- One requirement of every user is that the network should be resource efficient. There should be efficient energy consumption, storage consumption and time consumption. One of the most important factors on which they depend is the amount of data being transferred from the sender to the receiver. If huge data is being sent to the destination then it will consume more energy, more storage and even create more delay at the time of sending the data. So, for efficient transmission of data in the IoT network, data should be in compressed form. The benefit of sending the data in compressed form is that it allows huge data to be sent in

an efficient manner without any deduction in that data. Compressed data will consume less energy, less delay and storage due to its reduced form.

These are therefore some of the requirements which should be considered when designing a novel data collection method in the IoT.

11.7   Conclusion

This chapter described an overview which should be considered when designing a data collection method using fog computing. It explained why a user should use fog computing. It also described the features, threats and applications of fog computing. Fog computing architecture is also explained in this chapter to analyse how data transmission takes place while considering fog-enhanced systems. Later, the requirements for designing data collection methods were described. This chapter concludes that an efficient fog-assisted data collection method can be designed.

Bibliography

Alduais, N. A. M., Abdullah, I., & Jamil, A. (2019). An Efficient Data Collection Algorithm for Wearable / Mobile Tracking System in IoT /WSN. *2018 Electrical Power, Electronics, Communications, Controls and Informatics Seminar, EECCIS 2018*, 250–254. https://doi.org/10.1109/EECCIS.2018.8692815

Cheng, C. T., Ganganath, N., & Fok, K. Y. (2017). Concurrent data collection trees for IoT applications. *IEEE Transactions on Industrial Informatics*, 13(2), 793–799. https://doi.org/10.1109/TII.2016.2610139

Cheng, S., Li, Y., Tian, Z., Cheng, W., & Cheng, X. (2019). A model for integrating heterogeneous sensory data in IoT systems. *Computer Networks*, 150, 1–14. https://doi.org/10.1016/j.comnet.2018.11.032

Cherradi, G. (2016). Smart Data Collection Based on IoT Protocols, (November 2017).

Dabhi, A. A., Raval, T. J., & Chaudhary, K. (2017). Fog computing: A review and conceptual architecture, issues, applications and its challenges. *Pdfs.Semanticscholar.Org*, (5), 717–722. Retrieved from https://pdfs.semanticscholar.org/2f7e/41e41764f97cff39503118d0c2f94fd75e77.pdf

Desai, S., & Thakkar, A. (n.d.). *The Fog Computing Paradigm : A Rising Need of IoT World*. Springer Singapore. https://doi.org/10.1007/978-981-13-1610-4

Guan, Z., Zhang, Y., Wu, L., Wu, J., Li, J., Ma, Y., & Hu, J. (2019). APPA: An anonymous and privacy preserving data aggregation scheme for fog-enhanced IoT. *Journal of Network and Computer Applications*, 125, 82–92. https://doi.org/10.1016/j.jnca.2018.09.019

Hayajneh, T., Abdalla, A. N., Bhuiyan, M. Z. A., Zain, J. M., Hassan, M. M., & Tao, H. (2018). Secured Data Collection with Hardware-based Ciphers for IoT-based Healthcare. *IEEE Internet of Things Journal, PP*(X), 1–1. https://doi.org/10.1109/jiot.2018.2854714

Hossain, M. S., & Muhammad, G. (2016). Cloud-assisted Industrial Internet of Things (IIoT) - Enabled framework for health monitoring. *Computer Networks*, 101, 192–202. https://doi.org/10.1016/j.comnet.2016.01.009

Jang, J., Jung, I. Y., & Park, J. H. (2018). An effective handling of secure data stream in IoT. *Applied Soft Computing Journal*, 68, 811–820. https://doi.org/10.1016/j.asoc.2017.05.020

Li, G., He, J., Peng, S., Jia, W., Wang, C., Niu, J., & Yu, S. (2019). Energy efficient data collection in large-scale internet of things via computation offloading. *IEEE Internet of Things Journal*, 6(3), 4176–4187. https://doi.org/10.1109/JIOT.2018.2875244

Li, Z., Liu, Y., Liu, A., Wang, S., & Liu, H. (2018). Minimizing Convergecast Time and Energy Consumption in Green Internet of Things. *IEEE Transactions on Emerging Topics in Computing*, 6750(c), 1–16. https://doi.org/10.1109/TETC.2018.2844282

Liu, A., Liu, X., Wei, T., Yang, L. T., Rho, S. C., & Paul, A. (2017). Distributed multi-representative re-fusion approach for heterogeneous sensing data collection. *ACM Transactions on Emnedded Computing Systems (TECS)*, 16(3), 73.

Liu, Y. N., Wang, Y. P., Wang, X. F., Xia, Z., & Xu, J. F. (2019). Privacy-preserving raw data collection without a trusted authority for IoT. *Computer Networks*, 148, 340–348. https://doi.org/10.1016/j.comnet.2018.11.028

Lu, R., Member, S., Heung, K., Lashkari, A. H., Ghorbani, A. A., & Member, S. (2017). A Lightweight Privacy-Preserving Data Aggregation Scheme for Fog Computing-Enhanced IoT, *XX*(X), 1–10. https://doi.org/10.1109/ACCESS.2017.2677520

Luo, E., Bhuiyan, M. Z. A., Wang, G., Rahman, M. A., Wu, J., & Atiquzzaman, M. (2018). PrivacyProtector: Privacy-Protected Patient Data Collection in IoT-Based Healthcare Systems. *IEEE Communications Magazine*, 56(2), 163–168. https://doi.org/10.1109/MCOM.2018.1700364

Rahman, G., & Wen, C. C. (2018). Fog Computing , Applications , Security and Challenges , Review, 7(3), 1615–1621. https://doi.org/10.14419/ijet.v7i3.12612

Shi, F., Adeel, U., Theodoridis, E., Haghighi, M., & Mccann, J. (2016). OppNet : Enabling Citizen-Centric Urban IoT Data Collection Through Opportunistic Connectivity Service, 723–728.

Thareja, C., & Singh, N. P. (n.d.). *Role of Fog Computing in IoT-Based Applications*. Springer Singapore. https://doi.org/10.1007/978-981-13-5953-8

Ullah, A., Said, G., Sher, M., & Ning, H. (2019). Fog-assisted secure healthcare data aggregation scheme in IoT-enabled WSN. *Peer-to-Peer Networking and Applications*. https://doi.org/10.1007/s12083-019-00745-z

Wang, W., Xu, P., & Yang, L. T. (2018). Secure data collection, storage, and access in cloud-assisted Iot. *IEEE Cloud Computing*, 5(4), 77–88. https://doi.org/10.1109/MCC.2018.111122026

Zhang, J., Hu, P., & Long, J. (2019). A hybrid transmission based data collection scheme with delay and reliability guaranteed for lossy WSNs. *IEEE Access*, 7(c), 70474–70485. https://doi.org/10.1109/ACCESS.2019.2919355

12

Role of Fog Computing Platform in Analytics of Internet of Things- Issues, Challenges and Opportunities

Mamoon Rashid * *and Umer Iqbal Wani*

Assistant Professor, School of Computer Science and Engineering, Lovely Professional University, Jalandhar, India

Abstract

A huge amount of data is being generated in Internet of Things (IoT) devices used in smart homes, traffic sensors, smart cities and various connected appliances. Fog computing is one area which is quite popular in processing this huge amount of IoT data. However, there are challenges in these models for performing real time analytics in such data for quick analytics and insights. Fog analytics pipeline is one such area which can be a possible solution to address these challenges. In this chapter, an overview of using Fog computing platform for analyzing the data generated by IoT devices is provided. Fog computing platforms will be compared with the state-of-the-art to differentiate its impact in terms of analytics.

Keywords   *Internet of Things; Edge Computing; Real Time Analytics; Sensors*

12.1   Introduction to Fog Computing

Fog computing acts in between cloud-datacenters and various IoT devices/sensors. Its functioning is based on making connections between embedded devices and cloud server. It mostly provisions computational services as well as storing data, and networking close to the IoT devices/sensors (Mahmud, 2018). A large amount of data is generated by IoT-based applications. The data has to be analyzed for various decision-making processes. Transporting this data to be stored in the

*Corresponding Author: mamoon873@gmail.com, umer.23368@lpu.co.in

Fog, Edge, and Pervasive Computing in Intelligent IoT Driven Applications, First Edition.
Edited by Deepak Gupta and Aditya Khamparia.

cloud gives rise to a number of problems which include data lagging, more bandwidth usage, real-time response delay, data centralization *etc.* (Dastjerdi, 2016). To avoid the challenges encountered by IoT applications while communicating with the cloud, the term fog computing was coined (Cisco ,2012.)

Fog computing provides capabilities of processing data and storing capabilities locally (Atlam, 2018), without sending data to the cloud for permanent storage and processing. Fog computing lessens the amount of traffic and latency, by sending only portion of the data for cloud storage, which needs permanent storage, increasing both efficiency and reliability (Wen, 2017). With fog computing, services can be guaranteed to be delivered on time with consistent performance while overcoming problems that are found in cloud computing *e.g.*, performance overheads and communication delay.

Fog computing is based on the concept of edge computing. With edge computing, computation of data is done closer to the data generation source. Edge devices can't provide IoT application handling due to limited resources giving rise to contention of resources and increased latency. Fog computing binds devices located at edge and cloud-based resource, overshadowing edge computing's associated limitations. Apart from edge devices' processing, fog computing features and functionality can also be extended to the core of the network (Bonomi, 2012). The infrastructure related to computing in fog computing includes the edge as well as networking components that belong to the core of the network. Fog computing expands cloud services like PaaS, IaaS, and SaaS, and takes them nearer to the edge device.

Currently, too many devices are connected with one another and the number is increasing manifold, as time passes. According to estimations, there will be 6.58 devices per person (on average) by 2020. This calls for more processing and more storage growth exponentially. An integration mechanism has to be devised between devices and associated data services. Fog computing eliminates a number of problems which arise due to the centralized nature of the cloud computing paradigm.

The proliferation of IoT devices in our day-to-day lives have led to the generation of huge amounts of data, thus the need for storage, computation and processing has arisen. The data is processed in servers which are located at a distance from the end devices (Rashid, 2020). The distance between the end devices and data-center servers leads to time-consuming in data processing, and also affects the performance of the applications.

12.1.1 Hierarchical Fog Computing Architecture

The hierarchical fog architecture comprises the following three layers and is shown in Figure 12.1.

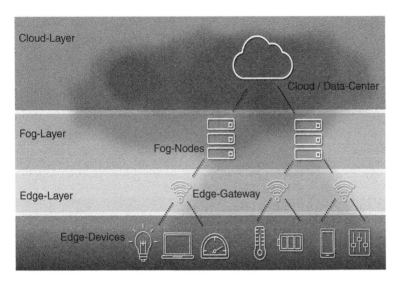

Figure 12.1 Hierarchical fog computing architecture.

Terminal Layer: One of the basic fog layers, it includes devices like, sensors, mobile phones, readers, smart vehicles, smartcards *etc*. This layer includes devices that can sense and also capture data. The devices are located at various places and distances from one another. The devices have the property of sensing events, data capturing and transferring it to higher layers for processing. Devices present in this layer span multiple platforms and architectures. Devices present in this layer are able to work in a heterogeneous environment, with other devices from different technologies.

Fog Layer: This layer consists of routers, gateways, access-points, base stations, and specific fog servers, all known as fog nodes. A fog node is located at the network edge, between end devices and cloud datacenters. A fog node can be located in a bus terminal or a coffee shop, known as a static node or can be fitted inside a moving vehicle, known as a moving node. Fog nodes deal with storing and performing computing on a temporary basis. Fog nodes are connected to cloud data centers through IP core network.

Cloud Layer: The cloud layer includes devices which possess high storage capabilities like high storage servers located in datacenters. The cloud layer provides unchanged and permanent data storage for a huge amount of data. The cloud layer includes datacenters that span multiple locations throughout the globe, comprising machines with both high processing capabilities and storage. The cloud layer provides all the basic cloud features to the users like scalability of service resources, ubiquitous access of services, on demand usage. The cloud layer is located at the far end of the fog architecture, at a larger distance from the end devices. It provides data back-up and permanent data storage. The data which is usually stored in the cloud layer isn't required for real time processing purposes.

12.1.2   Layered Fog Computing Architecture

Fog computing takes the computing power of large datacenters nearer to the edge devices. The computing and storage offered by fog is limited, these properties are offered in the middle layer, between the end devices and the cloud layer. The key motivation behind fog computing is to tackle issues related to latency in IoT-based applications (Mukherjee, 2020). The layered fog architecture of six layers is shown in Figure 12.2.

Physical and Virtualization Layer: Nodes are found in this layer – both physical and virtual. The nodes are distributed at a global level\and collect data. Sensing devices that are present in nodes perform sensing functions in the surrounding environment, capture data and transfer it to upper layers for more processing and filtering, *via* gateways. A node can be an independent device like a mobile phone or it can be located inside another device like a temperature sensor fitted inside a vehicle.

Monitoring Layer: A number of tasks related to nodes are taken into consideration here such as what type of tasks are being performed by the nodes, at what time is a node performing the task, *etc*. The performance, as well as the current status of applications, is also checked constantly. The amount of energy consumed by the fog node is also taken into consideration.

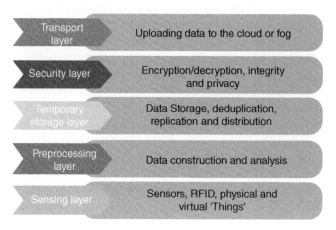

Figure 12.2   Layered fog computing architecture.

Pre-processing Layer: The pre-processing layer is used for the purpose of data-analysis. This involves cleaning data and filtering any unwanted data. The filtering of data is to assure that only important and useful data is left and all unimportant and less important data is removed.

Temporary Storage: The data is replicated and distributed on a temporary basis in this layer. The various storage devices used in this layer are FC, NAS, ISCSI *etc.* Once transferred to the cloud for permanent storage, data is no longer stored here.

Security Layer: The security layer mainly takes care of encryption and decryption of data. As well as this, the security layer deals with privacy and integrity. Privacy includes: privacy of data, privacy of user, privacy related to location of the user.

Transport Layer: The main function of the transport layer is to upload pre-processed and finely grained data in a secure way to the cloud layer. The data is uploaded for permanent storage purposes. The data is collected for efficiency purposes also. The use of smart gateways is to filter data before storing onto the cloud. Because of meagre fog related resources, the use of lightweight and efficient protocols is made.

12.1.3 Comparison of Fog and Cloud Computing

Cloud computing is a technology that has the features of being on-demand and scalable for a number of applications. It consists of a group of resources that are virtualized and are shared among a number of users (*e.g.* communication, computation, application, storage, and service). The resources are pooled in a centralized manner in big data centers. These resources can be allocated and de-allocated involving minimal management techniques (Cortés, 2015). The cloud provides more computing power than the computing power that can be made available near end devices. The cloud computing model assures an interference-free environment for computing, storage and other processing related activities.

With the generation of a large amount of data because of IoT enabled devices, it is not feasible or efficient to send data to and from cloud data-centers. Sending data to the cloud centers leads to increase in latency and gives rise to bottlenecks. Thus, for applications which are based on real time data manipulation like heath monitoring, emergency responses *etc.*, the option of storing data in cloud data-centers is not a viable and feasible option. The establishment of cloud data centers requires a large infrastructure investment, increasing the cost as well as energy consumption (Rashid, 2019).

The fog computing model is designed to address the issues which arise in the cloud computing paradigm. Fog computing promises to solve the problem of latency and thus an increase in bandwidth. The data will be processed and stored near to the end device. When data is processed nearer to the end devices, it leads

Table 12.1 Comparison of fog computing and cloud computing.

Parameters	Cloud Computing	Fog Computing
Real time interactions	Provides support	Provides support
Latency	High	Low
Geographical Distribution	Centrally based	Distributed in nature and thus decentralized
Mobility	Limited support	Supported
Mode of communication	IP-based network	Wireless communication: WiFi, 3G,4G, ZigBee, WLAN etc.
Bandwidth costs	High	Low
Energy consumption	High	Low
Working environment	Data center building	Outdoors (base stations), Indoors (cafes, houses)
Distance from end devices	Far	Nearer to the end devices
Number of server nodes	Few	Large number

to a decrease in response time. Fog computing makes the data processing and data storage available around the proximity of an end device. Fog computing is also considered to be a viable option when devices require frequent off-loading (Saharan, 2015). A comparison between cloud computing and fog computing is provided as shown in Table 12.1.

12.2 Introduction to Internet of Things

The IoT is a collection of a number of technologies like sensors, RFID, communication-based technologies and internet protocols. The basic idea is to have a network made up of physical objects with electronics, software, sensors and network connectivity that enables collection and exchange of data between these objects (Agarwal, 2016). Year-on-year, the number of IoT devices has increased by 31% totaling 8.4 billion in 2017 and by 2020 is estimated to reach 30 billion (Singh, 2015).

12.2.1 Overview of Internet of Things

Realizing the idea of the IoT, a large number of physical devices are being connected to the internet at a faster rate. The IoT, with the help of the Internet, enables

Figure 12.3 Representation of the Internet of Things.

objects to share data. An example is a smart home-based monitoring and control system. The devices act as smart objects in an IoT environment. IoT enabled objects are smart due to omnipresent and prevalent computing, the embedded nature of IoT devices, various communication technologies used in IoT devices, sensor networks, various internet applications and protocols (Atzori, 2010).

In future, the IoT will have a considerable impact on home and business applications, to enhance the quality of life and to boost the economy of the world (López, 2013). With smart homes, people will not have to get out of their cars to open their garages when reaching home, coffee can be prepared automatically, TVs and other appliances can be controlled remotely *etc*. In order to achieve such potential growth, market demands have to be met by emerging technologies and innovations and by the growth of service applications. In addition to this, devices that fit customer requirements, in terms of availability, need to be manufactured. Also, for compatibility between heterogeneous groups, new protocols are required. For companies to deliver quality products, a standard architecture for the IoT is the need of the hour. In addition, to match the challenges of the IoT, revised Internet architecture needs to be developed. For example, the vast collection of objects connecting to the Internet needs to be taken into consideration (Gantz, 2012). Security and privacy becomes another important requirement for the IoT due to the ability of IoT objects to monitor and supervise physical objects. To deliver high quality service at an efficient cost, management as well as monitoring of the IoT needs to take place.

12.3 Conceptual Architecture of Internet of Things

Due to the wide range of applications provided by the IoT numerous organizations want its products to be included in their business. However, practically this seems difficult because of the number of devices and conditions required to make it work. In other words, creating a stable architecture of the IoT becomes necessary. There are different approaches to IoT architecture. Conceptual architecture of IoT defines the basic constituents of IoT system, how they interact with each other and how data is processed and stored. This architecture is shown in Figure 12.4.

The different components that constitute this architecture are listed below.

Things: Things in the IoT include objects equipped with sensors and actuators. Sensors are devices used to collect data from an external environment which is later transferred to the network. Sensors may or may not be physically attached to the objects. Actuators are the devices that convert the command into actions. For example, an actuator may enable us to increase or decrease the rotational speed of an engine (Floyer, 2013).

Gateways: Gateways are used to connect the things with the cloud. These can be physical devices or software programs. All the data coming from or going to the cloud moves through the gateways. They pre-process the data going into the cloud to reduce its volume for easy processing. This also helps in controlling and transmitting the commands to the actuators (Krčo, 2014).

Cloud Gateways: Cloud gateways act as interfaces between the applications or things in the IoT system and the cloud-based storage. These gateways ensure compatibility with various protocols and communicate with field gateways using the protocol followed by that particular gateway (Yang, 2011).

Streaming Processor: The main aim of a streaming processor is to ensure the safe and efficient transmission of data to the data lake and applications.

Data Lake: The data lake is used to store the data coming from different objects in the same form as it is generated. The data is extracted from the data lake and loaded into the data warehouse for usable analysis (Tan, 2010).

Data Warehouse: The data warehouse is used to store the pre-processed and structured data only. This data is analyzed to gain meaningful insights. Data warehouse also stores the context information of sensors and things.

Data Analytics: Data analytics is used to analyze the data present in a data warehouse to find trends and make conclusions. It helps us know the performance of the devices or identifies inefficiencies in our IoT systems by analyzing the data.

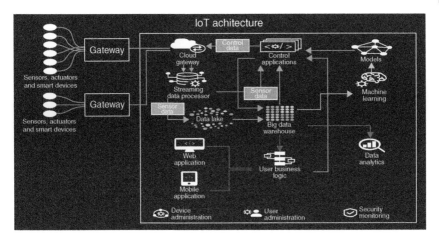

Figure 12.4 Conceptual architecture of IoT.

Machine Learning and Its Models: Machine learning enables us to create more accurate and effective models for controlling applications. These models are updated depending upon the data collected in data warehouses. These models are used for controlling applications once checked and analyzed by the data analysts (Koshizuka, 2010).

Control Apps: Control apps are used to send automated updates and instructions to the actuators. For example the windows in a smart house can open or close automatically by receiving the commands based upon weather updates, or in, the case of smart farming, soil is watered automatically by the watering system after receiving a particular command. Control apps are either controlled by machine learning models or by rules given by a specialist.

User Apps: User apps are used as an interface by the user to communicate with the IoT system. These apps help a user in controlling and monitoring the smart things. User apps can be web applications or mobile applications. They help the user in setting the automatic behaviour of smart things.

12.4 Relationship between Internet of Things and Fog Computing

Fog computing provides a connecting path between IoT devices and high-scale cloud computing services. Fog computing extends cloud-based services and takes them nearer to the end devices. The efficiency of IoT applications can be increased if the data generated is processed on a real-time basis with very low latency. Fog computing resolves a number of difficulties faced by IoT applications and provides a secure environment at the same time. The integration between fog and the IoT will bring a number of advantages like low latency, easy access, secure access *etc.* Fog will also provide efficient support for large sensor-based networks, which is the main problem currently faced by IoT applications.

12.5   Use of Fog Analytics in Internet of Things

Fog computing provides a number of advantages to the IoT applications and helps solve a number of problems faced by the IoT applications' environment.

Latency Constraints: One of the biggest problems faced by IoT applications is long time responses from the server units. Fog stores and analyzes data at a nearer location to the end devices, leading to decrease in latency.

Resource Constrained Devices: For devices where there is constraint on resources, fog computing performs operations that need huge resources. Saving cost in low power consumption and extending life cycle.

Uninterrupted Services: Fog computing can be made to run independently even when there is no connection for communicating with the cloud.

Security in IoT: For devices with limited resources, fog computing acts a proxy and updates software for the device and security credentials.

Bandwidth in IoT: Fog computing reduces the data which is meant for uploading in the cloud data-center. Depending on the application's demands, processing can be carried out nearer to the devices, reducing the bandwidth usage.

12.6   Conclusion

In this chapter, the authors have outlined an architecture of fog computing with emphasis on handling IoT-based huge data. Further the authors have tried to highlight the role of fog computing in handling IoT-based data. Comparison has been drawn between fog computing and cloud computing and key points were highlighted for the usage of fog computing replacing cloud computing. The authors believe that fog computing can have an important role in terms of analytics for IoT data.

Bibliography

Agarwal, S., Yadav, S., & Yadav, A. K. (2016). An efficient architecture and algorithm for resource provisioning in fog computing. International Journal of Information Engineering and Electronic Business, 8(1), 48.

Atlam, H. F., Walters, R. J., & Wills, G. B. (2018). Fog computing and the internet of things: a review. big data and cognitive computing, 2(2), 10.

Atzori, L., Iera, A., & Morabito, G. (2010). The internet of things: A survey. Computer networks, 54(15), 2787–2805.

Bonomi, F., Milito, R., Zhu, J., & Addepalli, S. (2012, August). Fog computing and its role in the internet of things. In Proceedings of the first edition of the MCC workshop on Mobile cloud computing (pp. 13-16). ACM.

Cortés, R., Bonnaire, X., Marin, O., & Sens, P. (2015). Stream processing of healthcare sensor data: studying user traces to identify challenges from a big data perspective. Procedia Computer Science, 52, 1004–1009.

Dastjerdi, A. V., Gupta, H., Calheiros, R. N., Ghosh, S. K., & Buyya, R. (2016). Fog computing: Principles, architectures, and applications. In Internet of Things (pp. 61–75). Morgan Kaufmann.

Floyer, D. (2013). Defining and sizing the industrial Internet. Wikibon, June, 27.

Gantz, J., & Reinsel, D. (2012). The digital universe in 2020: Big data, bigger digital shadows, and biggest growth in the Far East. IDC iView: IDC Analyze the future, 2007(2012), 1–16.

Koshizuka, N., & Sakamura, K. (2010). Ubiquitous ID: standards for ubiquitous computing and the internet of things. IEEE Pervasive Computing, (4), 98–101.

Krčo, S., Pokrić, B., & Carrez, F. (2014, March). Designing IoT architecture (s): A European perspective. In 2014 IEEE World Forum on Internet of Things (WF-IoT) (pp. 79–84). IEEE.

López, P., Fernández, D., Jara, A. J., & Skarmeta, A. F. (2013, March). Survey of internet of things technologies for clinical environments. In 2013 27th International Conference on Advanced Information Networking and Applications Workshops (pp. 1349–1354). IEEE.

Mahmud, R., Kotagiri, R., & Buyya, R. (2018). Fog computing: A taxonomy, survey and future directions. In Internet of everything (pp. 103–130). Springer, Singapore.

Mukherjee, M., Shu, L., & Wang, D. (2018). Survey of fog computing: Fundamental, network applications, and research challenges. IEEE Communications Surveys & Tutorials, 20(3), 1826–1857.

Rashid, M., Singh, H., & Goyal, V. (2019). Cloud Storage Privacy in Health Care Systems Based on IP and Geo-Location Validation Using K-Mean Clustering Technique. International Journal of E-Health and Medical Communications (IJEHMC), 10(4), 54–65.

Rashid, M., Singh, H., Goyal, V., Ahmad, N., & Mogla, N. (2020). Efficient Big Data-Based Storage and Processing Model in Internet of Things for Improving Accuracy Fault Detection in Industrial Processes. In Security and Privacy Issues in Sensor Networks and IoT (pp. 215–230). IGI Global.

Saharan, K. P., & Kumar, A. (2015). Fog in comparison to cloud: A survey. International Journal of Computer Applications, 122(3).

Singh, P., & Rashid, E. (2015). Smart home automation deployment on third party cloud using internet of things. Journal of Bioinformatics and Intelligent Control, 4(1), 31–34.

Tan, L., & Wang, N. (2010, August). Future internet: The internet of things. In 2010 3rd international conference on advanced computer theory and engineering (ICACTE) (Vol. 5, pp. V5–376). IEEE.

Wen, Z., Yang, R., Garraghan, P., Lin, T., Xu, J., & Rovatsos, M. (2017). Fog orchestration for internet of things services. IEEE Internet Computing, 21(2), 16–24.

Yang, Z., Yue, Y., Yang, Y., Peng, Y., Wang, X., & Liu, W. (2011, July). Study and application on the architecture and key technologies for IOT. In 2011 International Conference on Multimedia Technology (pp. 747–751). IEEE.

13

A Medical Diagnosis of Urethral Stricture Using Intuitionistic Fuzzy Sets

Prabjot Kaur* and Maria Jamal

Department of Mathematics, Birla Institute of Technology, India

Abstract

Medical diagnosis is a tedious, time consuming and costly procedure, however medical diagnosis with respect to the symptoms is not specific but can be done mathematically. To use the symptoms specifically for diagnosis we use linguistic variables. An Intuitionistic Fuzzy Method is used for diagnosing a disease named urethral stricture in patients most commonly males in the age group 20-40. To accomplish our purpose, first the medical data of different patients was collected from RIMS (Rajendra Institute of Medical Sciences), Ranchi under the supervision of a medical practitioner. The method of diagnosis is based on the relationship between patients and symptoms and the relationship between symptoms and diagnosis using linguistic variables by Intuitionistic Fuzzy Sets. The state of some patients after knowing the results of their medical tests is described by the degree of membership and degree of non-membership based on the relationship between patients and symptoms and the relationship between symptoms and diagnosis. Later we apply max-min-max composition and formula for calculating the Hamming distance in order to identify the disease with the least Hamming distance for various patients. We also apply revised max-min average composition to identify disease with the highest score. Finally, the diagnosis of urethral stricture of respective patients is mathematically shown.

Keywords *Fuzzy sets; Intuitionistic fuzzy sets; linguistic variable; hamming distance*

13.1 Introduction

The urethra is a tube which is used to carry urine from the bladder so that it can be expelled from the body. Usually the urethra is sufficiently wide for urine to flow

Fog, Edge, and Pervasive Computing in Intelligent IoT Driven Applications, First Edition.
Edited by Deepak Gupta and Aditya Khamparia.

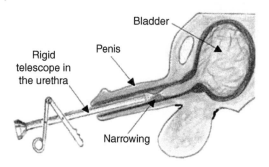

Figure 13.1 Urethral stricture.

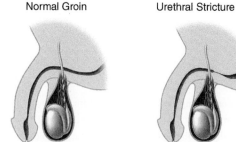

Figure 13.2 Normal groin *vs* urethral stricture.

easily through it. When the urethra narrows, it can restrict the flow of urine. This is known as a urethral stricture. Urethral stricture is a medical condition that mainly affects men. Urethral stricture involves constriction of the urethra. The narrowing of the urethra presents a multitude of symptoms and signs but most patients are unaware of any change in the force and calibre of their urinary stream. A doctor makes conclusions about the disease a patient is suffering after performing physical examinations and carrying out laboratory tests which makes diagnosis difficult and complex.

Zadeh (1965) proposed the fuzzy set theory. Fuzzy logic has found a wide application in medical sciences. Fuzzy sets let us better model imperfect information which is omnipresent in any conscious decision making. **Das and Borgohain (2010)** applied fuzzy soft set technology through the well-known Sanchez approach for medical diagnosis using arithmetic operations on fuzzy numbers. **Samuel-et-al (2012)** used fuzzy max-min composition to study the Sanchez approach for medical diagnosis (1976). **Sanchez** proposed an approach for medical diagnosis. **Juan-Carlor-Guz-Main, Patricia-Melin, German Prado (2016)** used a neuro fuzzy hybrid model for the diagnosis of blood pressure and **Sagir and Saratha** implemented an adaptive neuro fuzzy inference system to predict the degree of heart disease in patients. In decision making problems, especially in the case of medical diagnosis, there may exist a non-null uncertainty at all times of the evaluation of an unknown object. **Kim, Kim, Sorsich, Atanassov,**

Georgiev (1997) used intuitionistic fuzzy logic in decision making in medicine. **Titani** introduced 'intuitionistic fuzzy set theory' independently in 1984. The concept of intuitionistic fuzzy sets was introduced by **Atanassov (1986)**. **De et al. (2001)** gave an intuitionistic fuzzy sets approach in medical diagnosis. **Szmidt and Kacprzyk (2005)** started a new measure for intuitionistic fuzzy sets and its use in supporting medical diagnosis. **Samuel and Balamurugan (2013)** presented a study of Sanchez method of medical diagnosis using the notion of IFS (Intuitionistic Fuzzy Set) with fuzzy logic method. **Vassilev, Todorova and Surchev (2014)** used intuitionistic fuzzy estimates for a decision-making process in medicine. **Beg and Rashid (2015)** used trapezoidal valued intuitionistic fuzzy relations to solve a medical diagnosis decision making problem. To be more precise, intuitionistic fuzzy sets let us express *e.g.* the fact that the temperature of a patient changes, and other symptoms are not quite clear. It acts as a tool for a more human consistent reasoning under imperfectly defined facts and imprecise knowledge. This is the reason why we use IFS for medical diagnosis over other methods. We describe an attempt to diagnose urethral stricture using intuitionistic fuzzy set theory.

The organization of the paper is as follows: Section 1 gives a brief introduction of our problem. In Section 2, preliminaries of intuitionistic fuzzy sets are discussed. Section 3 elaborates the algorithm of medical diagnosis. Section 4 introduces a case study. Sections 5, 6 and 7 describe the results, conclusions and references.

13.2 Preliminaries

We give some basic definitions here which are used further in this paper.

13.2.1 Introduction

Fuzzy set theory defines set membership as a possibility distribution introduced in 1965 by L.A Zadeh. It was taken as a formal approach to deal with medical issues. But fuzzy sets lack the additional degree of freedom called non-membership function. IFS is an extension of fuzzy sets introduced by Krassimir Atanassov in 1983. The main advantage of using an IFS is its two degrees of freedom to handle uncertainty.

13.2.2 Fuzzy Sets

If Z is a collection of objects denoted by x, then a fuzzy set F in Z is defined by $F = \{x, \mu_F(x), \nu_F(x) \mid x \, \varepsilon \, Z\}, \mu_F : Z \rightarrow [0, 1]$

13.2.3 Intuitionistic Fuzzy Sets

An IFS F in Z is defined by

$F = \{(x, \mu_F(x), \nu_F(x) / x \varepsilon Z\}$

$\mu_F : Z \to [0, 1]$ and

$\nu_F : Z \to [0, 1]$

defines the degree of membership and the degree of non-membership of the elements $x \varepsilon Z$

$0 \le \mu_F(x) + \nu_F(x) \le 1$

13.2.4 Intuitionistic Fuzzy Relation

Let A and B be two sets. An intuitionistic fuzzy relation (IFR) R from A to B is an IFS of A × B characterized by the membership function μ_R and non-membership function ν_R.

An IFR R from A to B will be denoted by $R(A \to B)$.

13.2.5 Max-Min-Max Composition

Let $X(A \to B)$ and $Y(B \to C)$ be two IFR. The max-min-max composition YoX is the intuitionistic fuzzy relation from A to C, defined by membership function

$\mu_{YoX} (a,c) = \vee (\mu_X (a,b) \wedge \mu_Y (b,c))$

and the non-membership function

$\nu_{YoX} (a,c) = \wedge (\nu_X (a,b) \vee \nu_y (b,c) / \forall (a,c) \varepsilon A \times C$ and $\forall b \varepsilon B$

where,

$\vee$ = maximum value

$\wedge$ = minimum value

13.2.6 Linguistic Variable

A linguistic variable is a collection of words or sentences in a natural or artificial language which is an attribute having its value called linguistic values.

13.2.7 Distance Measure In Intuitionistic Fuzzy Sets

13.2.7.1 The Hamming Distance:

$$D_H(A, B) = \frac{1}{2} \sum_{i=1}^{n} \left(|\mu_A(x_i) - \mu_B(x_i)| + | \nu_A(x_i) - \nu_B(x_i) | + |\pi_A(x_i) - \pi_B(x_i)| \right)$$

13.2.7.2 Normalized Hamming Distance:

$$D_{n-H}(A, B) = \frac{1}{2n} \sum_{i=1}^{n} \left(|\mu_A(x_i) - \mu_B(x_i)| + | \nu_A(x_i) - \nu_B(x_i)| \right)$$

Table 13.1 Linguistic value and intuitionistic fuzzy values.

Linguistic Value	Intuitionistic Fuzzy Value
Low (L)	(0.15, 0.25)
Medium Low (ML)	(0.2, 0.3)
Medium (M)	(0.4, 0.4)
Medium High (MH)	(0.6, 0.3)
High (H)	(0.8, 0.2)
Very High (VH)	(0.9, 0.1)

13.2.7.3 Compliment of an Intuitionistic Fuzzy Set Matrix

Let $A = [a_{ij}] \ \varepsilon \ \text{IFSM}_{m \times n}$, where $a_{ij} = (\mu_j \ (C_i), v_j \ (C_i)) \ \forall \ i, j$. Then A^c is called an IFSCM if $A^c = [d_{ij}]_{m \times n}$, where $d_{ij} = (v_j \ (C_i), \mu_j \ (C_i)) \ \forall \ i, j$.

13.2.7.4 Revised Max-Min Average Composition of A and B (A Φ B)

$$A \ \Phi \ B = \text{Max} \ \frac{\mu_A(a_{ij}) + \mu_B(b_{jk})}{2}, \text{Min} \ \frac{v_A(a_{ij}) + v_B(b_{jk})}{2}, \forall \ i.j$$

13.3 Max-Min-Max Algorithm for Disease Diagnosis

In this section we present an application of IFS for medical diagnosis of a disease.

Let S be a set of symptoms, D a set of diagnosis and P a set of patients.

Now we come to intuitionistic fuzzy medical diagnosis. This methodology involves the following three steps:

1) The first step is to determine the symptoms.
2) The second step is to formulate the medical knowledge based on intuitionistic fuzzy relation.
3) The third step is to determine diagnosis on the basis of composition of intuitionistic fuzzy relations.

We apply IFS technology to develop a technique which helps in diagnosis of disease a particular patient is suffering from.

For this, we construct a mapping A: P → S. This gives us a matrix called patient-symptom matrix. Then we construct another mapping B: S → D. This gives us a matrix called symptom-diagnosis matrix. Now applying max-min-max composition we get another matrix called patient-diagnosis matrix. Lastly applying the Hamming distance formula we get crisp diagnosis matrix from where we can conclude which patient is suffering from which disease on the basis of least Hamming distance value.

13.4 Case Study

Consider a set of patients P = {Riya, Neha, Rahul, Sam} and set of symptoms S = {difficulty urinating, loss of bladder control, frequent urination, slow urine stream} and diseases having matching symptoms F = {urethral stricture, urinary tract infection (UTI), hematuria, diabetes}

The questionnaire used while collecting data from hospital is given below:

1) After urination did you feel that your bladder is not completely empty?
2) In 2 hours did you feel like urinating again?
3) Is it difficult to control urination?
4) Was the flow of urination very weak?
5) Did you pressurize yourself while urinating?
6) How many times do you get up to urinate at night?

Using max-min-max composition between Table 13.2 and Table 13.3 we get a relation between patients and diagnosis.

Using the Hamming distance formula between Table 13.2 and Table 13.3, we get the data given in Table 13.5.

Table 13.2 IFS relation between patients and symptoms.

Patients	Difficulty urinating	Loss of bladder control	Frequent urination	Slow urine stream
Riya	(0.9, 0.1)	(0.6, 0.3)	(0.9, 0.1)	(0.8, 0.2)
Neha	(0.8, 0.2)	(0.6, 0.3)	(0.8, 0.2)	(0.4, 0.4)
Rahul	(0.9, 0.1)	(0.8, 0.2)	(0.6, 0.3)	(0.6, 0.3)
Sam	(0.9, 0.1)	(0.6, 0.3)	(0.9, 0.1)	(0.8, 0.2)

Table 13.3 IFS relation between symptoms and diagnosis.

Symptoms	Urethral Stricture	Urinary Tract Infection (UTI)	Hematuria	Diabetes
Difficulty urinating	(0.9, 0.1)	(0.8, 0.2)	(0.8, 0.2)	(0.6, 0.3)
Loss of bladder control	(0.8, 0.2)	(0.4, 0.4)	(0.6, 0.3)	(0.2, 0.3)
Frequent urination	(0.9, 0.1)	(0.9, 0.1)	(0.8, 0.2)	(0.9, 0.1)
Slow urine stream	(0.9, 0.1)	(0.2, 0.3)	(0.15, 0.25)	(0.6, 0.3)

Table 13.4 IFS relation between patients and diagnosis.

Patients	Urethral Stricture	Urinary Tract Infection (UTI)	Hematuria	Diabetes
Riya	(0.9, 0.2)	(0.9, 0.3)	(0.8, 0.3)	(0.9, 0.3)
Neha	(0.8, 0.2)	(0.8, 0.3)	(0.8, 0.3)	(0.8, 0.3)
Rahul	(0.9, 0.2)	(0.8, 0.3)	(0.8, 0.25)	(0.6, 0.3)
Sam	(0.9, 0.2)	(0.9, 0.3)	(0.8, 0.3)	(0.9, 0.3)

Table 13.5

Patients	Urethral Stricture	Urinary Tract Infection (UTI)	Hematuria	Diabetes
Riya	0.0	0.05	0.025	0.05
Neha	0.0	0.05	0.05	0.075
Rahul	0.0	0.025	0.05	0.075
Sam	0.0	0.1	0.1875	0.05

13.5 Intuitionistic Fuzzy Max-Min Average Algorithm for Disease Diagnosis

Methodology:

1) Construct a mapping A: P → S called a patient-symptom matrix. Next, construct another mapping B: S → D called a symptom diagnosis matrix.
2) Obtain A^c, B^c.
3) Compute the intuitionistic fuzzy max-min average compositions A Φ B and A^c Φ B^c
4) Compute the matrices V, W and with the help of them obtain the score matrix S(A, B) where

$$S(A, B) = \frac{V + W}{2}$$

$$V = [(\mu_{A \; \Phi \; B} - \nu_{A^c \; \Phi \; B^c})]$$

$$W = \lfloor (\mu_{A^c \; \Phi \; B^c} - \nu_{A \; \Phi \; B}) \rfloor$$

Table 13.6 A Φ B.

Patients	Urethral Stricture	Urinary Tract Infection (UTI)	Hematuria	Diabetes
Riya	(0.9, 0.1)	(0.85, 0.1)	(0.85, 0.15)	(0.9, 0.1)
Neha	(0.85, 0.15)	(0.85, 0.15)	(0.8, 0.2)	(0.85, 0.15)
Rahul	(0.9, 0.1)	(0.85, 0.15)	(0.85, 0.15)	(0.75, 0.2)
Sam	(0.9, 0.1)	(0.9, 0.1)	(0.85, 0.15)	(0.9, 0.1)

Table 13.7 A^c Φ B^c.

Patients	Urethral Stricture	Urinary Tract Infection (UTI)	Hematuria	Diabetes
Riya	(0.25, 0.7)	(0.35, 0.5)	(0.3, 0.475)	(0.3, 0.4)
Neha	(0.25, 0.65)	(0.35, 0.3)	(0.325, 0.275)	(0.35, 0.4)
Rahul	(0.2, 0.75)	(0.3, 0.4)	(0.275, 0.375)	(0.3, 0.5)
Sam	(0.25, 0.7)	(0.35, 0.5)	(0.3, 0.475)	(0.3, 0.4)

5) Identify the maximum score for each patient for selection of appropriate diseases.

By using compliment of an IFS matrix we get A^c and B^c.
Now we find A Φ B from Tables 13.2 and 13.3.
Now we find A^c Φ B^c
Then we find S(A, B) with the help of V and W

13.6 Result

Here the Hamming distance approach is considered for each patient and the lowest Hamming distance provides us with a proper diagnosis. Patients 1, 2, 3 and 4 all are suffering from urethral stricture. The least Hamming distance is associated with urethral stricture which compliments the diagnosis of urethral stricture while the highest Hamming distance is associated with urinary tract infection in 2 cases, hematuria in 1 case and diabetes in 1 case. In case of revised max-min average composition also, all patients are suffering from urethral stricture shown by the maximum value in each row of Table 13.8 which provides us with correct diagnosis.

Table 13.8

Patients	Urethral Stricture	Urinary Tract Infection (UTI)	Hematuria	Diabetes
Riya	0.35	0.3	0.2625	0.175
Neha	0.375	0.15	0.325	0.325
Rahul	0.3	0.125	0.3	0.175
Sam	0.35	0.325	0.2625	0.175

13.7 Code for Calculation

```
C Code for finding Max-Min-Max composition
#include <stdio.h>
#include <stdlib.h>
#include < math.h>
float maxoffour (float x, float y, float z, float t)
{
float max;
if (x > = y)
{
if (x > = z)
{
if (x > = t)
max = x;
}
}
   if  (y > = x)
            {
                    if  (y > = z)   .
                    {
                            if  (y > = t)
                            max = y;
                    }
            }
            if  (z > = x)
            {
                    if  (z > = y)
                    {
                            if  (z > = t)
```

```c
                                        max = z;
                            }
                    }
                    if  (t > = x)
                    {
                            if  (t > = y)
                            {
                                    if  (t > = z)
                                    max = t;
                            }
                    }
                    return max;
            }
float min (float g, float h)
            {
            float min1;
            if  ( g < = h)
            min1 = g;
            else
            min1 = h;
            return min1;
            }
            void main ( )
            {
            float a[5][5], b[5][5], c[5][5][5], d[5][5];
            int i, j, k;
            printf ("Enter the membership function of
patient cross symptom matrix =")
            for ( i = 1; i < = 4; i++)
            {
                    for ( j = 1; j < = 4; j++)
            {
            printf ("Enter the value of a[%d][%d]=", i, j)
            scanf ("%f", &a[i][j]);
            }
            }
            printf ("Enter the membership function of
patient cross diseases matrix =")

                    for ( i = 1; i < = 4; i++)
            {
```

```
                for ( j = 1; j < = 4; j++)
    {
            printf ("Enter the value of b[%d][%d]=", i, j)
            scanf ("%f", &b[i][j]);
            }
            }
            for ( i = 1; i < = 4; i++)
            {
            for ( j = 1; j < = 4; j++)
            {
            for ( k = 1; k < = 4; k++)
            {
            c[i][j][k] = min(a[i][k], b[k][j]);
            printf ("%f", c[i][j][k]);
            }
            }
            }
                    for ( i = 1; i < = 4; i++)
            {
            for ( j = 1; j < = 4; j++)
            {
            k = 1;
            d[i][j] = maxoffour (c[i][j][k], c[i][j]
[k + 1], c[i][j][k + 2], c[i][j][k + 3]);
            printf ("d[%d][%d] = %f", i, j, d[i][j]);

            printf (" /n" );
            }
            }
```

C code for finding the Hamming distance

```
#include <stdio.h>
#include <stdlib.h>
#include < math.h>
void main ( )
{
float a[4][5], b[4][5], d[4][5], e[4][5], t[4][5],
g[4][5], p[4][5], z[4][5], m[4];
int r, c, i, j, n;
printf ("Enter the no of rows =");
```

```
scanf ("\%d", &r)
printf ("Enter the no of columns =");
scanf ("%d", &c)
printf ("Enter the no of symptoms =");
scanf ("%d", &n)
printf ("Enter membership function for matrix A =");
for ( i = 1; i <= r; i++)
{
for ( j = 1; j <= c; j++ )
{
printf ("Enter a[%d][%d] = ", i, j);
scanf ("%f", &a[i][j]);
}
}
printf ("Enter non-membership function for matrix A =");
           for ( i = 1; i <= r; i++)
           {
           for ( j = 1; j <= c; j++)
           {
           printf ("Enter b[%d][%d] = ", i, j);
           scanf ("%f", &b[i][j]);
           }
           }
           printf ("Enter membership function for matrix
B =");
           for ( i = 1; i <= r; i++)
           {
           for ( j = 1; j <= c; j++)
           {
           printf ("Enter d[%d][%d] = ", i, j);
           scanf ("%f", &d[i][j]);
           }
           }
           printf ("Enter non-membership function for
matrix B =");
           for ( i = 1; i <= r; i++)
           {
           for ( j = 1; j <= c; j++)
           {
           printf ("Enter e[%d][%d] = ", i, j);
           scanf ("%f", &e[i][j]);
           }
```

```
            }
    for (i = 1; i <= r; i++)
            {
            for ( j = 1; j <= c; j++)
            {
            z[i][j] = fabs( ( 1 - (a[i][j] + b[i][j]) ) -
(1 - (d[i][j] + e[i][j]) ) );
            g[i][j] = fabs( (a[i][j] - d[i][j]) );
            m[i][j] = ( ( b[i][j] - e[i][j]) );
            p[i][j] = z[i][j] + g[i][j] + m[i][j];
            t[i][j] = p[i][j] / 2 X n ;
            printf ("t[%d][%d] = %f", i, j, t[i][j]);
            printf ("/n");
            }
            }
            printf ("Matrix of Hamming Distance =");
                printf ( "/n");
                for ( i = 1; i <= r; i++)
                {
                for ( j = 1; j <= c; j++)
                {
                printf ("%f", t[i][j]);
                }
                printf ("/n");
                }
                getch ( );
                }
```

13.8 Conclusion

In this chapter we have applied IFR, max-min-max composition and the Hamming distance as well as revised max-min average composition formula along with Sanchez method of medical diagnosis. A case study has been given to simplify the whole procedure or technique. IFS can process imprecise data as well with the help of the degree of membership and non-membership. The non-membership function has a much more important role compared to the membership function because in medical diagnosis there is an absolute chance of the existence of a non-zero hesitation part. Using max-min-max composition, revised max-min average composition and Hamming distance formula it was easy to diagnose a disease.

13.9 Acknowledgement

This work was partially supported by Dr. Arshad Jamal, Department of Urology, Rajendra Institute of Medical Sciences, Bariatu, Ranchi, Jharkhand and Dr. S.R Sharma, Assistant Professor, Department of Bioengineering, BIT Mesra(Ranchi). We express our deep gratitude to them.

Bibliography

1 K. Atanassov, Intuitionistic Fuzzy Sets, *Fuzzy Sets and Systems* 20(1986) 87–96.
2 K. Atanassov, More on Intuitionistic Fuzzy Sets, *Fuzzy Sets and Systems* 33(1989) 37–46.
3 Anthony Shannon, Soon Ki Kim, Young Hyun Kim, Joseph Sorsich, Krassimir Atanassov, Peter Georgiev, 1997, *A possibility for implementation of elements of the intuitionistic fuzzy logic in decision making in medicine*, vol 3(4),40–43.
4 Gary T, Anderson et al, 1999, A rough set/fuzzy logic based decision making system for medical application, vol 29, 879–896.
5 Supriya Kumar De, Ranjit Biswas and Akhil Ranjan Roy, 2001, An application of Intuitionistic Fuzzy Sets in Medical diagnosis, vol 117(2), 209–213.
6 Anjan Mukherjee, Sadhan Sarkar, 2002, Distance based similarity measure for interval valued *Intuitionistic Fuzzy Set and its application*, vol 3(4), 34–42.
7 Eulalia Szmidt and Janusz Kacprzyk, 2004, *Medical Diagnostic Reasoning using a similarity measure for Intuitionistic Fuzzy Sets*, vol 10, 61–69.
8 Eulalia Szmidt, Janusz Kacprzyk, Mikhail Matveev, 2005, A new similarity measure for *Intuitionistic Fuzzy Sets and its use in supporting a medical diagnosis*, vol 11(4), 130–138.
9 P.K.Das and R.Borgohain, 2010, *An application of Fuzzy Soft Set in medical diagnosis using fuzzy arithmetic operations of Fuzzy Number*, vol 5, 107–116.
10 Jeong Yong Ahn, Kyung Soo Han, Young Oh and Chong Deuk Lee, 2010, *An application of interval valued Intuitionistic Fuzzy Sets for medical diagnosis of headache*, vol 7, 2755–2762.
11 A. Edward Samuel et al and M.Balamurugan, 2012, Fuzzy max-min composition technique in medical diagnosis, vol 6, 1741–1746.
12 Vijay Kumar, Isha Bharti, Y.K Sharma, 2012, Fuzzy diagnosis procedure of types of glaucoma, vol 1(6), 42–45.
13 A. Edward Samuel and M.Balamurugan, 2013, Intuitionistic fuzzy set with ordering technique in medical diagnosis, vol 7(6), 751–753.
14 S.Elizabeth and L.Sujatha, 2013, *Application of fuzzy membership matrix in medical diagnosis and decision making*, vol 7(127), 6297–6307.

15 Nur Alom Talukdart, Daizy Deb, Sudipta Roy, 2014, *Automated blood cancer detection using image processing based on fuzzy system*, vol 4(8), 467–472.

16 P.A Ejegwa et al, 2014, *A note n some models of intuitionistic fuzzy sets in real life situations*, vol 2(5), 42–50.

17 U Dev, A Sultana, D Saha and N.K Mitra, 2014, Application in fuzzy logic in medical data *interpretation*, vol 49(3), 137–146.

18 Peter Vassilev, Lyudmila Todorova and Jivko Surchev, 2014, *Determining intuitionistic fuzzy estimates for decision making in medical tasks*, vol 20(5), 62–68.

19 Ismat Beg and Tabasam Rashid, 2015, A system for medical diagnosis based on intuitionistic fuzzy relation, vol 21(3), 80–89.

20 Ejiofor C.I and Okengwu U.A, 2015, *Simplified fuzzy approach for cataract identification*, vol 2, 626–630.

21 Juan Carlor Guzmain, Patricia Melin, German Prado-Arechiga, 2016, *Neuro fuzzy hybrid model for the diagnosis of blood pressure*, vol 667, 573–582.

22 Tran Manh Tuan, Nguyen Thanh Duc, Pham Van Hai and Le Hoang Son, 2017(in press), *Dental diagnosis from X-ray images using fuzzy rule based systems*, vol 6(1), 1–16.

23 Sagir, Abdu Masanawa, Sathasivam, Saratha, 2017, *A novel adaptive neuro fuzzy inference system based classification model for heart disease prediction*, vol 25(1), 43–56.

14

Security Attacks in Internet of Things

Rajit Nair[1,], Preeti Sharma[2], and Dileep Kumar Singh[3]*

[1] *Jagran Lakecity University*
[2] *Bansal College of Engineering*
[3] *Jagran Lakecity University*

Abstract

An Internet of Things (IoT) environment involves different types of devices such as IoT devices/end nodes, computing devices, gateways/edges and cloud servers. Initially IoT devices or end nodes such as smart home appliances, smart cars, computers, etc., pass information to the cloud, after that information is processed through different machine learning or artificial intelligence approaches through which the client can gain valuable information. It is the capability of the device that makes the IoT brilliant and this has been achieved by putting the intelligence into the devices. The intelligence in the sensors is developed by adding sensors and actuators which can collect information and pass it to the cloud through Wi-Fi, Bluetooth, ZigBee, and so on. But these IoT networks are vulnerable to different types of attack such as physical attacks, network attacks, software attacks, and encryption attacks, these attacks divert the IoT devices from performing their normal operations. So, it is very important that we overcome these attacks. This chapter will discuss all types of attack that occur in the IoT network with their counter measures, and will also cover the different layers, protocols, and the security challenges related to the IoT.

Keywords *sensors; layers; protocols; Radio frequency (RF); encryption; edges; wireless network*

*Corresponding Author: rajitnitbpl@gmail.com

Fog, Edge, and Pervasive Computing in Intelligent IoT Driven Applications, First Edition.
Edited by Deepak Gupta and Aditya Khamparia.
© 2021 John Wiley & Sons, Inc. Published 2021 by John Wiley & Sons, Inc.

14.1 Introduction

Today the IoT (Internet of Things) is one of the emerging fields in which different objects are connected together with some sensors and electronic devices through communication over wireless networks [1]. Basically, the IoT connects smart things such as intelligent devices and sensors. Most of the data collected by smart things are sent to cloud-based services so that they can process all the data collected and share this information with multiple users. But this IoT environment is vulnerable to different types of attack [2]. Due to these attacks or threats there has been an increase in interest related to understanding the emerging cyberthreats. Vulnerability in IoT devices occurs due to a combination of heterogeneous components, weak security protocols, naïve security configuration, etc. In dedicated and diversified communication protocols there can be a limited number of nodes in the IoT due to which their ability to protect themselves get weaker. In this chapter the discussion will be based on different types of attacks in the IoT, we already know that already many types of attacks have existed for a very long time. So, we are interested in what is new in the attacks in the IoT? Millions of devices are connected together in the IoT; almost all of them are potential victims to traditional attacks or threats. The core concept behind the IoT is to connect different devices through a network so that they can perform tasks but at the same time when new devices are connected they create a new entry path in the network which will impose a risk to security and privacy.

Attacks in the IoT follow the same approach as in the past but the impact of these attacks may be different depending on various factors such as the architecture, devices, and the environment in which the system has been built. IoT frameworks can be assaulted effectively from various perspectives.

This chapter will aim to find all the possible issues in the IoT, cover the classification of common types of cyber-attacks and how these attacks have given rise to an abnormality in the working of the IoT. Solutions will also be discussed that will explore the potential of technologies like block chain attacks, software defined networks and many others. The chapter will start from the IoT reference model with its communication protocol and also cover the security issues in each layer with their existing solutions.

14.2 Reference Model of Internet of Things (IoT)

Initially during the emergence of the IoT there were three layers in the architecture, the perception layer, transport layer, and application layer but some researchers have further divided the application layer into two layers, i.e., the processing layer and application layer. This is done because the processing layer is the one which makes strategic decisions in the cloud so it has to be dealt with

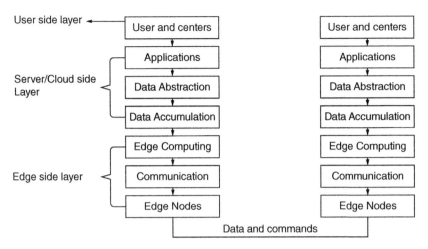

Figure 14.1 Reference Model of the Internet of Things (IoT).

separately rather than merging with the application layer. This reference model therefore becomes a four layer architecture [3].

Later there have been some work done by Cisco in which they have defined a seven layer IoT reference model in which communication is bidirectional and the flow of data and commands will pass from application layer to edge node layer in a control system, and in case of a monitoring system the flow will be from node layer to application layer.

Descriptions of each layer are as follows:

1. Edge side layer – This layer consist of computing nodes like sensors [4], RFID readers, controllers and actuators. This layer is basically used to provide integrity and confidentiality of the data that is sent across the network. It is further subdivided into three layers which are defined below:
   - Edge nodes – This layer consists of different sensors and various other devices that collect information from different sources and monitors the network. These are the physical devices and controllers that can control various devices. Mainly these are the things in the IoT and they have different end points which can send and receive information. In today's scenario the list of IoT devices is broad and sometimes it can become difficult to add more equipment to the IoT. This happens because they are diverse in nature and they have no protocols about size, location, origin, etc. The IoT has to support the entire range of devices even when some of the devices are, such as a silicon chip or large, such as a vehicle. There are many manufacturers who are producing IoT devices at large scale and to maintain compatibility and manufacturability, IoT reference model will explain the processing level needed from level 1 devices.

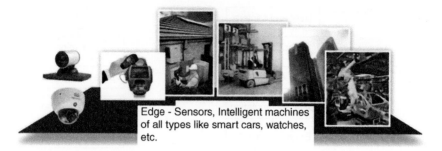

Edge - Sensors, Intelligent machines of all types like smart cars, watches, etc.

Figure 14.2 Edge nodes.

- Communication – Although it is a communication level, connectivity is also concentrated in the same level, i.e., level 2. The communication layer consists of objects that can be used for communication in the other levels, that is, the first level, second level and third level. The most important function of this layer is the reliability and timely transmission of information. Transmission is done from devices, i.e., level 1 to the network and it passes across the network. Through this layer the flow of information is done from level 2 to low level information processing at level 3. Generally, the IoT reference model does not require or indicate creation of a different network, it mostly relies on the existing network. It has been observed that some of the legacy devices are not Internet Protocol (IP) enabled in that case they need communication gateways. It might also be possible that connectivity devices in level 1 can change at level 2 during communication. Regardless of the details, level 1 devices communicate through the IoT system by interacting with level 2 connectivity equipment, as shown in Figure 14.3. In this layer functionality focuses on east–west communication. Some of key tasks performed by layer 2 are as follows:
  a) Communicating with and between the level 1 devices.
  b) Implementation of various protocols.
  c) Security at the network level.
  d) Switching and routing.
  e) Reliable delivery across the networks.
  f) Translation between protocols.
  g) Self-learning or networking analytics.

Figure 14.3 Communication layer.

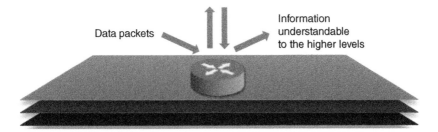

Figure 14.4 Fog computing layer.

- Edge or fog computing – Edge computing is very similar to fog computing in that it mainly initiates data processing and reduces the computation load at next level for fast response. This layer consists of data preprocessing methods and algorithms so that real time applications can perform computation with edge nodes close to the network. The function of this layer is to convert the network data into information that is suitable for storage and processing by higher level, i.e., level 4. The main focus will be on high volume data analysis and transformation. Let us take an example: a sensor device at level 1 will generate data samples multiple times per second, full day and a whole year. A general ideology of the IoT reference model is to initiate information processing as early and as close to the edge of the network as possible in the intelligent system. Layer 3 performs this task and is sometimes referred to as fog computing. The data which is usually submitted at level 2 by devices in small units are processed later on at level 3 on a packet-by-packet basis. The processing done at this level is limited as they have only awareness of data but not about sessions or transactions. Level 3 processing can encompass many examples as follows:
    a) Evaluation: Data are evaluated according to criteria so that they could be processed at a higher level.
    b) Formatting: Data are reformatted for consistent higher-level processing.
    c) Expanding/decoding: Cryptographic details are handled with additional context.
    d) Distillation/reduction: Data is reduced or summarized to impact on the traffic of the network and higher level processing system.
    e) Assessment: Estimate whether the data represents an alert or a threshold so that it can redirect the data to other destinations.

The functionality in this layer focus on north–south communication.

Some of the highlighted features of the fog computing layer are given below:

1. Data filtering, Cleanup, aggregation.
2. Packet content inspection.

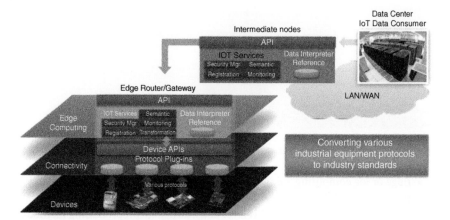

Figure 14.5 Functioning of three layers.

3. Combination of network and data level analytics.
4. Thresholding.
5. Event generation.

2. Server/Cloud side layer – In this layer we discuss the collection of data onto the servers and clouds and how they are processed further during communication.

 • Data Accumulation – Networking systems are used to move data reliably. Prior to this level, data is in motion which moves it across the network at a rapid rate. It has already been discussed that level 1 devices do not include the computing capabilities themselves. Some computational activities could be done at level 2 like protocol translation or application of network security policy. Additionally, computing tasks like packet inspection can be performed at level 3. Fog computing refers to computational tasks that are performed as close to the edge of the IoT as possible with heterogeneous systems distributed across multiple domains [5]. It is computing and services that are the distinguishing characteristics of the IoT. Data accumulation, i.e., level 4 converts the data in motion into data at rest. It is a level that determines some of the characteristics which are as follows:

 a) Passing of data of interest to higher levels – This layer is basically configured to serve the specific data which is required at higher levels.
 b) Persistence of data – It also decides whether the data should be kept in a non-volatile state or in an accumulated form in the disk for short-term use.
 c) Type of Storage – Different type of file system such as big data or a relational database for persistence.

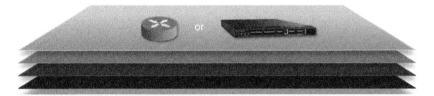

Figure 14.6 Data Accumulation layer.

d) Recombination or recomputation of data – Data is integrated, recomputed or assembled with previously stored information that might be not coming from IoT sources.

So, it is clear that level 4 captures data and puts it at rest, it will be used by different types of applications in a non-real-time basis and will access the data whenever needed. It also converts the event-based data to query processing that is a crucial step in managing the dissimilarities between the real-time networking and non-real-time application world. Figure 14.6 summarizes the activities at level 4.

Some of the highlighted features of data accumulation are as follows:

1. Aggregation of events.
2. Comparison of events.
3. Event persistence in storage.
4. Filtering and sampling of events.
5. Event based rule evaluation.
6. North and south bound alerting.

- Data Abstraction

In the real world application IoT systems have to scale themselves to a corporate or global level and they require multiple storage systems to store the data generated from IoT devices and data taken from traditional enterprises like CRM, HRMS, ERP and other systems [6]. Level 5, i.e., data abstraction is able to perform this function and it focusses on rendering and storing data in a way which enable it to develop simple performance improved applications. There are multiple reasons why data generated through multiple devices are not kept at the single place:

a) Large amount of data becomes accumulated in one place.
b) Lots of power is consumed during processing of moving data, so it is better to retain separation between data retrieving and data generation. This concept is implemented by OLTP (onlinetransaction processing) and data warehouses [7].
c) Geographically separated devices are processed locally.

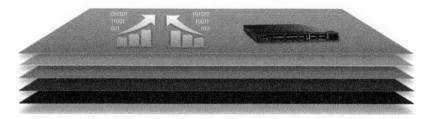

Figure 14.7 Data Abstraction Layer.

d) Level 3 and level 4 might have separate data that represent a continuous stream of data. For example Hadoop can store continuous streaming of data [8], i.e., big data and a relation for faster query processing [9].

To perform the above steps, a data abstraction layer has to process different things, as follows:

1. Processing might be different for different kinds of data, e.g., in store data processing is different from all store processing because at abstraction level there is processing of different things.
2. Integration of multiple data formats from different sources and assure the semantics of data across the different sources.
3. Also confirms that the data is complete for higher level application.
4. Data is consolidated in one place through extract, transform, and load (ETL) [10], extract, load, and transform (ELT) or data replication and access is provided to multiple data stores through virtualization [11].
5. Data is protected with proper authorization and authentication.
6. Fast application is done through normalizing or denormalizing and indexing of data.

Some of the highlighted features of level 5 are as follows:

i. Data is viewed in the manner in which application demands.
ii. Applications are simplified by combining the data from multiple sources.
iii. Data is filtered, selected, projected, and reformatted to serve the need of client applications.
iv. Adapt the differences in data format, shapes, access protocols, semantics and security.

- Application

This is another important level in which information interpretation occurs. In this level software interacts with data which are at rest and is formed by level 5, that is why it does not have to process at network speeds. Applications are based on

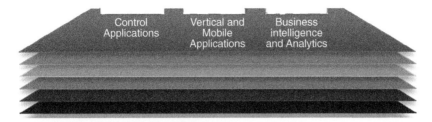

Figure 14.8 Application layer in the IoT.

device nature, vertical markets, and the requirement of the business. For example, due to this the IoT reference model does not strictly define an application. Some applications are based on monitoring the device data, some focus on controlling devices, and some integrate the IoT device and non-device data. Different application models, programming structure and software stacks are based on monitoring and control applications that lead to the emergence of application servers, operating systems, mobility, multi-threading, hypervisors [12], etc. The complexity of the application may vary widely and some of the applications are as follows:

a) Applications based on analysis that can interpret the data for business decisions.
b) Reports generated through business intelligence server.
c) Handling simple interactions through mobile application.
d) Business critical applications like generalized ERP or some specialized industry solutions.
e) System management and control center applications can control the IoT system itself but don't put any impact on the data generated by it.

It has been analyzed that if levels 1–5 are constructed properly then the amount of work will be reduced, i.e., required by level 6 and if level 6 is designed properly the user can perform the work in a better way.

Some of the highlighted features are:

i. Reporting.
ii. Analytics.
iii. Control.

3. User side layer – This layer represent the user interaction with IoT devices, basically through this layer one can learn how people and processes interact during the usage of IoT devices.

• Collaboration and processes

One of the important properties of the IoT is that it includes people and processes. To represent this property there is the last level, i.e., collaboration

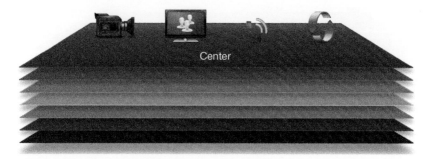

Figure 14.9 Collaboration and processes layer in the IoT.

and processes. It will be of little value if the IoT system and information does not involve people and processes because through the execution of business logic one can empower the people. Specific needs of people are fulfilled by the use of different applications and associated data. The same applications can be used by multiple people for different purposes, so the main objective is to work better through empowering the people. The bottom layer, i.e., level 6 provides accurate data to business people but requires multiple people to perform actions because it cannot be handle by a single person properly. It is required that people communicate and collaborate through the Internet to achieve fulfillment of the IoT. This level also requires multiple steps and handles multiple applications at the same time.

14.3 IoT Communication Protocol

The IoT network protocol stack is different from the traditional OSI protocol model as resources in the IoT environment are more constrained compared with the traditional model. The protocols are considered to be lightweight in the case of the IoT. Figure 14.10 shows the IoT protocol stack.

We now discuss the communication protocol of the IoT which is as follows:

6LoWPAN – This is the lowest protocol termed as IPv6 over low power wireless personal area that helps to make the resources constrained device compatible with IP world and that enables the internet to access the sensor devices [13].

UDP – UDP (user datagram protocol) are connectionless datagrams that enables smaller packets and cycles to be transmitted with some overhead and faster wake-up [14].

CoAP – CoAP (constrained application protocol) is a transfer protocol that is very similar to hypertext transfer protocol (HTTP) and is designed in such a way that it can help in communication for resource constrained devices and resource

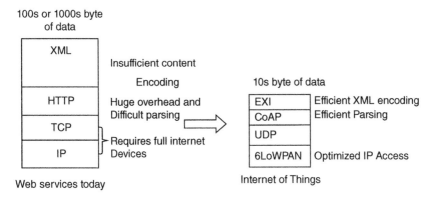

Figure 14.10 IoT Protocol suite.

rich devices through the internet. Semantic of this protocol is designed to be similar to HTTP [15]. It is a binary protocol that is transported over UDP and through the use of UDP it increases the flexibility in communication models and its capability to minimize latency. One of the benefits of using HTTP over CoAP and UDP is that it uses the same protocol code while talking to cloud as well as to other devices on the local network [16].

EXI – EXI (efficient XML interchange) is a compact XML representation and is defined to provide techniques that are needed by resource constrained devices to support XML applications as it requires low bandwidth and at the same time it enhances the encoding or decoding performance. On the basis of the present XML schema, processing stage, and context, EXI compression helps to minimize the document structure content by producing small tags internally. It also ensures that the tags are used to optimize the data representation. Given a document that is in binary format which has data tags encoded using event code, these events codes are the binary tags that will keep their value only in their specific position within the ESXI stream [17].

14.4 IoT Security

Security has to be maintained at each level for the movement of data between the levels. To implement a proper IoT reference model we must ensure some security measures which are as follows:

a. Each device or system must be secured.
b. Each level process must have security.
c. Movement and communication between levels must be secure whether they are north or south bound.

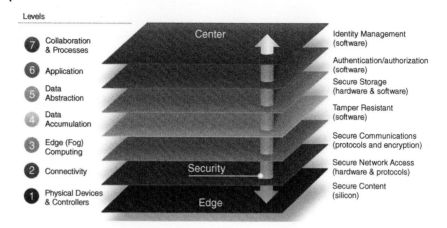

Figure 14.11 Security layer in IoT.

In the traditional network system the expected requirement for secure networks are confidentiality, integrity, availability, non-repudiation, and privacy [18]. This are still the foremost requirements of any network system and IoT also works on the same requirements. Violation of any of these requirements can cause a disaster in the system.

Confidentiality in a network will stop unauthorized access which could steal critical or sensitive information such as money transaction details, patient records, etc.

Integrity property ensures reliable communication between the sender and receiver. It checks whether the participants in the network are legitimate or not. This type of attack takes place on devices such as pacemakers, insulin pumps, etc.

Availability of data is another major requirement as most analytics and strategic decisions are made on the basis of real-time data generated through the IoT domain.

Non-repudiation property ensures that an event which has occurred in the network will either lead to reliable networking of network or not. There can be different vulnerabilities in individual layers of the IoT reference model that can susceptible to different attacks.

14.4.1 Physical Attack

We trust and put cash into banks because we realize that they have physical assurance and surveillance systems appropriately inside their premises to guarantee the safety of our assets. In the same way, for IoT frameworks to develop and the environment to thrive, security must be set up. We hear a great deal about IoT and how huge it is. IoT is truly happening now and such is its capacity that something

Table 14.1 Difference between traditional network security and IoT security.

Traditional network security	IoT network security
Added security features are reactive in nature.	Proactive built in security.
Requires more computation and processing time due to complex algorithms.	Algorithms are for resource constrained devices so they are lightweight and need less computation time.
Most often it is placed in a closed environment.	It is placed in both open and closed environments depend on the IoT application.
Technological heterogeneity is less.	It communicates with different devices so has large technological heterogenity and also has a large attack surface.

as basic as power devices would now be able to be associated with the cloud and controlled remotely.

- Requirement for assurance against physical assaults

Physical assaults are regularly used to distinguish new IoT vulnerabilities. Before a genuine assault is carried out, the attacker will initially attempt to get physical access to devices by purchasing a duplicate of the targeted IoT device from the market. Later, through reverse engineering, they will generate a false attack to "test" to perceive what kind of output can be acquired from it.

These physical attacks uncover the vulnerabilities of the framework. Some of the physical attacks are:

a. Attackers can disassemble the device and can study the flash memory to analyze the software.
b. Attackers can also tamper with the microcontroller and can acquire sensitive information. It can also cause unintended behavior.

Just by performing the above two steps, an attacker can easily gain understanding of the complete system and can find out how he can benefit from performing a physical attack remotely. Some of the common ways to perform this are as follows:

1. Node tampering – In the case of node tampering an attacker physically alters the compromised nodes and can steal sensitive information such as encryption keys, passwords, etc. [30].
2. RF interference – Denial of service is performed by sending noise signals over radio frequency signals. RFID communication uses these signals.
3. Node jamming – An attacker can disturb wireless communication by using a jammer.

Table 14.2 Summarizing the attacks on different layers of IoT with counter measures.

User side layer or application layer	Code injection, data access and authentication	IDS diglossia [19]
	Virus, worms, malware attacks, phishing attacks, spyware	Anti-virus, firewall, IDS [20], secure application code, educating users to use complex passwords, access control mechanisms, key agreement, log monitoring, file and database monitoring tools, anti-malwares to protect applications against malwares.
Support layer security	DoS, wormhole, black hole, interoperability and portability, business continuity and disaster recovery, cloud audit, virtualization security	IDS designed for IoT [21] Lightweight encryption techniques like CLEFIA [22] and PRESENT [23], need for continuous cloud audits, implementation of cloud security alliance standards, secure virtualization technologies, tenants separation, storage encryption for users data confidentiality and integrity
Network layer security	Side-channel attacks: Sybil	Malicious firmware/software detection [24], randomized delay [25], intentionally-generated noise [26], balancing Hamming weights [27].
	Battery-draining, sleep deprivation attack, routing attacks, node jamming in WSN	Policy-based mechanisms and intrusion detection systems (IDSs)
	RFID tag	Personal RFID firewall [28], anonymous tag, lightweight cryptographic protocol [29]
Physical layer or edge layer security	Node tampering, fake node, malicious code injection, side channel attack,	Authentication with encryption techniques. Physical security in nodes vicinity, need for lightweight encryption algorithms for constrained nodes, sensor data privacy, effective
	Mass node authentication, protecting sensor data	authentication and access control mechanisms for devices, anti DoS attacks mechanisms.
	Physical famage	It is better to keep some physical protection for the devices.

4. Physical damage – An attacker can physically attack the device by breaking components of the IoT devices.
5. Social engineering – Social engineering is the act of manipulating people so that they send confidential information. The kinds of information that attackers are looking for can be changed from time to time, however when individuals are targeted then attackers attempt to steal banking information such as user id and passwords. Sometimes these attackers execute a malicious program that provides access to valuable information about clients even though they can control the complete system of the user. Nowadays social engineering is carried out in the form of phishing attacks through mails in which links can automatically transfers the target to a web page which can ask you for banking details [31].
6. Malicious node injection – During malicious code injection attackers physically insert a new node between two or more nodes. Using this they can pass the wrong information to the remaining nodes, the attacker also uses the multiple nodes to perform malicious node injection attacks. Imagine there are two nodes A and B then a replica of one of the nodes will be created first and will be inserted before inserting a malicious node. Execution of the attack is carried out by both the replicated node and the malicious one. Due to this, the victim node cannot send and receive any packet. Even this can make the watchdog nodes perform incorrectly by announcing the legitimate node as malicious one. A MOVE (monitoring verification) scheme has been introduced to prevent this kind of attack by monitoring nodes and detecting malicious behavior of the nodes. Later a verifier will decide which nodes are legitimate and which are malicious [32].
7. Sleep deprivation attack: In this attack greater power is consumed by the attacker which causes nodes to be shut down [33].

- Overcoming physical attacks

IoT devices consist of application processors, connectivity chips, sensors, actuators, and a power supply. So, it is required that proper security measures should be taken after understanding all these vulnerabilities to attack. To counter this, use of the correct safety efforts on IoT devices is essential. This is best done by utilizing equipment-based security.

From Table 14.3 it can be clearly stated that hardware-based security measures are the most helpful approach for overcoming physical attacks. If no security measure is taken then the application processor, consisting of sensors, actuators, a power supply, and connectivity is at risk. To overcome this risk an application processor should be implemented in a secure environment through tamper-resistant hardware. Physical security is required to protect components of IOT devices from destruction. Hardware security can also be used to authenticate device ID. This

Table 14.3 Security measures on different types of attacks.

	Physical attacks on MCU	Micro-architectural attacks	Software attacks
Software-based security measures	✕	✕	✕
Hardware based security measures	✓	✓	✓
Isolated IP security measures	✕	✓	✓

means that a series of security measures can be put between the server and the device itself to establish the authenticity of that device.

14.4.2 Network Attack

1. Traffic analysis attack – In this attack intruders intercept and examine messages passing through the network so that they can obtain network information.
2. Man-in-the-middle attack – In this attack, an attacker or hacker will interrupt and breach the communication between two separate systems. This attack tends to be dangerous because in this attack an intruder secretly intercepts a message and transmits it to two different parties making them believe that they are communicating directly and being unaware that the message they received is sent by illegitimate users. These attackers can change the original message and send it to the receiver, the receiver will think that they are receiving a genuine message.
3. RFID spoofing – An attacker can spoof the RFID signal which means they can capture the original information that is transmitted from a RFID tag. This attack alters the real information and sends it to the recipient. The recipient believes that the message which the system accepts is the original one [34].
4. Radio frequency (RF) jamming: Radio jamming is basically used to block wireless IoT devices such as alarm security devices. This jamming, blocking or interference is carried out with the help of wireless communication. Jamming can be carried out by an illegal RF jammer device and can disconnect the IoT devices, and reduce its ability to communicate with the network. In the case of home and commercial alarm security systems, which are usually connected over the cellular network, they can be jammed by enabling and blocking a signal that could be send for security [35].
5. Denial of service and many more – This type of attack can be done in many ways, one example is flooding the traffic with many users so that services become unavailable to the intended users.

- Overcoming from the network

First and the foremost it is necessary to make sure that only required ports are exposed and available. After that prepare the services that must not be vulnerable to butter overflow and fuzzing attacks.

i. Name your router – Do not keep the name provided by the manufacturer, change the name of the router as soon as possible.

ii. Strong encryption method for Wi-Fi router setting – In the router settings it is good practice to use a strong encryption method, like WPA2, when you set up Wi-Fi network access. This will help keep your network and communications secure [36].

iii. Alter the default username and passwords – In many cases it has been observed that attacker already knows the default ID and passwords, this make it easy for them to access the IoT devices. So, it is very important for the user name and password are changed.

iv. Set up a guest – It is always be preferable that you set up a guest network in Wi-Fi access this will prevent your network from intruders accessing your private data.

v. Disable the features – IoT devices come with many features that are already enabled, so it is better that if a feature is not needed you disable it. It has been seen that a remote access feature is enable in many IoT devices, if it is not required then this feature should be disabled.

vi. Change the setting of the devices – It is very important to change the default privacy and security setting which are provided by the manufacturer otherwise this can create a big problems for the network.

vii. Software update – Software must be updated from time to time because if it is not updated at regular intervals this can cause security flaws in security devices.

viii. Avoid public Wi-Fi networks – Try to avoid connecting to public Wi-Fi networks, sometimes public Wi-Fi networks contain trap for us.

ix. Two step verification – This verification enables us to keep our network secure, in this verification during the initial process a one time is code is sent to the receiver so that one can easily understand whether you are communicating with the genuine sender [37].

x. Use strong, unique passwords for Wi-Fi networks and device accounts – Strong and unique passwords are essential for Wi-Fi networks and devices.

xi. Disable Telnet login [38] and use SSH [39] where possible.

xii. Auditing of the IoT devices on the network is also needed at regular intervals.

Another important prevention technique is connecting the IoT devices to the 0G network because this is a dedicated, low power wireless network specially designed for sending small and critical messages from any IoT device to the Internet [40]. This type of network is created to save power and it does not depend on the traditional communication protocol that is a constant and two-way communication protocol between the source and destination. The main advantage in using the 0G network is that once the IoT device wakes up and sends data through the 0G network asynchronously it then goes back into sleep mode. Due to this an extremely slow window is formed for attackers which is very difficult for them to break and take the control of the network and the devices. The 0G network not support network-initiated downlinks, it only support device-initiated downlinks. These properties make the 0G network not susceptible to network attackers. It is almost impossible to jam the 0G network because of its robustness signaling scheme and the pseudo-randomness of its data transmissions. The 0G network can also serve as a back-up network for cellular devices that are susceptible to RF jamming. Through this network industries can realize the limitless potential of the IoT without any security issues.

14.4.3 Software Attack

1. Viruses and worms – This is another popular method of attack in which systems are damaged by malicious code such as attaching the code to emails, during the downloading of files from the internet, etc. A worm has the ability to replicate itself without any human input and can be spread across the system [41].
2. Trojan social horse – This is another way of attacking a system by entering through an incorrect path [42].
3. Malicious scripts – Using scripts one can gain information about a system [43].
4. Spyware – This has become a major problem for IoT gadgets as it is very difficult to detect them, they can execute in the background with any Internet capable device. Most the hackers use spyware for tracking activity through an operating system and send information to the cybercriminal who is the mastermind behind the attack [44].

- **Overcoming software attacks**
  i. Regular updating and installation of antivirus and antispyware software – The update and installation of antivirus and antispyware software in every system is required regularly.
  ii. Firewall protection is needed in Internet connections.
  iii. Install updates that are required by your operating system and application software.
  iv. Always maintain a backup copy of business data and information.
  v. Physical access to your system and network components must be controlled.
  vi. Keep your Wi-Fi network secure and it would be better for it to be hidden.

vii. Have an individual user account for each employee.

viii. Limited accessibility of data and information should be provided to employees with limited authority for them to install software.

ix. Passwords should be changed at regular intervals.

x. Trained cyber security professionals are needed.

xi. Use proper certification for accessing the internet.

14.4.4 Encryption Attack

1. Cryptanalysis attacks – An attacker tries to obtain the encryption key either by using plain text or cipher text. There are different types of cryptanalysis attacks, some attacks are just based on a cipher test in which the attacker accesses the cipher text and decrypts it to discover the plain text. In the case of a plain text attack the attacker has some information about the cipher text and later decrypts the remaining cipher text to obtain the complete information [45].

2. Side channel attacks – In this attack, the attacker utilizes side channel information that is generated with the help of encrypting devices. This information cannot be considered as plain text or cipher text, this is information related to power, time and fault frequency to perform the operation. Attackers can predict the encryption key through this side channel information which is generated by encrypting devices. There are different types of side channel attacks, e.g., timing attacks, differential fault analysis attacks, and simple and differential power attacks. Timing attacks provide information related to secret keys [46].

- **Overcoming encryption attack**

  The public key infrastructure (PKI) is "a set of policies, software/hardware, and procedures that is needed for the creation, management, and distribution of the digital certificates". This is one of the processes which has proved to be an effective solution over the years for security related to the IoT.

  PKI has ensured that the encryption of data must be done through asymmetric and symmetric encryption processes. Previously, data encryption and decryption are both done through the same key, later both are done through different keys. This encryption and decryption are carried out to maintain data privacy and to minimize the possibilities of the data being stolen.

  Security measures include using digital certificates to verify the identity of all the devices connected together in the IoT. This also ensures that information should be secure and kept from potential attackers. A cryptographic key and X509 digital certificate are some of the IoT PKI security methods that can be used as well as public or private key management, distribution, and revocation.

14.5 Security Challenges in IoT

Security is the fundamental prerequisite of any client of computerized media. A web client will not share their classified and significant information on the system unless the system is trusted. With the rise of distributed computing security requests of its client also expand as they need to trust a cloud owned by a third party. For cloud sellers to attract more clients to utilize their administrations they have to create client trust through cloud reviews and confirmation of consistency to CSA security measures or other measures of security. In spite of the fact that network security arrangements are sufficiently experienced, however, it isn't possible to apply it in the setting of the IoT because of the size of IoT systems, heterogeneity in its design and asset compelled IoT end hubs.

14.5.1 Cryptographic Strategies

Currently available cryptographic algorithms such as symmetric key cryptographic algorithms, and advance encryption standard (AES) [47] are utilized to safeguard information secrecy, which is to be a secure algorithm. The most frequently used algorithm for digital signature exchange is Rivest Shamir Adelman (RSA) [48] which is additionally secure. Secure hash calculations (SHA) [49] are used for data integrity and Diffie Hellman (DH) [50] is used for key agreement. Elliptic curve cryptography (ECC) is an additional, effective cryptographic strategy which is not used very much at present [51]. The above discussed algorithms are secure and efficient but the main issue is that they require more CPU power and consume more battery power. For this reason, they are not a feasible way to verify IoT devices, so there has been an emergence of new cryptographic calculations or advances the existing ones for battery operated IoT devices.

14.5.2 Key Administration

Key administration is a significant and most referenced issue in every cryptographic algorithm. Research has already implemented many ways to solve this issue [52]. These solutions do not completely fit to the IoT framework because of the huge scale associated nodes at device layer of IoT design. Therefore, key administration in the IoT framework is a significant research challenge and needs more thought regarding finding a perfect solution.

14.5.3 Denial of Service

Denial of service attack can be more overpowering in the IoT as it can cause loss of life if it is launched successfully on a smart car IoT application [53]. DDoS detection and mitigation solutions that are used in traditional network systems may not be feasible for the IoT system due to the fact that in the IoT we cannot permit even 10 attack messages to sensor nodes before identifying the DoS attack and blocking occurs due to battery operated resource constrain sensor nodes. Still the solutions provided to overcome these attacks are not very satisfactory [54].

14.5.4 Authentication and Access Control

The IoT concentrates on a machine to machine (M2M) method of correspondence [55]. To perform these verifications of communication nodes is very important and through this security and privacy can be insured. During any common task when two or more nodes are communicating with each other, authentication between them is required otherwise a fake node attack can be done. Up to now there has been no efficient authentication mechanism for the network that consists of a large number of IoT devices that can provide security at large level.

14.6 Conclusion

This chapter has covered all attacks related to the IoT in detail with full justification and also discusses some of the security measures to overcome these attacks. The security of the IoT is a research area within industry and the academic world, but there is still a need for further consideration and to concentrate on investigating distinct security issues in the IoT. This chapter has examined significant security issues in the IoT with their counter measures. We have shown a complete investigation of each layer with their security issues. We likewise present brief countermeasures to various security difficulties to verify IoT frameworks. This chapter has discussed challenges to legacy security solutions in the IoT and has also presented the authentication and access control mechanism in the IoT. We have discussed that traditional security solutions are not feasible for the IoT network that consist of large number of nodes. Therefore a new system is required to validate IoT devices in M2M communication. We present an investigation of the best in class confirmation and access control mechanism for IoT. This study will guide researchers in dealing with security issues in the IoT and where they should put their efforts to develop better security solutions for the IoT.

Bibliography

1 Nair R, Sharma P, Bhagat A, Dwivedi VK. A Survey on IoT (Internet of Things) Emerging Technologies and Its Application. *Int J End-User Comput Dev.* 2019.

2 Abomhara M, Køien GM. Cyber security and the internet of things: Vulnerabilities, threats, intruders and attacks. *J Cyber Secur Mobil.* 2015.

3 Green J. The Internet of Things Reference Model. *Internet of Things World Forum.* 2014.

4 International Electrotechnical Commission, Yinbiao S, Lee K, Lanctot P, Juanbin F, Hao H, et al. Internet of Things: Wireless Sensor Networks. Int Electron Commision. 2014.

5 Samuel F, Chowdhury M, Boutaba R. PolyViNE: Policy-based virtual network embedding across multiple domains. J Internet Serv Appl. 2013.

6 Chen SL, Chen YY, Hsu C. A new approach to integrate internet-of-things and software-as-a-service model for logistic systems: A case study. Sensors (Switzerland). 2014.

7 Parekh A. Introduction on Data Warehouse with OLTP and OLAP. Int J Eng Comput Sci. 2013.

8 Freiknecht J, Papp S, Freiknecht J, Papp S. Hadoop. In: Big Data in der Praxis. 2018.

9 Zafar R, Yafi E, Zuhairi MF, Dao H. Big Data: The NoSQL and RDBMS review. In: ICICTM 2016 - Proceedings of the 1st International Conference on Information and Communication Technology. 2017.

10 Bergamaschi S, Guerra F, Orsini M, Sartori C, Vincini M. A semantic approach to ETL technologies. Data Knowl Eng. 2011.

11 Nanda S, Chiueh T. A Survey on Virtualization Technologies. RPE Rep. 2005.

12 Muditha Perera P, Keppitiyagama C. A performance comparison of hypervisors. In: Proceedings of International Conference on Advances in ICT for Emerging Regions, ICTer 2011. 2011.

13 Shelby Z, Bormann C. 6LoWPAN: The Wireless Embedded Internet. 6LoWPAN: The Wireless Embedded Internet. 2009.

14 Loshin P. User Datagram Protocol. In: TCP/IP Clearly Explained. 2007.

15 Kumar A, Kumar A. Hypertext Transfer Protocol (HTTP). In: Web Technology. 2019.

16 Shelby Z, Hartke K, Bormann C. The Constrained Application Protocol (CoAP). Rfc 7252. 2014.

17 W3C. Efficient XML Interchange (EXI) Format 1.0. W3C Recomm. 2011.

18 Kumar M, Meena J, Singh R, Vardhan M. Data outsourcing: A threat to confidentiality, integrity, and availability. In: Proceedings of the 2015 International Conference on Green Computing and Internet of Things, ICGCIoT 2015. 2016.

19 Son S, McKinley KS, Shmatikov V. Diglossia: Detecting code injection attacks with precision and efficiency. In: Proceedings of the ACM Conference on Computer and Communications Security. 2013.

20 Azmandian F, Moffie M, Alshawabkeh M, Dy J, Aslam J, Kaeli D. Virtual machine monitor-based lightweight intrusion detection. ACM SIGOPS Oper Syst Rev. 2011.

21 Raza S, Wallgren L, Voigt T. SVELTE: Real-time intrusion detection in the Internet of Things. Ad Hoc Networks. 2013.

22 Shirai T, Shibutani K, Akishita T, Moriai S, Iwata T. The 128-bit blockcipher CLEFIA. In: Lecture Notes in Computer Science (including subseries Lecture Notes in Artificial Intelligence and Lecture Notes in Bioinformatics). 2007.

23 Bogdanov A, Knudsen LR, Leander G, Paar C, Poschmann A, Robshaw MJB, et al. PRESENT: An ultra-lightweight block cipher. In: Lecture Notes in Computer Science (including subseries Lecture Notes in Artificial Intelligence and Lecture Notes in Bioinformatics). 2007.

24 Msgna M, Markantonakis K, Mayes K. The B-side of side channel leakage: Control flow security in embedded systems. In: Lecture Notes of the Institute for Computer Sciences, Social-Informatics and Telecommunications Engineering, LNICST. 2013.

25 Spreitzer R, Moonsamy V, Korak T, Mangard S. Systematic Classification of Side-Channel Attacks: A Case Study for Mobile Devices. IEEE Commun Surv Tutorials. 2018.

26 Meng Z, Jha NK. FinFET-based power management for improved DPA resistance with low overhead. ACM J Emerg Technol Comput Syst. 2011.

27 Sundaresan V, Rammohan S, Vemuri R. Defense against side-channel power analysis attacks on microelectronic systems. In: National Aerospace and Electronics Conference, Proceedings of the IEEE. 2008.

28 Rieback MR, Crispo B, Tanenbaum AS. RFID guardian: A battery-powered mobile device for RFID privacy management. In: Lecture Notes in Computer Science. 2005.

29 Dimitriou T. A lightweight RFID protocol to protect against traceability and cloning attacks. In: Proceedings - First International Conference on Security and Privacy for Emerging Areas in Communications Networks, SecureComm 2005. 2005.

30 Potrino G, Rango F De, Fazio P. A Distributed Mitigation Strategy against DoS attacks in Edge Computing. In: Wireless Telecommunications Symposium. 2019.

31 Lin Z, Dong L. Clarifying Trust in Social Internet of Things. IEEE Trans Knowl Data Eng. 2018.

32 Ho JW, Wright M, Das SK. Distributed detection of mobile malicious node attacks in wireless sensor networks. Ad Hoc Networks. 2012.

33 Bhattasali T, Chaki R, Sanyal S. Sleep Deprivation Attack Detection in Wireless Sensor Network. Int J Comput Appl. 2012.

34 Shang-Ping W, Qiao-Mei M, Ya-Ling Z, You-Sheng L. An Authentication Protocol for RFID Tag and its Simulation. J Networks. 2011.

35 Wilhelm M, Martinovic I, Schmitt JB, Lenders V. Short paper: Reactive jamming in wireless networks - How realistic is the threat? In: WiSec'11 - Proceedings of the 4th ACM Conference on Wireless Network Security. 2011.

36 Nazeh M, Wahid A, Ali A, Marwan M. A Comparison of Cryptographic Algorithms: DES, 3DES, AES, RSA and Blowfish for Guessing Attacks Prevention. J Comp Sci Appl Inf Technol. 2018.

37 Van Oorschot PC, Wan T. TwoStep: An authentication method combining text and graphical passwords. In: Lecture Notes in Business Information Processing. 2009.

38 Cisco Systems. Configuring SSH and Telnet This. Cisco Support. 2017.

39 SSH Communications Security. SSH Protocol – Secure Remote Login and File Transfer | SSH.COM. ssh.com. 2017.

40 Mir MM ud in, Kumar S. Evolution of Mobile Wireless Technology from 0G to 5G. Int J Comput Sci Inf Technol. 2015.

41 Elements of computer security. Choice Rev Online. 2011.

42 Gupta S, Singhal A, Kapoor A. A literature survey on social engineering attacks: Phishing attack. In: Proceeding - IEEE International Conference on Computing, Communication and Automation, ICCCA 2016. 2017.

43 Canali D, Cova M, Vigna G, Kruegel C. Prophiler : A Fast Filter for the Large-Scale Detection of Malicious Web Pages Categories and Subject Descriptors. Proc Int World Wide Web Conf. 2011.

44 Elmalaki S, Ho BJ, Alzantot M, Shoukry Y, Srivastava M. SpyCon: Adaptation based spyware in human-in-the-loop IoT. In: Proceedings - 2019 IEEE Symposium on Security and Privacy Workshops, SPW 2019. 2019.

45 Limbasiya T, Karati A. Cryptanalysis and improvement of a mutual user authentication scheme for the Internet of Things. In: International Conference on Information Networking. 2018.

46 Lyu Y, Mishra P. A Survey of Side-Channel Attacks on Caches and Countermeasures. J Hardw Syst Secur. 2018.

47 Heron S. Advanced Encryption Standard (AES). Netw Secur. 2009.

48 RSA Public Key Cryptography Algorithm A Review. Int J Sci Technol Res. 2017.

49 Gueron S, Johnson S, Walker J. SHA-512/256. In: Proceedings - 2011 8th International Conference on Information Technology: New Generations, ITNG 2011. 2011.

50 McCurley KS. A key distribution system equivalent to factoring. J Cryptol. 1988.

51 Rahman H, Azad S. Elliptic curve cryptography. In: Practical Cryptography: Algorithms and Implementations Using C++. 2014.

52 Wu CK. Key management. In: Network Security, Administration and Management: Advancing Technology and Practice. 2011.

53 Kasinathan P, Pastrone C, Spirito MA, Vinkovits M. Denial-of-Service detection in 6LoWPAN based Internet of Things. In: International Conference on Wireless and Mobile Computing, Networking and Communications. 2013.

54 Bawany NZ, Shamsi JA, Salah K. DDoS Attack Detection and Mitigation Using SDN: Methods, Practices, and Solutions. Arabian Journal for Science and Engineering. 2017.

55 Ali A, Shah GA, Farooq MO, Ghani U. Technologies and challenges in developing Machine-to-Machine applications: A survey. Journal of Network and Computer Applications. 2017.

15

Fog Integrated Novel Architecture for Telehealth Services with Swift Medical Delivery

Inderpreet Kaur[1], Kamaljit Singh Saini[2], and Jaiteg Singh Khaira[3]

[1] *Reseach Scholar, CU Gharaun, Dept. of Computer Applications, CGC Landran*
[2] *Dept. of Computer Applications, Chandigarh University, Gharuan*
[3] *Dept. of Computer Applications, Chitkara University, Punjab Campus*

Abstract

Nowadays, the implementation of the Internet of Things (IoT) is very prominent and used widely for an enormous number of applications with automation using sensor devices and advanced wireless technologies. In current scenarios, wireless technology loaded devices are based on internet protocols (IP) and these are mapped with their identification modules. These modules are accessed by servers and satellites for monitoring so that real time analytics can be done. Telehealth or telemedicine is one of the diverse applications for which the IoT is used. As shown in the research reports and analytics from Allied Market Research, the revenue from telemedicine will soon exceed 13 billion dollars. From the research reports from Statista, they expect IoT based deployments to be used significantly and with increasing frequency because of the usage patterns in many domains. In another research report, the use and number of tablets in many locations from 2014 to 2019 are increasing. Because of these data, it is necessary to enforce security mechanisms for the IoT and wireless based environment. In the domain of IoT based telehealth services, there are various sections where there is the need for work to be carried out on the advanced technologies so that a higher degree of accuracy and performance can be achieved. In this chapter the multiple dimensions of IoT in health services are explored and presented in association with the fog computing-based environment.

Keywords *Fog Computing; Fog Integrated Novel Architecture; Telehealth Services; Medical Delivery*

Fog, Edge, and Pervasive Computing in Intelligent IoT Driven Applications, First Edition.
Edited by Deepak Gupta and Aditya Khamparia.
© 2021 John Wiley & Sons, Inc. Published 2021 by John Wiley & Sons, Inc.

15.1 Introduction

The domain of telehealth or telemedicine focuses on the delivery of medical services using advanced communication technologies including wireless communications, remote medical access, video-conferencing and many others [1]. In telehealth, advanced wireless technologies are used so that patients can be provided with remote medical services without delay using real time communication by the medical practitioners [2]. As shown in the research reports, in Singapore more than 25% of the population will be 65 or older by 2030 and these people will be in need of rapid medical services.

Telemedicine is one of the key areas of research in the area of social insurance and remote monitoring of patients with accurate medical conditions [3]. The present state of medicinal sciences and human organizations requires more accuracy so that real time medical access is available to patients [4]. In telemedicine, patients have Internet of Things (IoT) based medical devices and gadgets which are controlled by the remote medical experts [5]. For example, when an ambulance is moving on a highway, the patient can be in contact with doctors with IoT based medical devices [6].

In many situations, patients do not have direct access to medical experts or doctors. Using telemedicine-based devices and gadgets, doctors can be in contact of patients remotely. In this way, lives can be saved with greater frequency and minimum delay [7, 8]

Further key points about telemedicine [9, 10] include

- Telerehabilitation
- Telepharmacy
- Teleneuropsychology
- Telenursing
- Emergency services handling
- Teletrauma care

  Specialist care delivery

- Teleophthalmology
- Teledermatology
- Telepsychiatry
- Teleaudiology
- Telepathology
- Teleradiology
- Teletransmission of ECG
- Telecardiology
- Teledentistry

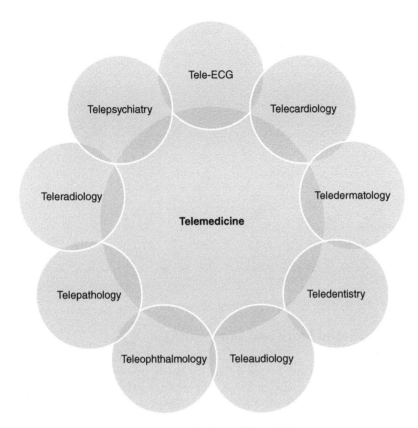

Figure 15.1 Key Perspectives of Telehealth or Telemedicine.

Figure 15.2 Taxonomy of Telemedical or Telehealth Services.

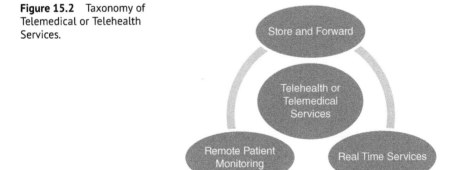

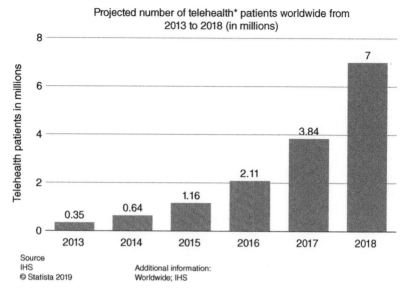

Figure 15.3 Global Scenario of Telemedicine.

15.2 Associated Work and Dimensions

Alaba, F. A. *et al.* [8] presented work on the IoT-based network environment on real time scenarios with an effective view on performance-based services. Their work underlines the application areas of IoT in different situations where the IoT will be more efficient.

De Cremer *et al.* [9] presented work on the integrity and privacy challenges of the IoT with multiple views so that improvements in the network environment can be carried out.

Trappe, W. *et al.* [10] emphasized the work on low energy and power-based devices which are widely used in the IoT which need high performance algorithms in order to be efficient. The low energy and low resource environments need to incorporate the approaches so that resource optimization with high performance is possible.

Telehealth services have multiple dimensions and its taxonomy includes store and forward, remote patient monitoring and real time interactive services [11, 12]

Telemedicine technologies [13–15] include the following

- Mobile OS technology
- Mobile apps
- Mobile telemedicine/telecare devices
- Chat bots and spiders
- Microcomputers

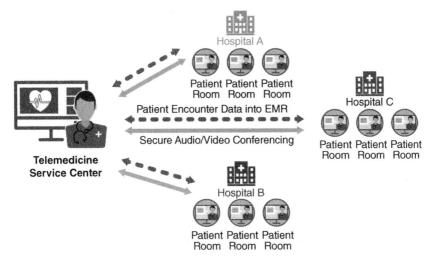

Figure 15.4 Architecture associated with Telemedicine.

- Microprocessor based medical gadgets
- Patient monitoring devices
- Data collection software

15.3 Need of Security in Telemedicine Domain and Internet of Things (IoT)

The IoT is also not immune to cyber-attacks where millions of smart devices, including smart watches, webcams, e-health gadgets and many others, are connected with each other. A number of assaults implemented in IoT infrastructure and these require specialized frameworks to maintain security and integrity.

Nowadays, most people are dependent on smartphones and technology-based devices for routine work whether it is official communication or social media. The huge usage of smart devices and Internet connected gadgets attract cyber-attacks and the whole world is affected with this problem. In India also, cyber-attacks are very common and occur frequently [16–18].

As reported in the Annual Threat Report 2019 by Quick Heal, the major cities of India, including Delhi, Bangalore, Mumbai and Kolkata, are most affected by cyber-attacks. The reports underline that there were more than 950 million attacks instances registered last year.

Another report from inc42.com states that India is the second most affected country in terms of cyber-attacks of different types. The report presents that the average cost per affected record by cyber-attack was around 4552 Indian Rupees. In 2018, more than 50% of cyber-attacks affected the revenue of more than 5 lac dollars. These figures are worrying, and even the general public need more awareness of cyber-security and digital forensics.

15.3.1 Analytics Reports

Reports from the Economic Times suggests that there is 22% growth in cyber-attacks on IoT deployments in India. The research analytics show that there were more than 2500 unique malware impressions which affected the IoT environment and deployments.

These figures and research reports are alarming and must be investigated by law enforcement and forensic investigation agencies. The cyber investigating teams and the government officials are required to be equipped the advanced tools and programming languages so that in any malicious attempt the root cause of the crime can be identified [19–21].

A number of software libraries, frameworks, tools and programming languages are available for cyber security and digital forensics and all citizens should be aware of these. With the awareness of these tools and technologies, cyber-attacks on privacy can be reduced.

With the use of IoT based devices in association with Cloud technologies, the implementation achieves a higher degree of performance and minimum delay. With such performance-integrated approaches the overall effectiveness of cloud-based services can be elevated for multiple real-world applications [22–24].

15.4 Fog Integrated Architecture for Telehealth Delivery

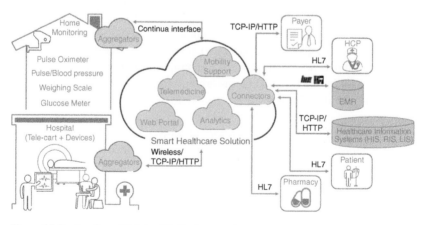

Figure 15.5 Fog Based Telehealth Environment.

15.5 Research Dimensions

Research work has a multi-dimensional scope where medical services can be delivered using advanced technologies.

There is a need to integrate the higher performance-based platforms and technologies so that the overall security and integrity in IoT based telemedicine can be escalated [25–27]. With the use of software-defined networking and a fog computing-based environment, the overall performance can be elevated on many parameters.

As telemedicine is more dependent on wireless technologies and advanced architectures, it is necessary to associate the performance-aware tools and algorithms [28–30]. In addition, the benchmark datasets can be used to integrate with the machine learning and deep learning models so that predictive analytics with more accuracy will be available.

The following are the benchmark datasets which can be used for the training and learning of the models for predictive analytics and thereby the overall performance of telemedicine and telehealth segment can be elevated.

15.5.1 Benchmark Datasets

A number of research repositories are available with the benchmark datasets whereby the enormous data sources exist for research and development. Table 15.1 provides a few of the primary datasets in benchmark taxonomy.

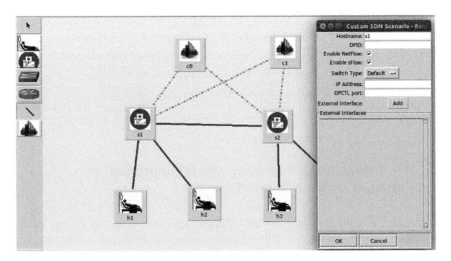

Figure 15.6 Telemedicine Scenario using Software Defined Networking.

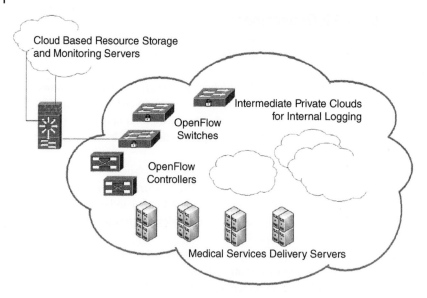

Figure 15.7 Telemedicine Scenario using Virtualization based Environment.

Table 15.1 Benchmark Datasets for Research in Telemedicine.

Dataset Library	Link
BrainSignals	http://www.brainsignals.de/
Health Data	https://www.healthdata.gov
EEG Dataset	http://www.bsp.brain.riken.jp/~qibin/homepage/Datasets.html
ECG Dataset	https://www.physionet.org/physiobank/database/ptbdb/
Physionet	https://www.physionet.org/pn6/chbmit/
UCI Machine Learning Repository	https://archive.ics.uci.edu/ml/datasets.html

15.6 Research Methodology and Implementation on Software Defined Networking

Advanced technologies can provide elevated performance to the telemedicine-based environment [31–34]. The integration of software defined networking is quite performance aware and has greater accuracy and minimum downtime [35, 36].

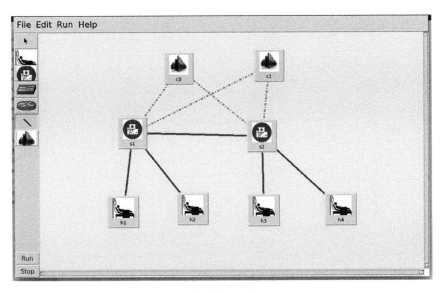

Figure 15.8 SDN based Environment for Telemedicine.

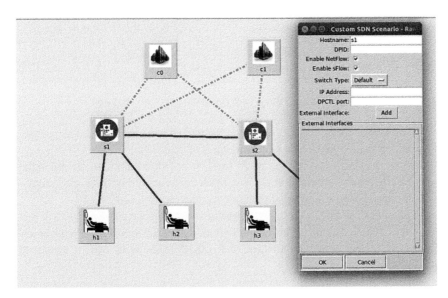

Figure 15.9 Delivery of Services to Patients.

Figure 15.10 Controller Panel for Telemedicine.

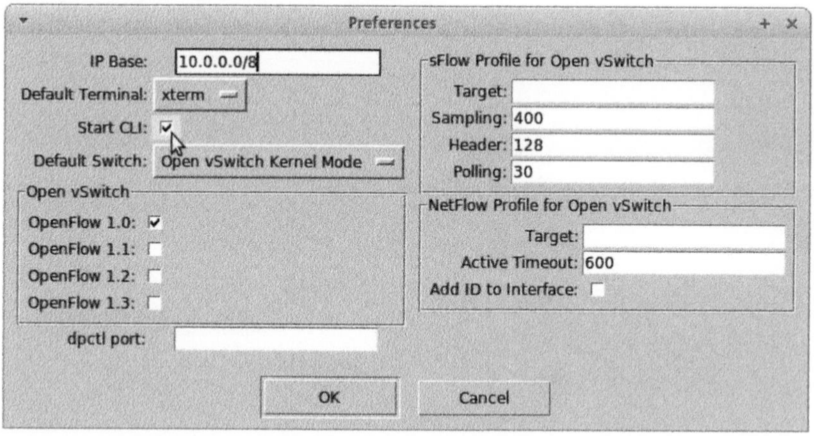

Figure 15.11 Fog Integrated Settings with OpenFlow.

The cloud-based environment for patients in the telemedicine-based infrastructure can be created to have a greater degree of effectiveness.

Advanced algorithms provide more uptime and throughput to telemedicine and the IoT based environment [37, 38]. With the advanced settings and integration of recent approaches higher effectiveness can be achieved [39, 40].

As OpenFlow is one of the most powerful and performance-aware SDN based tools, it can be used as the advanced integration for the delivery of health resources. Using these tools and technologies, the overall throughput can be escalated [41, 42, 43].

Once the integration of fog and advanced SDN is completed, Wireshark is added to have the detailed logs of the data capturing and signals under transmission.

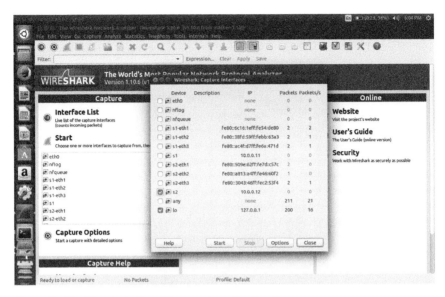

Figure 15.12 Wireshark Based Packets Capturing Panel.

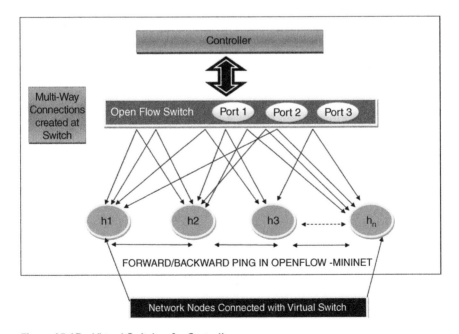

Figure 15.13 Virtual Switches for Controller.

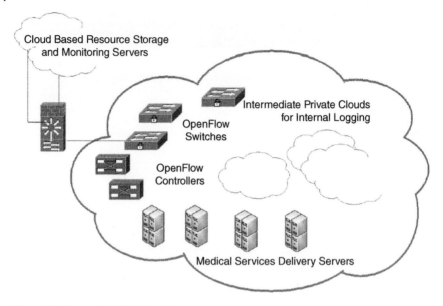

Figure 15.14 Medical Services Framework.

The software defined network and fog computing-based environment can be associated with the virtual switches so that the overall environment can be implemented with software defined network.

15.6.1 Key Tools and Frameworks for IoT, Fog Computing and Edge Computing

Title	URL
iFogSim	github.com/Cloudslab/iFogSim
DSA	iot-dsa.org
Cooja	contiki-os.org
IMUNES	imunes.net
CupCarbon	cupcarbon.com
EmuFog	github.com/emufog/emufog
Mininet	mininet.org
Cloonix	clownix.net
FogTorch	github.com/di-unipi-socc/FogTorch

Title	URL
NS3	nsnam.org
Node-RED	nodered.org
Unetlab	routereflector.com/unetlab
IoTivity	iotivity.org
Zetta	zettajs.org
Shadow	shadow.github.io
GNS3	gns3.com/
Netkit	netkit.org
OpenIoT	openiot.eu
KAA	kaaproject.org
CORE	nrl.navy.mil/itd/ncs/products/core

The following are the classes of iFogSim which are required to simulate the fog network

- Sensor
- Tuple
- Fog application
- Fog device
- Monitoring edge
- Actuator
- Resource management service

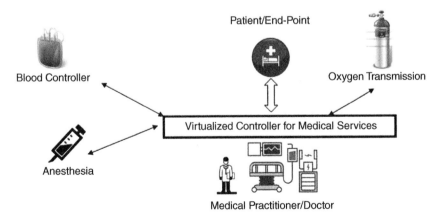

Figure 15.15 Delivery of Medical Services using Telemedicine.

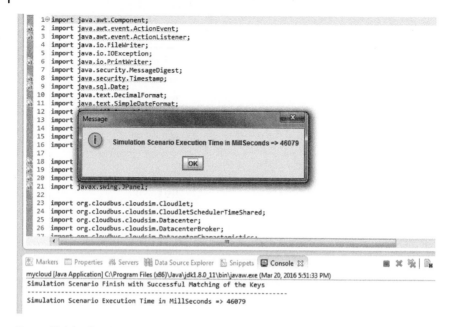

```
 1  import java.awt.Component;
 2  import java.awt.event.ActionEvent;
 3  import java.awt.event.ActionListener;
 4  import java.io.FileWriter;
 5  import java.io.IOException;
 6  import java.io.PrintWriter;
 7  import java.security.MessageDigest;
 8  import java.security.Timestamp;
 9  import java.sql.Date;
10  import java.text.DecimalFormat;
11  import java.text.SimpleDateFormat;
12  import
13  import      Message
14  import
15  import        (i)   Simulation Scenario Execution Time in MillSeconds => 46079
16  import
17
18  import                        OK
19  import
20  import
21  import javax.swing.JPanel;
22
23  import org.cloudbus.cloudsim.Cloudlet;
24  import org.cloudbus.cloudsim.CloudletSchedulerTimeShared;
25  import org.cloudbus.cloudsim.Datacenter;
26  import org.cloudbus.cloudsim.DatacenterBroker;
```

Markers   Properties   Servers   Data Source Explorer   Snippets   Console ☒

mycloud [Java Application] C:\Program Files (x86)\Java\jdk1.8.0_11\bin\javaw.exe (Mar 20, 2016 5:51:33 PM)
Simulation Scenario Finish with Successful Matching of the Keys
----
Simulation Scenario Execution Time in MillSeconds => 46079

Figure 15.16   Showing Execution Time of Simulation in milliseconds.

15.6.2   Simulation Analysis

The integration of IoT based implementation for Telemedicine is very effective for the real time scenarios of health care and delivery of medical services. The presented section gives a scenario for dynamic key exchange for secured delivery of telehealth services using the IoT and the cloud. With these methodologies, the general execution and viability can be improved to a higher level of precision.

Static Key Approach without SDN for Static Security Key Transmitted =>
1(@@1$@1$*@@(#

Static Key Approach without SDN for Static Security Key Transmitted from FogBroker=> (*)(*@#*97119733

Randomly Selected Number of Processors int in a range: 19

MemoryBlocks => 1114

Bandwidth =>1111

Processors =>19

Virtual Machine Manager =>XenFVM

InitiatingCloudSim version 3.1

FogArchitecture-1 is initiating…

FogArchitecture-1 is initiating…

FogBroker isinitiating…

Entities started.

1.1:FogBroker: Cloud Resource List captured with 1 resource(s)

1.1:FogBroker: Trying to Create FVM #1 in FogArchitecture-1

1.1:FogBroker: Trying to Create FVM #1 in FogArchitecture-1

[FVMScheduler.FVMCreate] Allocation of FVM #1 to Host #1 failed by MIPS

1.1:FogBroker: FVM #1 has been created in Datacenter #1, Host #1

1.1:FogBroker: Creation of FVM #1 failed in Datacenter #1

1.1:FogBroker: Trying to Create FVM #1 in FogArchitecture-1

1.1:FogBroker: FVM #1 has been created in Datacenter #3, Host #1

1.1:FogBroker: Sending MyFogBlock1 to FVM #1

1.1:FogBroker: Sending MyFogBlock1 to FVM #1

1.1:FogBroker: Sending MyFogBlock1 to FVM #1

161.1:FogBroker: MyFogBlock1 captured

311.1:FogBroker: MyFogBlock1 captured

311.1:FogBroker: MyFogBlock1 captured

311.1:FogBroker: All FogMods executed. Finishing...

311.1:FogBroker: Destroying FVM #1

311.1:FogBroker: Destroying FVM #1

FogBroker is shutting down...

Simulation: No more future events

CloudInformationService: Notify all CloudSim entities for shutting down.

FogArchitecture-1 is shutting down...

FogArchitecture-1 is shutting down...

FogBroker is shutting down...

Simulation run completed.

Simulation run completed.

================== OUTPUT ==================

MyFogModID STATUS Data center ID FVM ID Time Start Time Finish Time

===

1	SUCCESS	3	1	161	1.1	161.1
1	SUCCESS	1	1	311	1.1	311.1
1	SUCCESS	1	1	311	1.1	311.1

Fog Simulation Finish

Simulation Scenario Finish with MATCHING (AUTHENTICATION SUCCESS-FUL) of the Keys

----

Simulation Scenario Processing Time in MillSeconds => 313

Security Parameter => 11.419715571313161

1114-16-31 13: 53: 31.197

InitiatingFog Simulation with Dynamic and Hybrid Secured Key

Fog Simulation Starts for Security in Data Centers

Initialising...

MultiLayeredSecurityDigest(in Hex. format):: f7614f374fd73e4cdc111c1bd1c58 619

Secured Approach (SecuredKey) Hash Hex format : 8f81897ef4c8118d6fdab5c5 531383e1337eca1d6b5a8e651ce91fd8b7ae6119

Secured Approach Based (SecuredKey) Security Key Transmitted => Hskgh@*(@@*#(@#

Secured Approach Secured Mode Transmission (SecuredKey) from FogBroker- with Key => (*$@(@B@JK@J#

Randomly Selected Number of Processors int in a range : 8

MemoryBlocks =>511

Bandwidth =>1111

Processors =>8

and

Virtual Machine Manager =>Xen

Initialized

InitiatingCloudSim version 3.1

FogArchitecture-1 is initiating...

FogArchitecture-1 is initiating...

FogBroker isinitiating...

Entities started.

1.1:FogBroker: Cloud Resource List captured with 1 resource(s)

1.1:FogBroker: Trying to Create FVM #1 in FogArchitecture-1

1.1:FogBroker: Trying to Create FVM #1 in FogArchitecture-1

[FVMScheduler.FVMCreate] Allocation of FVM #1 to Host #1 failed by MIPS

1.1:FogBroker: FVM #1 has been created in Datacenter #1, Host #1

1.1:FogBroker: Creation of FVM #1 failed in Datacenter #1

1.1:FogBroker: Trying to Create FVM #1 in FogArchitecture-1

1.1:FogBroker: FVM #1 has been created in Datacenter #3, Host #1

1.1:FogBroker: Sending MyFogBlock1 to FVM #1

1.1:FogBroker: Sending MyFogBlock1 to FVM #1

1.1:FogBroker: Sending MyFogBlock1 to FVM #1

161.1:FogBroker: MyFogBlock1 captured

311.1:FogBroker: MyFogBlock1 captured

311.1:FogBroker: MyFogBlock1 captured

311.1:FogBroker: All FogMods executed. Finishing...

311.1:FogBroker: Destroying FVM #1

311.1:FogBroker: Destroying FVM #1

FogBroker is shutting down...
Simulation: No more future events
CloudInformationService: Notify all CloudSim entities for shutting down.
FogArchitecture-1 is shutting down...
FogArchitecture-1 is shutting down...
FogBroker is shutting down...
Simulation run completed.
Simulation run completed.
MyFogModID STATUS Data center ID FVM ID Time Start Time Finish Time
1   SUCCESS   1   1   311   1.1   311.1
1   SUCCESS   3   1   161   1.1   161.1
1   SUCCESS   1   1   311   1.1   311.1

Fog Simulation Finish
Simulation Scenario Finish with MATCHING (AUTHENTICATION SUCCESS-FUL) of the Keys
----

Simulation Scenario Processing Time in MillSeconds => 141
Security Parameter => 31.73876581994184

InitiatingFog Simulation with Dynamic and Hybrid Secured Key

Fog Simulation Starts for Security in Data Centers
Initialising...
MultiLayeredSecurityDigest(in Hex. format):: 0#@Z@#@0#FFFFZJEJFJJ
Secured Approach (SecuredKey) Hash Hex format : i7987917!*473
Secured Approach Based (SecuredKey) Security Key Transmitted =>
okvwnpdkgjc4©^{32}¬¦¦^{2}«±
Secured Approach Secured Mode Transmission (SecuredKey) from FogBroker-with Key => okvwnpdkgjc4©^{32}¬¦¦^{2}«±
Randomly Selected Number of Processors int in a range : 18
MemoryBlocks =>511
Bandwidth =>1111
Processors =>18
and
Virtual Machine Manager =>Xen
Initialized
Security Key Not Matched... Operation Terminated
Simulation Scenario Finish with NON-MATCHING (AUTHENTICATION FAILED) of the Keys
Operation Terminated

Simulation Scenario Processing Time in MillSeconds => 118
Security Parameter => 31.47111431637181
Static Key Approach without SDN for Static Security Key Transmitted =>
@!(!*10$#@$1)(@
Static Key Approach without SDN for Static Security Key Transmitted from
FogBroker=> pwsthvgkayc4|||¡¥ª¦²|¡®
Randomly Selected Number of Processors int in a range : 9
MemoryBlocks =>511
Bandwidth =>1111
Processors =>9
and
Virtual Machine Manager =>Xen
Initialized
Security Key Not Matched... Operation Terminated
Simulation Scenario Finish with NON-MATCHING (AUTHENTICATION
FAILED) of the Keys
Operation Terminated
Simulation Scenario Processing Time in MillSeconds => 139
Security Parameter => 11.859743354619565
Static Key Approach without SDN for Static Security Key Transmitted =>
@!(!*10$#@$1)(@
Static Key Approach without SDN for Static Security Key Transmitted from
FogBroker=> jdtpptwpqqu4¤|«|¢«¬¥§|¦
Randomly Selected Number of Processors int in a range : 17
MemoryBlocks =>511
Bandwidth =>1111
Processors =>17
Virtual Machine Manager =>Xen
Initialized
InitiatingCloudSim version 3.1
FogArchitecture-1 is initiating...
FogArchitecture-1 is initiating...
FogBroker isinitiating...
Entities started.
1.1:FogBroker: Cloud Resource List captured with 1 resource(s)
1.1:FogBroker: Trying to Create FVM #1 in FogArchitecture-1
1.1:FogBroker: Trying to Create FVM #1 in FogArchitecture-1
[FVMScheduler.FVMCreate] Allocation of FVM #1 to Host #1 failed by MIPS
1.1:FogBroker: FVM #1 has been created in Datacenter #1, Host #1
1.1:FogBroker: Creation of FVM #1 failed in Datacenter #1
1.1:FogBroker: Trying to Create FVM #1 in FogArchitecture-1

1.1:FogBroker: FVM #1 has been created in Datacenter #3, Host #1
1.1:FogBroker: Sending MyFogBlock1 to FVM #1
1.1:FogBroker: Sending MyFogBlock1 to FVM #1
1.1:FogBroker: Sending MyFogBlock1 to FVM #1
161.1:FogBroker: MyFogBlock1 captured
311.1:FogBroker: MyFogBlock1 captured
311.1:FogBroker: MyFogBlock1 captured
311.1:FogBroker: All FogMods executed. Finishing...
311.1:FogBroker: Destroying FVM #1
311.1:FogBroker: Destroying FVM #1
FogBroker is shutting down...
Simulation: No more future events
CloudInformationService: Notify all CloudSim entities for shutting down.
FogArchitecture-1 is shutting down...
FogArchitecture-1 is shutting down...
FogBroker is shutting down...
Simulation run completed.
Simulation run completed.
================ OUTPUT =======================
MyFogModID STATUS Data center ID FVM ID Time Start Time Finish Time
==

1 SUCCESS 3 1 161 1.1 161.1
1 SUCCESS 1 1 311 1.1 311.1
1 SUCCESS 1 1 311 1.1 311.1
Fog Simulation Finish
Simulation Scenario Processing Time in MillSeconds => 181
Security Parameter => 11.858184353151471
Processing Time=> 3861 Fog Generated Dynamic Key for Secured Telemedicine: ajdwcxfubba4«® ©§ ¥± |-
Processing Time=> 5896 Fog Generated Dynamic Key for Secured Telemedicine: nrsieyvocfm4±|-£^{3} ||°|®
Processing Time=> 4681 DynamicSecurity Key cqjyyptkwpq4±$^{a®}$¨ ¢|^2£|®¬
Processing Time=> 5914 DynamicSecurity Key iyoqhefjwcn4||¨¡®||^2|¬
Processing Time=> 5767 Fog Generated Dynamic Key for Secured Telemedicine: ygcxsbyybpr4¢- ^{a}¢£|¡®-£
Processing Time=> 15417 Fog Generated Dynamic Key for Secured Telemedicine: xcyvqjijbyp4||¡® ^{a}|¬-|¡
Processing Time=> 14193 Fog Generated Dynamic Key for Secured Telemedicine: rojoeonhivv4©¥¨ ¥||¥|-$^{2®}$
Processing Time=> 18161 Fog Generated Dynamic Key for Secured Telemedicine: anfcjunqhct4§¨ |¡^{2}¡$^{®°22}$

The performance elevation aspects are quite astounding and can be integrated using a fog-based environment with advanced protocols and frameworks.

15.7 Conclusion

The domain of telemedicine is now quite prominent and needs high performance frameworks for integrity, privacy and security to work in the IoT-based environment. This chapter has presented the use of software-defined networking and advanced fog-based environments for the telemedicine and cumulative performance using secured implementations. For future work, the present approach can be elevated in the performance and related aspects using the integration of soft computing and meta-heuristics. With these approaches, the overall performance and effectiveness can be improved with a higher degree of accuracy. The future scope of work can be integrated with the multi-dimensional hybrid approaches of soft computing and fuzzy-based implementations to achieve the outcomes on multiple perspectives and an accuracy-aware environment.

Bibliography

1 P. Thongpanh and S. Choomchuay, "A telepathology system for small servicing units," in Information and Communication Technology, Electronic and Electrical Engineering (JICTEE), 2014 4th Joint International Conference on. *IEEE*, 2014, pp. 1–5.

2 M. Marchevsky, A. Relan, and S. Baillie, "Self-instructional virtual pathology laboratories using web-based technology enhance medical school teaching of pathology," *Human pathology*, vol. 34, no. 5, pp. 423–429, 2003.

3 W. H. Organization, *Application of the international classification of diseases to dentistry and stomatology*. World Health Organization, 1994.

4 Karatzanis, A. Iliopoulos, M. Tsiknakis, V. Sakkalis, and K. Marias, "A collaborative central reviewing platform for cancer detection in digital microscopy images," in In Silico Oncology and Cancer Investigation (IARWISOCI), 2014 6th International Advanced Research Workshop on. *IEEE*, 2014, pp. 1–5.

5 M. Alves, A. Savaris, C. G. von Wangenheim, and A. von Wangenheim, "Software quality evaluation of the laboratory information system used in the santacatarina state integrated telemedicine and telehealth system," in Computer-Based Medical Systems (CBMS), 2016 IEEE 29th International Symposium on. *IEEE*, 2016, pp. 76–81.

6 F. Nobre and A. von Wangenheim, "Development and implementation of a statewide telemedicine/telehealth system in the state of santacatarina, brazil," in Technology Enabled Knowledge Translation for eHealth, ser. Healthcare Delivery in the Information Age, K. Ho, S. JarvisSelinger, H. Novak Lauscher, J. Cordeiro, and R. Scott, Eds. *Springer New York*, 2012, pp. 379–400.

7 J. Wallauer, D. Macedo, R. Andrade, and A. von Wangenheim, "Creating a statewide public health record starting from a telemedicine network," *IT Professional*, vol. 10, no. 2, pp. 12–17, 2008.

8 Alaba, F. A., Othman, M., Hashem, I. A. T., & Alotaibi, F. (2017). Internet of Things security: A survey. *Journal of Network and Computer Applications*, 88, 10-28.

9 De Cremer, D., Nguyen, B., & Simkin, L. (2017). The integrity challenge of the Internet-of-Things (IoT): on understanding its dark side. *Journal of Marketing Management*, 33(1-2), 145-158.

10 Trappe, W., Howard, R., & Moore, R. S. (2015). Low-energy security: Limits and opportunities in the internet of things. *IEEE Security & Privacy*, 13(1), 14-21.

11 "Telemedicinatelemedicina." [Online]. Available: http://site. telemedicina.ufsc.br/telemedicina/

12 D. Clunie, D. Hosseinzadeh, M. Wintell, D. De Mena, N. Lajara, M. Garcia-Rojo, G. Bueno, K. Saligrama, A. Stearrett, D. Toomey et al., "Digital imaging and communications in medicine whole slide imaging connectathon at digital pathology association pathology visions 2017," Journal of Pathology Informatics, vol. 9, no. 1, p. 6, 2018.

13 J. M. Alves, D. B. Albino, M. C. Resener, M. Zannin, A. Savaris, C. G. von Wangenheim, and A. von Wangenheim, "Quality evaluation of poison control information systems: A case study of the datatox system," in Computer-Based Medical Systems (CBMS), 2016 IEEE 29th International Symposium on. *IEEE*, 2016, pp. 30–35.

14 F. Shafique, D. Al-Tamimi, and H. Kussaibi, "Virtual slides and instant sharing of medical diagnosis: Emerging telepathology practices at king fahd hospital," in Science and Information Conference (SAI), 2015. *IEEE*, 2015, pp. 298–304.

15 Pantanowitz, "Digital images and the future of digital pathology," Journal of pathology informatics, vol. 1, 2010.

16 W. Tolone, G.-J. Ahn, T. Pai, and S.-P. Hong, "Access control in collaborative systems," *ACM Computing Surveys (CSUR)*, vol. 37, no. 1, pp. 29–41, 2005.

17 J. M. Alves, C. Wangenheim, T. Lacerda, A. Savaris, and A. Wangenheim, "Adequate software quality evaluation model v1. 0," Instituto Nacional para Convergencia Digital–INCoD, Tech. Rep ˆ , 2015.

18 H. N. Boone and D. A. Boone, "Analyzing likert data," *Journal of extension*, vol. 50, no. 2, pp. 1–5, 2012.

19 V. R. B.-G. Caldiera and H. D. Rombach, "Goal question metric paradigm," *Encyclopedia of software engineering*, vol. 1, pp. 528–532, 1994.

20 Goode, B. Gilbert, J. Harkes, D. Jukic, and M. Satyanarayanan, "Openslide: A vendor-neutral software foundation for digital pathology," Journal of pathology informatics, vol. 4, 2013.

21 S. Jodogne, C. Bernard, M. Devillers, E. Lenaerts, and P. Coucke, "Orthanc-a lightweight, restful dicom server for healthcare and medical research," in Biomedical Imaging (ISBI), 2013 IEEE 10th International Symposium on. *IEEE*, 2013, pp. 190–193.

22 R. K. Yin, *Case study research and applications: Design and methods*. Sage publications, 2017.

23 K. Finstad, "The usability metric for user experience," *Interacting with Computers*, vol. 22, no. 5, pp. 323–327, 2010

24 S. Dubey and A. Gambhir and S. K. Jain and A. V. Jha and A. Jain and S. Sharma, "IoT application for the design of digital drug administration interface" 2017 International Conference on Information, Communication, Instrumentation and Control (ICICIC), pp. 1-5, 2017.

25 Dickman, J. Schneider and J. Varga "The Syringe Driver: Continuous Subcutaneous Infusions in Palliative Care," Published by Oxford University, Second edition, 2005.

26 K. Liu, R. Wu, Y. Chuang, H. S. Khoo, S. Huang, and F. Tseng "Microfluidic Systems for Biosensing," *Sensors (Basel).* 2010; 10(7): pp. 6623–6661.

27 Juarez et al. "AutoSyP: A Low-Cost, Low-Power Syringe Pump for Use in Low-Resource Settings." The American Journal of Tropical Medicine and Hygiene 95.4 (2016): 964–969. PMC. Web. 29 Sept. 2018.

28 Saidi, L. A. Ouni and M. Benrejeb, "Design of an Electrical Syringe Pump Using a Linear Tubular Step Actuator," International Journal of Sciences and Technologies of Automatic control & computer engineering IJ-STA, Volume 4, No-2, December 2010, pp. 1388-1401.

29 Imed, R. Habib and A. Mahfoudh "Design and Modelling of a Linear Switched Reluctance Actuator for Biomedical Applications" International Journal of the Physical Sciences Vol. 6(22), pp. 5171- 5180, 2 October, 2011.

30 Rashid, "Power Electronics Handbook," 3rd Edition, Published by Butterworth-Heinemann, Elsevier Corporate Drive, Suite 400, Burlington, MA 01803, USA.

31 Akash, M. P. Kumar and N. Venkatesan, "A Single Acting Syringe Pump Based on Raspberry Pi –SOC," 2015 IEEE International Conference on Computational Intelligence and Computing Research (ICCIC), 10-12 Dec. 2015.

32 M. A. Mazidi, J. Gillspie, Mckinlay and D. Rolin, " The Microcontroller in Embedded System: Using Assembly and C," 2nd edition published by Pearson Education.

33 T. Nihtila et al., "System Performance of LTE and IEEE 802.11 Coexisting on A Shared Frequency Band," *IEEE Wirel. Commun. Netw. Conf. WCNC*, pp. 1038–1043, 2013.

34 H. Zhang, X. Chu, W. Guo and S. Wang, "Coexistence of Wi-Fi and Heterogeneous Small Cell Networks Sharing Unlicensed Spectrum," *IEEE Commun. Mag.*, vol. 53, no. 3, pp. 158–164, 2015.

35 W. Wu, Y. Gong and K. Yang, "On Constructing A LTE-based Experimental Network in Power Distribution Systems," 8th International Conference on Communications and Networking in China (CHINACOM), pp 241-245, 2013.

36 Alshamali, εGSM Based Remote Ionized Radiation Monitoring System,ε Electronics and Micro-electronics, International Conference on Advances in(ENICS), pp. 155-158, 2008.

37 S. Gaonkar, J. Li, R.R. Choudhury, L. Cox and A. Schmidt. "MicroBlog: Sharing and Querying Content Through Mobile Phones and Social Participation". MobySys'08, 17-20 June 2008.

38 Health Resources and Services Administration. https://www.hrsa.gov/ rural-health/telehealth/index.html. Accessed on 02-03-2019.

39 American Telemedicine Association. https://thesource.americantelemed.org/ resources/telemedicine-glossary. Accessed on 02-03-2019.

40 Chryssanthou, I. Varlamis, and C. Latsiou, "Security and trust in virtual healthcare communities," Proceedings of the 2nd International Conference on PErvasive Technologies Related to Assistive Environments, p. 72, 2009.

41 G. S. Lee and B. Thuraisingham, "Cyberphysical systems security applied to telesurgical robotics," *Computer Standards Interfaces*, vol. 34, no. 1, pp. 225–229, 2012.

42 M. Hussain, A. Al-Haiqi, A. A. Zaidan, B. B. Zaidan, M. Kiah, S. Iqbal, and M. Abdulnabi, "A security framework for mhealth apps on android platform," Computers Security, vol. 75, no. 191-217, 2018.

43 M. Hossain, S. R. Islam, F. Ali, K. S. Kwak, and R. Hasan, "An internet of things-based health prescription assistant and its security system design," *Future Generation Computer Systems*, vol. 82, pp. 422–439, 2018.

16

Fruit Fly Optimization Algorithm for Intelligent IoT Applications

Satinder Singh Mohar, Sonia Goyal, and Ranjit Kaur

Department of Electronics and Communication Engineering, Punjabi University Patiala.

Abstract

The Internet of things (IoT) plays s vital role in the day to day life of human beings, for example managing the flow of heat in a room by measuring the room's temperature with the help of sensors. The IoT is basically a network that consists of sensor nodes, software and has some connectivity to the network in order to sense, collect and exchange information between the devices. Optimization in the IoT is used to improve the performance of the network by enhancing the efficiency of the network, reducing overheads and energy consumption, and increasing the rate of deployment of various devices in the IoT. The applications of the IoT are smart cities, augmented maps, the IoT in health care etc. and various issues in the IoT, for example, security, addressing schemes etc. are discussed. Various optimization techniques such as heuristic and bio-inspired algorithms, evolutionary algorithms and their applicability in the IoT are described. Further the fruit fly optimization algorithm (FOA) and flow chart of FOA is explored in detail. Finally, the applications of FOA in the IoT and node deployment using FOA are explained. On the basis of observation FOA can be used to increase the coverage rate of sensor nodes.

Keywords *IoT; Optimization techniques; Applications of IoT; Fruit Fly Optimization algorithm; sensor nodes*

16.1 An Introduction to the Internet of Things

The Internet of Things **(IoT)** is also known as the Internet of Everything, basically the term IoT refers to the large networks that make up a number of devices which communicate with other devices with the help of wireless protocols without the

Fog, Edge, and Pervasive Computing in Intelligent IoT Driven Applications, First Edition.
Edited by Deepak Gupta and Aditya Khamparia.
© 2021 John Wiley & Sons, Inc. Published 2021 by John Wiley & Sons, Inc.

requirement of direct human interface. The efficient and fast transfer of data is provided by links made through IoT devices which are required to support various applications. The applications based on the IoT provide major improvements in terms of efficiency and security [1, 2].

The IoT is used in various fields such as wireless sensor network (WSN), machine learning, sensor technologies, artificial intelligence (AI) and analysis of big data, etc. The major area of the IoT is WSNs which is made up of a large number of sensors which are deployed in the sensing field to monitor and record some information [3]. In pervasive network configuration, all heterogeneous devices are connected with other devices to sense, collect and investigate data of a dissimilar nature. All these activities are performed without human interface [2, 4].

16.2   Background of the IoT

In this section evolution of the IoT and different elements which are involved in IoT communication are discussed in detail.

16.2.1   Evolution of the IoT

The term IoT was first invented by Kevin Ashton [5] in 1999 to draw the attention of members of the organization where the author was employed to stock sequence optimization using radio frequency identification. In order to record and calculate the number of items which were used in stock sequencing without requiring human involvement the author planned to use the Internet together with radio frequency identification. To accomplish the above target, Kevin Ashton proposed a novel idea called the IoT. The term IoT did not attract the attention of researchers over the world until 2010 [6].

However, in mid-2010 the IoT had recovered its attractiveness. Moreover, in 2011 China had declared that the IoT was a significant topic in its five-year plan. In the same year, the IoT was seen as an evolving tool in the field of hype rotation for emergent technology developments in Gartner research. In addition to this, Europe's main Internet conference on IoT was conducted by LeWeb in 2012. The International Data Corporation has predicted that USD market for the IoT will be 8.9 trillion in 2020. According to Cisco prediction the things which are connected to the internet will be 50 billon by 2020 [6].

16.2.2   Elements Involved in IoT Communication

Figure 16.1 shows the three elements which are involved in the communication of IoT applications which are: machine to machine (M2M) connection, people to people (P2P) connection and machine to people (M2P) connection [2, 7].

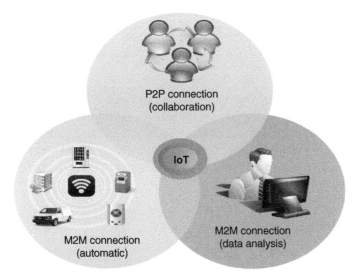

Figure 16.1 Internet of Things (IoT) elements [7].

People to People (P2P) connection: In P2P connection the transfer of information takes place between two people, this usually occurs during a video call, social communication or a telephone call. It is also called collaboration connection [7].

Machine to People (M2P) connection: In M2P connection the data is transferred from machines such as sensors, mobile phones and computers to the user to investigate that information. For example: in a weather forecasting application, sensors are used to collect information from the environment and data is send back to the control room for advance examination of data [2].

Machine to Machine (M2M) connection: In M2M connection information is exchanged between two devices without human communication. For example: a car talking to another car about changing lanes, braking intentions, etc. [2, 7].

The communication of IoT systems consisting three major categories can be summarized as shown below:

IoT = Human + Physical Objects (sensors, devices, etc.) + Internet

16.3 Applications of the IoT

Applications of the IoT are increasing day by day in various fields. The IoT is used in various domains such as smart parking, smart homes and offices, agriculture, healthcare, environmental monitoring, augmented maps, industry etc. The demands of IoT applications are increasing in day to day life as people changed their requirements as they use the Internet according to their needs [8].

16.3.1 Industrial

The IoT is used in industries to monitor the levels of temperature, oxygen and also various gases such as toxic gas for the safety of the people working in the industry. In food industries the IoT is used during meat production to monitor the level of ozone. The information which is collected from a CAN bus is used to send warning messages to the driver in emergency situations so that driver can take necessary actions [1, 9–11].

16.3.2 Smart Parking

Wireless sensors are used in parking systems to sense vehicles entering and leaving the parking area. It is an efficient system that gives information about the total number of cars in the parking area and the number of empty spaces for new vehicles which can be parked in the parking area. It also allows people to people to book parking spaces online directly from the cars. It helps lessen traffic jams in the parking area, helps to reduce the consumption of fuels such as petrol, diesel and also reduce the emission of carbon dioxide [8, 12–14].

16.3.3 Health Care

The IoT has an important role in monitoring the health of patients. It has all of the records for the patient's health condition and sends this information to the monitoring room. The doctors analyze the data and advice is sent back to the patient [8, 15]. The IoT health monitoring system can also monitor people suffering from a number of diseases such as diabetes etc. The IoT based health monitoring system reduces costs and also improve the quality of care [1, 16].

16.3.4 Smart Offices and Homes

In day-to-day life human beings use electronic devices such as microwave ovens, air conditioners, refrigerators, fan and lights etc. Sensors can be installed to consume power in an efficient way. Wireless sensors will monitor the temperature outside the room and also regulate the amount of cooling and heating in the room. Smart homes and offices help to reduce power consumption [8, 17]. Modern smart homes can also be controlled by smart mobile phones and computers. For example: the user can turn air conditioners and lights on and off using their smart phones [1, 18, 19].

16.3.5 Augment Maps

Near-field communications play an important role in tourist's augmented maps. These maps allow users to search the data about nearby places such as hotels, tourist attractions, restaurants, malls etc. by connecting smart phones to the Internet. The augmented maps help visitors to find the best hotels for their needs in that particular area. It will display all the tourist attraction and hotels near to the user's location on the screen of the user's smart phone [8, 20].

16.3.6 Environment Monitoring

The IoT is used in environment applications to create real-time maps of noise pollution, air pollution, water pollution, temperature and dangerous gases etc. The IoT can be used to gather data and this information can be stored for environmental records and can send alarm messages to residents and control centres. Government agencies can also take decisions from the collected data for the healthy and safe environment for residents [21–23].

16.3.7 Agriculture

The IoT plays a significant role at various levels in agricultural production. The IoT can be used to estimate field parameters such as atmospheric conditions, biomass of animals or plants and soil state. The IoT can also be used to evaluate different variables, for example, humidity and temperature. Moreover, it also provides data about properties and origin of the product to the final customer. The IoT is a vital tool for people working with agricultural industrial systems such as farmers, distributors, consumers, businessmen and government agencies [21, 24, 25].

16.4 Challenges in the IoT

The IoT has applications in various fields but it has also some challenging areas such as security, addressing schemes, energy consumption etc. The wireless sensor network (WSN) is a major part of the underlying technology for the IoT. Some of the design issues for the IoT are discussed below [2].

16.4.1 Addressing Schemes

The major issue in the IoT is to discover objects for the successful completion of IoT applications. It requires to individually categorize hundreds of devices and controls the devices through the internet. The major features of generating unique addresses are scalability, uniqueness, and reliability [26]. Smart devices require unique addresses that allow them to communicate with each other and become a part of the Internet. A unique addressing scheme for the IoT is a great challenge and a popular research field [2].

16.4.2 Energy Consumption

The IoT makes a large network which consists of several devices which are connected to the Internet. Energy is a critical issue for IoT devices since many applications of the IoT are driven by battery. Since the underlying network of the IoT is WSN which consist of thousands of sensor nodes which are constrained in terms of energy and storage capabilities [27]. It is not possible to change the batteries of sensor nodes because WSN consists of millions of nodes [2]. Therefore, energy efficient network architecture and quick routing mechanism must be designed for various IoT applications.

16.4.3 Transmission Media

Transmission medium is the physical route that makes the link and transmits the information from the source to the destination. The IoT uses various kinds of technologies to transfer the information such as radio frequency identification, Zigbee and Bluetooth etc. The issues which are linked with transmission medium are bandwidth, inference, high error rate and fading etc. Every transmission medium needs dedicated energy, bandwidth and system hardware [2, 28]. So, enhancing the transmission media is a big issue in IoT applications to extend the life of networks.

16.4.4 Security

The security issue is one of the main challenges of the IoT. When packets are sent from a sender they pass through various intermediate devices to reach the final destination on the Internet. The secrecy and reliability of the information must be maintained when it travels through different devices over the Internet. Many IoT devices are constrained in terms of power so existing cryptographic techniques cannot be used directly in the IoT [2]. Therefore, it is essential to protect the data collected by sensor nodes from unauthorized devices while transmitting the data to the final destination [29].

16.4.5 Quality of Service (QoS)

In various applications, the data which is collected by various sensor nodes must be delivered to the final destination within a certain period of time. Data that is delivered late is useless in certain applications [2]. Therefore, QoS necessities must meet with distinguished services and loss of packet, packet delay and bandwidth. These necessities are required for effective end-to-end service. So, QoS needs research for optimization and implementation in the IoT [30].

16.5 Introduction to Optimization

Optimization is the way of obtaining the supreme feasible outcomes under given situations. Engineers have to make choices in designing, constructing and maintaining a network. The aim of all choices is either to reduce the effort or to exploit profit. The profit or effort is generally defined as a function of different variables which are used in designing a particular network [31]. Therefore, optimization is defined as a process of discovering the circumstances that will result in either maximizing or minimizing the value of the desired function.

A particular technique cannot be used for finding the solution to all optimization issues efficiently. Several optimization algorithms have been developed by various researchers to solve different optimization issues. The methods which are used to find the optimum solutions are also recognized as mathematical programming techniques. These techniques are part of operations research [32, 33].

16.6 Classification of Optimization Algorithms

The optimization problem consists of various input variables, output variables, one or more objective functions and some constraints. Therefore, a single optimization algorithm cannot be used to solve the different optimization problems in the IoT. Various researchers have proposed different optimization algorithms such as genetic algorithms, particle swarm optimization algorithms, evolutionary algorithms, heuristic algorithms etc. to solve different optimization problems in the IoT [6].

16.6.1 Particle Swarm Optimization (PSO) Algorithm

PSO is used to optimize certain difficulties in networks by refining aspirant solutions in an iterative manner in order to provide a standard quality of service. PSO was invented from the crowd behaviour of birds, animals and their training environment. The essential data is scattered among all people in a crowd [6]. For example, in [34], PSO is used to improve the performance of a network by decreasing message overhead, energy consumption, latency in delivering the messages and by enhancing network life. In [35] a multi-objective PSO algorithm is used to reduce response time and energy consumption in the IoT application.

16.6.2 Genetic Algorithms

A genetic algorithm (GA) uses normal progress action and also the existence of the fittest value in order to allocate the appropriate value to the solution for the issue in a particular problem. Both reserved and unrestricted optimized problems can be solved by using GA [6]. GA in [36] is used to enhance the speed and life of a network. In [37] GA is used to select the sensor nodes which have the maximum energy and storage capacity in order to extend the life of the network and the efficiency of the sensor nodes as well as the wireless sensor network.

16.6.3 Heuristic Algorithms

Heuristic algorithms are used to discover result out of various options and offers reasonably nearby answer to a complex difficulty in easy and quicker way [6]. Heuristic algorithm in [38] is used to reduce the energy consumption in IoT devices. In [39] heuristic algorithms are used to deliver the coverage for many IoT devices and effective communication between the devices.

16.6.4 Bio-inspired Algorithms

Bio-inspired algorithms are inspired from the activities of ants or swarms of bees and their behaviour in performing tasks in an efficient way. The bio-inspired algorithms are used to various optimization and mathematical problems [6]. Bio Inspired Artificial Bee Colony (ABC) algorithm is proposed in [40] to decrease the burden on networks while finding appropriate paths to deliver messages.

16.6.5 Evolutionary Algorithms (EA)

Evolutionary algorithms make use of population-based methods in which two or more solutions are involved in all iterations and develops a new set of solutions in all iterations. Evolutionary algorithms are more flexible and easier to implement as compared to other algorithms [6]. In [41] Bacterial Foraging Optimization (BFO) algorithm is used to provide secure transmission of data and to preserve the energy while choosing the cluster head for transmission of messages from the sender to receiver.

Various researchers have used many optimization techniques such as particle swarm optimization [42], bacterial foraging algorithm [43], artificial bee colony [44], fruit fly optimization algorithm [45], cuckoo algorithm [46], glow worm swarm optimization [47] etc. for finding the optimal solutions to different problems. In this chapter the fruit fly optimization algorithm is discussed in detail. The fruit fly optimization algorithm is a novel technique in which all fruit flies collect information about current best position for further search [45].

16.7 Network Optimization and IoT

Network optimization is defined as a methodology which is used to enhance the performance of the network in any situation. The network optimization improves the data rate, discards the redundant information and also enhances the reaction time of the network. Network optimization in the IoT has gained a lot of attention from researchers because of the probability of a high rise in traffic flow from IoT devices, as millions of IoT devices will be expected to join the network in the future. So efficient resource utilization and to decrease the traffic which is created by IoT devices is a big challenge for researchers [6].

The traffic made by IoT devices is dissimilar to the traffic caused by cellular networks due to different kinds of devices and applications. In addition to this the traffic caused by IoT driven applications is required to be controlled in order to monitor the services provided by IoT devices. IoT application produces a small amount of data but the combination of devices to the application creates a large amount of data which increases the traffic in the network. So, in order to overcome the traffic of data an effective optimization technique is required [6]. In this chapter, various optimization algorithms to deliver network optimization in IoT will be discussed in the following sections.

16.8 Network Parameters optimized by Different Optimization Algorithms

Various optimization techniques are used by researchers to optimize the various network parameters such as load balancing, management of network, maximizing the network lifetime etc. These parameters are discussed in detail as follows.

16.8.1 Load Balancing

In routing of data, load balancing plays a vital role in maximizing the lifetime of the network. Multipath routing will provide reliable end-to-end communication of information with fewer chances of sensor node failure in wireless networks [6]. Various optimization techniques such as, particle swarm optimization (PSO) [34], heuristic optimization algorithms [38], artificial bee colony optimization algorithms [40] and bio-inspired algorithms [41] are used to optimize the load balancing in the network.

16.8.2 Maximizing Network Lifetime

Various parameters such as load balancing, management for the failure of nodes and energy effective routing will help maximize the lifetime of a network. Since wireless sensor nodes are restricted in term of energy the optimization algorithms must be used for effective resource utilization and to increase the lifetime of the network [6]. PSO [34], genetic algorithm [37], heuristic optimization algorithms [38] and evolutionary algorithms [41] are used to improve the lifetime of the network.

16.8.3 Link Failure Management

Link failure occurs in sensor networks due to the failure of many nodes which affects various parameters such as signal strength and lifetime of the network. Therefore, optimization algorithms must be used to optimize the failures of the links to deliver consistent communication [6]. Various optimization algorithms such as PSO [34], genetic algorithm [37] and heuristic optimization algorithms [38] are used to optimize the failure of the links.

16.8.4 Quality of the Link

This factor provides quality of service to the message. In multipath routing, all routes are tested and information is sent by the shortest path which reduces delay and retransmission of messages [6]. The quality of the link maximizes the life of the network as it decreases retransmission of data. PSO [34], genetic algorithm [37] and evolutionary algorithms [48] are used to enhance the quality of links as well as enhancing the performance of the network.

16.8.5 Energy Efficiency

The underlying architecture of the IoT is WSN. WSN consists of a huge number of sensor nodes. The sensor nodes have a limited amount of energy. Energy consumption is a critical factor in the IoT and various optimization algorithms are used to reduce the energy consumption of sensor nodes [6]. Many algorithms such as PSO [34], genetic algorithms [37] and heuristic algorithms [38] offer energy management plans to exploit the lifetime of WSN.

16.8.6 Node Deployment

Sensor nodes are used in the IoT to sense and collect large amounts of data. The performance of the sensor network depends on the positioning of sensor nodes [6]. Various optimization algorithms such as artificial bee colony [44], fruit fly optimization algorithm [45], genetic algorithm [49], PSO [50], and bio-inspired algorithms [51] are used to provide efficient deployment schemes in order to increase the coverage rate of sensor nodes.

In this chapter fruit fly optimization algorithm is used to deploy the sensor nodes in the network and to increase the coverage rate of sensor nodes.

16.9 Fruit Fly Optimization Algorithm

The fruit fly optimization algorithm (FOA) is a swarm intelligence-based procedure for discovering the overall optimization solution. FOA is encouraged from the activities of searching for food of a fruit fly swarm. The studies which are linked with FOA state that fruit flies have a crowd hunting performance and they can also exchange data about food with other fruit flies. As the fruit fly is dominant over other kinds in terms of fragrance and visualization they can also rapidly find the location of food sources [45].

The searching activity of fruit flies consists of two phases. In the first phase the fruit flies are accumulating odours drifting in the air using osphresis tissues. The fruit fly group gets closer to the target in the second phase by flying in the direction of the target position accurately with the help of its searching visualization [52]. FOA have replicated the procedure of searching behaviour, every fruit fly has its own site and fragrance concentration decision value.

The fragrance concentration decision value is calculated using the fragrance concentration decision function. Then the fruit fly cluster will constantly track the location which has the present best fragrance concentration decision value and progressively they get closest to the optimum position [53].

Merits of FOA

The advantages of FOA are listed below [54]:

- FOA is simple structured algorithm.
- FOA resolves the problems fast compared to other optimization algorithms.
- It is easily adaptable to the applications.
- Program code of FOA is short and can be easily understand.
- FOA is a stable optimization algorithm and has only a few parameters to tune.

16.9.1 Steps Involved in FOA

The main steps which are involved in the fruit fly optimization algorithm are explained as follows:

Step 1: Initialize all the parameters including the group size (n), the maximum number of iterations and the initial positions of fruit fly group ($X_{initial}$, $Y_{initial}$).

Step 2: Every single fruit fly will be given an arbitrary direction and distance using osphresis for discovering food.

$$X(i) = X_{initial} + \text{Random value}$$

$$Y(i) = Y_{initial} + \text{Random value}$$

where i varies from 1 to n and n is the size of the fruit fly group.

Step 3: Evaluate the distance of food to the origin for each fruit fly which is denoted as (D).

$$D(i) = \sqrt{X(i)^2 + Y(i)^2}$$

Then calculate the fragrance concentration decision value (S) for every fruit fly which is obtained by taking the reciprocal of distance

$$S(i) = \frac{1}{D(i)}$$

Step 4: To calculate the fragrance concentration value (Fragrance) of each fruit fly the fragrance concentration decision value (S) is substituted in the fragrance concentration decision function.

$$\text{Fragrance}(i) = \text{Function}(S(i))$$

Step 5: The fruit fly which has the maximum fragrance concentration value compared to other fruit flies is selected. Store its value and index.

$$[\text{best Fragrance, best index}] = \max(\text{Fragrance})$$

Step 6: Keep the best fragrance concentration value and its coordinates, then the fruit fly cluster flies towards that best location using their searching visualization.

$$\text{Fragrance best} = \text{best Fragrance}$$

$$X_{axis} = X(\text{best index})$$

$$Y_{axis} = Y(\text{best index})$$

Step 7: Check whether it has met the stop criteria, if not then repeat the steps from 2 to 5. If the fragrance concentration value is higher than the previous iteration fragrance concentration value then implements step 6 [45, 52, 53].

16.9.2 Flow Chart of Fruit Fly Optimization Algorithm

The flow chart of the fruit fly optimization algorithm is shown below in Figure 16.2.

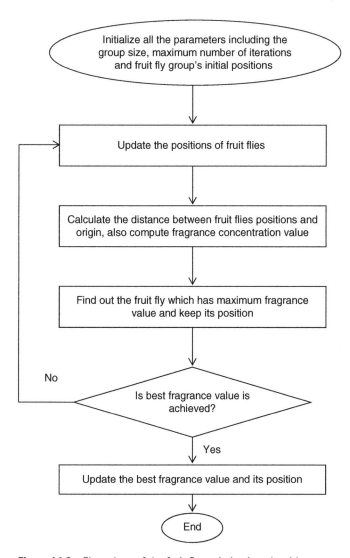

Figure 16.2 Flow chart of the fruit fly optimization algorithm.

16.10 Applicability of FOA in IoT Applications

In this section applications of fruit fly optimization algorithms in IoT are discussed in detail as follows.

16.10.1 Cloud Service Distribution in Fog Computing

In 2018, Atlam, Walter and Wills emphasized the integration of IoT computing and fog. The design of fog is flexible to the necessities of IoT. The fog can support the IoT in the field of smart traffic lights, wireless sensor networks, smart homes, cyber systems and health care. The cloud service distribution is a major issue in the optimization problem. The fruit fly optimization algorithm can be used to reduce the error rate and execution time [55].

16.10.2 Cluster Head Selection in IoT

The efficient broadcasting of information from sensor nodes to the base station is possible due to the IoT. Nodes include a battery with small size and low cost and they are used for sensing some parameters such as temperature, pressure etc. which consumes a small amount of power. The routing is a major issue in such kinds of applications. The fruit fly optimization-based cluster head selection technique is proposed in [56] to conserve energy. The cluster head selection is done on the basis of residual energy of node and length of the link to the base station. The performance of the proposed method is better compared to PSO and LEACH.

16.10.3 Load Balancing in IoT

Effective task scheduling is one of the major issues in cloud computing and it emphasizes balancing the load of tasks among machines. The preservation of energy is another issue in cloud surroundings which further decreases the operation costs. The fruit fly optimization algorithm (EFOA-LB) is proposed in [57] to solve load balancing issues between machines, to achieve fast processing times and to reduce energy consumption. The simulation results show that the proposed algorithm is more effective than the existing algorithms.

16.10.4 Quality of Service in Web Services

Web services are a fundamental concept of service-oriented computing. The current issue in research is how web services can be composed to provide quality of service (QoS). The increase in the number of web services across the cloud data centres has a significant impact on the service performance experienced by the user. In [58] a FOA based network-aware service composition approach is proposed to optimize the various web service QoS parameters such as response time, cost and execution time.

16.10.5 Electronics Health Records in Cloud Computing

Cloud computing is an exciting technology for revolutionizing and reorganizing healthcare associations to deliver the best facilities to the clients. The demand for healthcare facilities is increasing day by day in cloud computing which leads to inequality in usage of sources and the consumption of power increases which results in high operational cost. To optimize the utilization of resources and to attain fast execution times, the hybrid fruit fly optimization technique is proposed in [59] to increase the rate of convergence and to optimize accuracy. The FOA is used to attain the optimum use of resources and decreases the cost and consumption of energy in cloud computing scenarios. The simulation results show that the suggested FOA outperforms resourcefully as compared to other current load balancing procedures.

16.10.6 Intrusion Detection System in Network

For security reasons in cyberspace intrusion detection systems (IDS) are used to find abnormalities in the regular performance of the system. A model for finding the interference is used to appropriately categorize the receiving information as attack or normal data. The major aim is to reduce false positives and increase the exposure rate and accuracy. In [60] an IDS model based on optimized support vector machines (SVM) is proposed. The parameters of SVM are optimized by FOA. The proposed FOA increased the accuracy of classification, exposure rate and also reduced the false alarm rate. The result demonstrates that the proposed FOA achieved better performance compared to other existing models in terms of exposure rate and accuracy.

16.10.7 Node Capture Attack in WSN

Wireless sensor networks (WSN) consist of enormous numbers of sensor nodes which consume low power and have small size and ability to perform tasks. WSN is greatly exposed to various kinds of physical attack due to limited resource capabilities. The main attack in WSN is sensor node capture attack in which the invader physically captures the sensor node and deletes private data from the memory of the node. A multi-objective FOA is proposed in [61] to detect the node capture attack. The result shows that FOA has fewer attacking rounds and energy cost than genetic algorithm (GA) and has also increased attacking efficiency than GA by using the minimum number of sensor nodes in the network.

16.10.8 Node Deployment in WSN

The positions of the sensor nodes play an important role in WSN. A new sensor deployment method based on FOA is proposed in [45] to improve the coverage rate. The node deployment based on FOA shows that fruit fly groups collect the existing overall optimum position for advance exploration to achieve the best sensor deployment. The simulations are carried out in both ideal and obstacle environments. The simulation results show that the proposed algorithm has a higher coverage rate and fast convergence speed compared to other sensor deployment techniques.

16.11 Node Deployment Using Fruit Fly Optimization Algorithm

The sensor deployment is major issue in WSN, so the FOA is used for positioning the sensor nodes and to improve the coverage rate. The main objective is to attain the maximum coverage rate using a limited number of sensor nodes. The sensing area is divided into $M*N$ grid points and n sensors are randomly placed in the deployment area [45]. The distance between the sensor nodes and grid points is calculated using the following formula:

$$\text{Dist}_i = \sqrt{(X_i - x_1)^2 + (Y_i - y_1)^2}$$

where X_i and Y_i are initial positions of ith sensor node (fruit fly)

x_1 and y_1 are coordinates of the first grid point

The position of each fruit fly searching for food using osphresis is updated independently and is given by:

$$X(i, :) = X_{\text{axis}} + 2 * s * rand(1, d) - s$$

where i represent the ith fruit fly

$X\_$axis is the initial position of the ith sensor node

d represents the number of sensor nodes

s represents the search step length

The fragrance concentration values (Fragrance_i) for each fruit fly is calculated using the fragrance concentration judgement function as follows:

$$\text{Fragrance}_i = \text{Computecover}(X(i, :))$$

where Computecover is defined as the minimum distance between grid point and fruit fly which is selected after comparing all the distances among n fruit flies and grid points at each iteration [45].

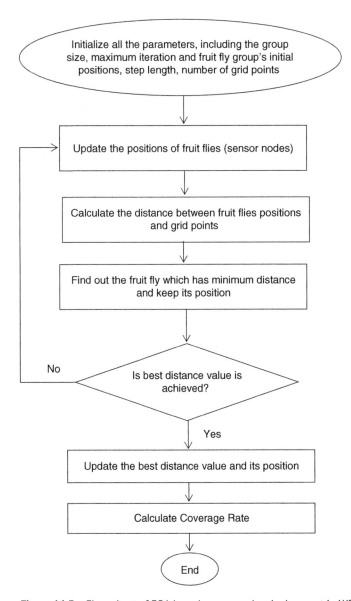

Figure 16.3 Flow chart of FOA based sensor nodes deployment in WSN.

The minimum distance value and coordinates of sensor nodes are stored and checked whether best distance value is achieved or not. If the best distance value is achieved then the distance value and its positions are updated and the coverage rate calculated, otherwise repeat the above process.

The flow chart of fruit fly-based sensor deployment in WSN is shown in Figure 16.3.

Binary detection model which expresses the coverage $C_{xy}(S_i)$ of a grid point p by sensor S_i. and is defined as follow:

$$C_{xy}(S_i) = \begin{cases} 1, \text{Dist}_i \leq r \\ 0, \text{otherwise} \end{cases}$$

where r is the sensing radius and Dist_i is the Euclidean distance

Coverage rate is the objective function of the sensor network coverage problem which requires to be improved. The coverage rate (CR) is defined as follows:

$$CR = \frac{N_{\text{effect}}}{M * N}$$

where N_{effect} is the number of grid points which are covered by sensor nodes [45].

16.12 Conclusion

In this chapter evolution of the Internet of Things (IoT) with its application and challenges has been discussed. Different optimization techniques and their applications in the IoT have been explored. Further, the fruit fly optimization algorithm and various steps involved in the fruit fly optimization algorithm are described in detail. In the last section, applications of FOA in the IoT and node deployment using FOA are explained briefly. It has been observed that FOA can be used to increase the coverage rate of sensor nodes. In future, various optimization techniques such as the bat algorithm and other nature inspired optimization algorithms can be used to deploy the sensor nodes and to increase the coverage rate.

Bibliography

1 Dudhe, P.V., Kadam, N.V., Hushangabade, R.M. et al. (2017). Internet of Things (IOT): An overview and its applications. *Proceedings of the International Conference on Energy, Communication, Data Analytics and Soft Computing (ICECDS)*, Chennai, India (1–2 August 2017).

2 Farhan, L., Kharel, R., Kaiwartya, O. et al. (2018). A Concise Review on Internet of Things (IoT) - Problems, Challenges and Opportunities. *Proceedings of* the 11*th* International Symposium on Communication Systems, Networks and Digital Signal Processing (CSNDSP), Budapest, Hungary (18–20 July 2018).

3 Fuqaha, A., Guizani, M., Mohammadi, M. et al. (2015). Internet of things: A survey on enabling technologies, protocols, and applications. *IEEE Communications Surveys Tutorials* 17 (4): 2347–2376.

4 Farhan, L., Kharel, R., Kaiwartya, O. et al. (2018). Towards green computing for Internet of Things: Energy oriented path and message scheduling approach. *Sustainable Cities and Society* 38: 195–204.

5 Ashton, K. (2009). That internet of things thing, *RFiD J.* 22 (7): 97–114.

6 Srinidhi, N.N., Dilip Kumar, S.M. and Venugopal, K.R. (2019). Network optimizations in the Internet of Things: A review. *Engineering Science and Technology, an International Journal* 22: 1–21.

7 Farhan, L., Alzubaidi, L., Abdulsalam, M. et al. (2018). An efficient data packet scheduling scheme for Internet of Things networks. *Proceedings of the 3ʳᵈ Scientific Conference of Engineering Sciences (ISCES) and 1st International Scientific Conference of Engineering Sciences*, Diyala, Iraq (10–11 January 2018).

8 Shah, S., Yaqoob, I. (2016). A survey: Internet of Things (IOT) technologies, applications and challenges. *Proceedings of the 4ᵗʰ IEEE International Conference on Smart Energy Grid Engineering*, Oshawa, ON, Canada (21–24 August 2016).

9 Xu, L.D., He, W. and Li, S. (2014). Internet of Things in Industries: A Survey. *IEEE Transactions on Industrial Informatics* 10 (4): 2233–2243.

10 Sisinni, E., Saifullah, A. et al. (2018). Industrial Internet of Things: Challenges, Opportunities, and Directions. *IEEE Transactions on Industrial Informatics* 14 (11): 4724–4734.

11 Yang, C., Shen, W., and Wang, X. (2016). Applications of Internet of Things in manufacturing. *Proceedings of the IEEE 20ᵗʰ International Conference on Computer Supported Cooperative Work in Design (CSCWD)*, Nanchang, China (4–6 May 2016).

12 Khanna, A., and Anand, R. (2016). IoT based smart parking system. *Proceedings of the 2016 International Conference on Internet of Things and Applications (IOTA)*, Pune, India (22–24 January 2016).

13 Lookmuang, R., Nambut, K. and Usanavasin, S. (2018). Smart parking using IoT technology. *Proceedings of the 2018 5th International Conference on Business and Industrial Research (ICBIR)*, Bangkok, Thailand (17–18 May 2018).

14 Alsafery, W., Alturki, B. et al. (2018). Smart Car Parking System Solution for the Internet of Things in Smart Cities. *Proceedings of the 2018 1st International Conference on Computer Applications & Information Security (ICCAIS)*, Riyadh, Saudi Arabia (4–6 April 2018).

15 Baker, S., Xiang, W. and Atkinson, I. (2017). Internet of Things for Smart Healthcare: Technologies, Challenges, and Opportunities. *in IEEE Access* 5: 26521–26544.

16 Kwak, D., Kabir, M. et al. (2015). The Internet of Things for Health Care: A Comprehensive Survey. *in IEEE Access* 3: 678–708.

17 Darianian, M., Michael, M.P. (2008). Smart Home Mobile RFID-Based Internet-of-Things Systems and Services. *Proceedings of the International Conference on Advanced Computer Theory and Engineering*, Phuket, Thailand (20–22 December 2008).

18 Rajab, H. and Cinkelr, T. (2018). IoT based Smart Cities. *Proceedings of the 2018 International Symposium on Networks, Computers and Communications (ISNCC)*, Rome, Italy (19–21 June 2018).

19 Dlodlo, N., Gcaba, O. and Smith, A. (2016). Internet of things technologies in smart cities. *Proceedings of the 2016 IST-Africa Week Conference*, Durban, South Africa (11–13 May 2016).

20 Rana, R.K., Chou, C.T. et al. (2010). *Ear-phone: an end to-end participatory urban noise mapping system, in: acm request permissions.*

21 Talavera, J.M., Tobon, L.E. et al. (2017). Review of IoT applications in agro-industrial and environmental fields. *Computers and Electronics in Agriculture* 142: 283–297.

22 Sastra, N.P. and Wiharta, D.M. (2016). Environmental monitoring as an IoT application in building smart campus of Universitas Udayana. *Proceedings of the 2016 International Conference on Smart Green Technology in Electrical and Information Systems (ICSGTEIS)*, Bali, Indonesia (6–8 October 2016).

23 Han, F.L., Drieberg, M. et al. (2018). An Internet of Things Environmental Monitoring in Campus. *Proceedings of the 2018 International Conference on Intelligent and Advanced System (ICIAS)*, Kuala Lumpur, Malaysia (13–14 August 2018).

24 Jaiganesh, S., Gunaseelan, K. and Ellappan, V. (2017). IOT agriculture to improve food and farming technology. *Proceedings of the 2017 Conference on Emerging Devices and Smart Systems (ICEDSS)*, Tiruchengode, India (3–4 March 2017).

25 Dagar, R., Som, S. and Khatri, S.K. (2018). Smart Farming – IoT in Agriculture. *Proceedings of the 2018 International Conference on Inventive Research in Computing Applications (ICIRCA)*, Coimbatore, India (11–12 July 2018).

26 Gubbi, J., Buyya, R. et al. (2013). Internet of Things (IoT): A vision, architectural elements, and future directions. *Future Generation Computer Systems* 29: 1645–1660.

27 Saha, D., Yousuf, M.R. et al. (2011). Energy Efficient Scheduling Algorithm for S-Mac Protocol In Wireless Sensor Network. *International Journal of Wireless and Mobile Networks* 3: 129–140.

28 Vejlgaard, B., Lauridsen, M. et al. (2017). Interference impact on coverage and capacity for low power wide area IoT networks. *Proceedings of the IEEE Wireless Communications and Networking Conference (WCNC)*, San Francisco, CA, USA (19–22 March 2017).

29 Liu, C., Zhang, Y. et al. (2013). A novel approach to IoT security based on immunology. *Proceedings of the 9th International Conference on Computational Intelligence and Security*, Leshan, China (14–15 Dec. 2013).

30 Brogi, A., Forti, S. (2017). QoS-aware deployment of IoT applications through the fog. *IEEE Internet of Things Journal* 4: 1–8.

31 Swisher, J.R., Hyden, P.D. et al. (2000). A survey of simulation optimization techniques and procedures. *Proceedings of the Winter Simulation Conference Proceedings*, Orlando, FL, USA, USA (10–13 December 2000).

32 Csiszar,S. (2007). Optimization Algorithms (survey and analysis). *Proceedings of the 2007 International Symposium on Logistics and Industrial Informatics*, Wildau, Germany (13–15 September 2007).

33 Tyagi, Kanika and Seth (2015). A Comparative Analysis of Optimization Techniques. *International Journal of Computer Application* 131(10): 6–12.

34 Hu, Y., Ding,Y. et al.(2014). An immune orthogonal learning particle swarm optimisation algorithm for routing recovery of wireless sensor networks with mobile sink. *Int. J. Syst. Sci.* 45 (3): 337–350.

35 Kumrai, T., Ota, K. et al. (2017). Multi-objective optimization in cloud brokering systems for connected internet of things. *IEEE Int. Things J.* 4 (2): 404–413.

36 Dhumane, A.V., Prasad, R.S. et al. (2017). An optimal routing algorithm for internet of things enabling technologies. *Int. J. Rough Sets Data Anal. (IJRSDA)* 4 (3): 1–16.

37 Martins, J., Mazayev, A. et al. (2017). Gacn: self-clustering genetic algorithm for constrained networks. *IEEE Commun. Lett.* 21 (3): 628–631.

38 Dhondge, K., Shorey, R. et al. (2016). Heuristic and opportunistic link selection algorithm for energy efficiency in industrial internet of things (IIoT) systems. *Proceedings of the 8th International Conference on Communication Systems and Networks (COMSNETS)*, Bangalore, India (5–10 January 2016).

39 Shailendra, S., Rao, A. et al. (2017). Power efficient RACH mechanism for dense IoT deployment. *Proceedings of the IEEE International Conference on Communications Workshops (ICC Workshops)*, Paris, France (21–25 May 2017).

40 Ismail, N.H.A., Hassan, R. (2013). Lowpan local repair using bio inspired artificial bee colony routing protocol. *Procedia Technol.* 11: 281–287.

41 Reddy, P.K., Babu, R. (2017). An evolutionary secure energy efficient routing protocol in internet of things. *Int. J. Intell. Eng. Syst.* 10 (3): 337–346.

42 Kulkarni, R.V., Kumar, G. et al. (2011). Particle Swarm Optimization in Wireless-Sensor Networks: A Brief Survey. *IEEE Transactions on Systems, Man and Cybernetics-Part C: Applications and Reviews*, 41(2): 262–267.

43 Kulkarni, R.V., Kumar, G. et al. (2009). Bio-Inspired Node Localization in Wireless Sensor Networks. *Proceeding of the IEEE International Conference on Systems, Man and Cybernetics*, San Antonio, TX, USA (11–14 Oct. 2009).

44 Yu, X., Zhang, J. et al. (2013). A Faster Convergence Artificial Bee Colony Algorithm in Sensor Deployment for Wireless Sensor Networks. *International Journal of Distributed Sensor Networks* 9(10): 1–9.

45 Zhao, H., Zhang,Q. et al. (2015). Novel Sensor Deployment Approach Using Fruit Fly Optimization Algorithm in Wireless Sensor Networks. *Proceedings of the IEEE Conference on Trustcom/BigDataSE/ISPA*, Helsinki, Finland (20–22 Aug. 2015).

46 Goyal, S., Patterh, M.S. (2014). Wireless Sensor Network Localization Based on Cuckoo Search Algorithm. *Wireless Personal Communication* 79(1): 223–234.

47 Liao, W., Kao Y., and Li, Y. (2011). A sensor deployment approach using glow-worm swarm optimization algorithm in wireless sensor networks. *Journal of Expert Systems with Applications* 38: 12180–12188.

48 Hamza, K.S., Amir, F. (2016). Evolutionary clustering for integrated WSN-RFID networks. *Proceedings of the 10th International Conference on Informatics and Systems*, Giza, Egypt (May 9–11, 2016).

49 Deif, D.S., Gadallah, Y. (2014). Wireless Sensor Network Deployment Using a Variable-Length Genetic Algorithm. *Proceedings of the IEEE Conference on Wireless Communications and Networking (WCNC)*, Istanbul, Turkey (6–9 April 2014).

50 Li, Z., Lei, L. (2009). Sensor Node Deployment in Wireless Sensor Networks Based on Improved Particle Swarm Optimization. *Proceedings of the IEEE International Conference on Applied Superconductivity and Electromagnetic Devices*, Chengdu, China (25–27 Sept. 2009).

51 Kulkarni, R.V., Kumar, G. et al. (2010). Bio-inspired Algorithms for Autonomous Deployment and Localization of Sensor Nodes. *IEEE Transactions on Systems, Man, and Cybernetics-PART C: Applications and Reviews* 40(6): 663–675.

52 Pan, W.T. (2012). A new fruit fly optimization algorithm: taking the financial distress model as an example. *Knowledge-Based Systems* 26: 69–74.

53 Yuan, X., Dai, X. and Zhao J. (2014). On a novel multi-swarm fruit fly optimization algorithm and its application. *Applied Mathematics and Computation* 233: 260–271.

54 Iscan, H., Gunduz, M. (2015). A Survey on Fruit Fly Optimization Algorithm. *Proceedings of the 11ᵗʰ International Conference on Signal-Image Technology and Internet-Based Systems (SITIS)*, Bangkok, Thailand (23–27 November 2015).

55 Atlam, H.F., Walters, R.J. et al. (2018). Fog Computing and the Internet of Things: A Review. *Big Data and Cognitive Computing* 2(10): 1–18.

56 Poluru, R.K., and Kumar, L. (2009). An Improved Fruit Fly Optimization (IFFOA) based Cluster Head Selection Algorithm for Internet of Things. *International Journal of Computers and Applications*: 1–9.

57 LawanyaShri, M., Subha, S. and Balusamy, B. (2017). Energy-Aware Fruitfly Optimisation Algorithm for Load balancing in Cloud Computing Environments. *International Journal of Intelligent Engineering and Systems* 10: 75–84, 2017.

58 Shehu, U., Safdar, G. et al. (2016). Fruit Fly Optimization Algorithm for Network-Aware Web Service Composition in the Cloud. *International Journal of Advanced Computer Science and Applications* 7: 1–11.

59 LawanyaShri, M., Subha, S. and Balusamy, B. (2017). Energy-aware hybrid fruitfly optimization for load balancing in cloud environments for EHR applications. *Informatics in Medicine Unlocked* 8: 42–50.

60 Haneef, F., Singh, S. (2017). Improved intrusion detection system based on optimized SVM using M-FOA. *International Journal of Advanced Research and Development* 2(2): 644–650.

61 Bhatt, R., Maheshwary, P. et al. (2019). Implementation of Fruit Fly Optimization Algorithm (FFOA) to escalate the attacking efficiency of node capture attack in Wireless Sensor Networks (WSN). *Computer Communications* 149: 134–145.

17

Optimization Techniques for Intelligent IoT Applications

Priyanka Pattnaik[1,], Subhashree Mishra[2], and Bhabani Shankar Prasad Mishra[1]*

[1] School of Computer Engineering, KIIT University, Bhubaneswar, Odisha
[2] School of Electronics Engineering, KIIT University, Bhubaneswar, Odisha

Abstract

In the current time, engineering problems are associated with multiple objectives. Based on the number of objectives an optimization problem can be classified as single-, multi-, and many-objective optimization. In the case of a single-objective optimization it is easy to find a single solution, however it becomes very difficult to get a single solution in the cases of multi- and many-objective optimization as the objectives are quite contradictory. Hence, evolutionary and swarm-based algorithms are widely used to address such problems where the search space is very large and the problem is associated with multiple contradictory objectives. Today optimization is a powerful tool of trade for the engineer in virtually every discipline. It provides them with a rigorous, systematic method for rapidly zeroing in on the most innovative, cost-effective solutions to some of today's most challenging engineering design problems. The Internet of Things (IoT) is the concept of connecting everyday devices to the Internet allowing the devices to send and receive data. With the IoT, devices can constantly report their status to a receiving computer that uses the information to optimize decision making. The IoT network optimization offers a lot of benefits for improving traffic management, operating efficiency, energy conservation, reduction in latency, higher throughput and faster rate in scaling up or deploying IoT services and devices in the network. This chapter presents an overall and in depth study of different optimization algorithms inspired from the behaviour of nature.

Keywords *Optimization; Evoltion; IoT; Search*

Fog, Edge, and Pervasive Computing in Intelligent IoT Driven Applications, First Edition.
Edited by Deepak Gupta and Aditya Khamparia.

Figure 17.1 Cuckoo bird.

17.1 Cuckoo Search

17.1.1 Introduction to Cuckoo

Computer science always works by making efficient algorithms and to make such algorithms it gains inspiration from various fields [3], and nature is always an inspiration for us. Cuckoo search is a type of innovation which is completely galvanized by the behaviour of the cuckoo (in Figure 17.1 and Figure 17.2).

17.1.2 Natural Cuckoo

Cuckoo Search is an innovation for optimization that was discovered in 2009 by Xin-she Yang and Suash Deb [21]. As it is inspired by the cuckoo it follows the behaviour of the cuckoo, i.e., its egg laying behaviour. Cuckoos have a natural behaviour of laying egg in other birds' nests. It chooses its target nest carefully and lays eggs, so that the host bird will not discover the outsider's egg and so takes care of the egg as if it were its own [1]. If the host bird discovers the outsider's egg then it then it will either dump the outsider egg or will leave the nest.

Figure 17.2 A nest.

17.1.3 Artificial Cuckoo Search

This is inspired from the natural search behaviour of the cuckoo and follows three conditions to make it a better search optimization technique. The first condition is that each cuckoo should lay one egg at a time and will choose the nest for its egg randomly [4]. The second condition is if the egg is not discovered by the host bird then the egg will progress and the nest will be marked and chosen as the best nest for the next generation. The third condition is if the host bird discovers the parasite egg than that nest is counted as the worst nest solution and discarded [2].

17.1.4 Cuckoo Search Algorithm

The Cuckoo search algorithm is based on three primitive operations: initialization, fitness evaluation and creation of a new nest.

As we know, in the first phase the cuckoo chooses its nest randomly which is actually followed up by the Levy flights mechanism. The Levy flight mechanism is characterized by a random walk of the animals or birds that they take while searching for the food. This is also followed by the step length of the animals or birds which obey the power law distribution.

$$y_{t+1} = y_t + sP_t \tag{17.1}$$

P_t: The standard normal distribution
S: The step size

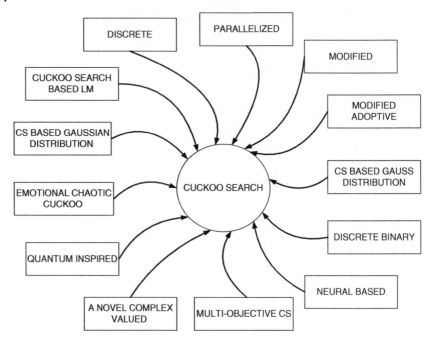

Figure 17.3 Variants of cuckoo search.

17.1.5 Cuckoo Search Variants

When the cuckoo search developed it gave us a new method for optimization. People have attempted to make some variants of it and and have got some good results from it. Such variants are shown in the figure below (in Figure 17.3).

17.1.6 Discrete Cuckoo Search

The objective of the discrete cuckoo is to enhance the comprehensive search for the best solutions for the population, by using Levy flights (Figure 17.4 presents a flowchart of a discrete cuckoo search) [6].

17.1.7 Binary Cuckoo Search

In a binary cuckoo search first the algorithm itself creates solutions at the initial stage with the operator (in Figure 17.5). When we initialize it then the task is to evaluate the compliance with the stopping criteria. So, for stopping we basically take two criteria, i.e., the maximum iteration number or if it gives the optimal value continuously [5]. When we get one of the stopping criteria then the k-means operator for transition is executed and it perform the binarization. If the transitions

Figure 17.4 Discrete cuckoo search.

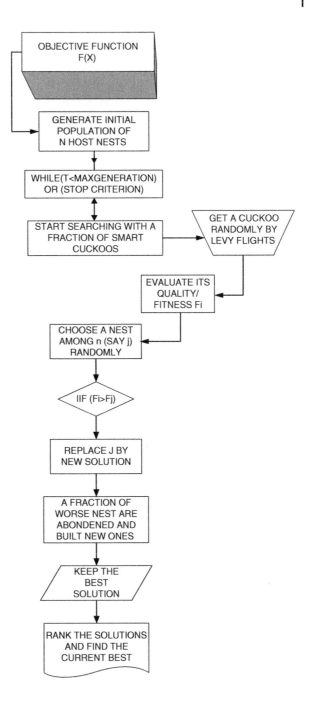

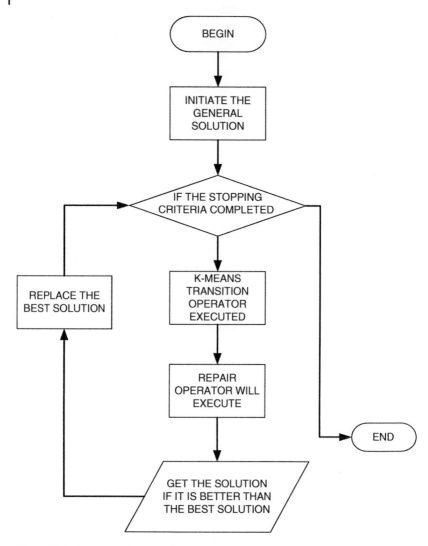

Figure 17.5 Binary cuckoo search.

are already obtained then it applies a repair operator for the solution which is not accomplished with the problem restriction.

17.1.8 Chaotic Cuckoo Search

In the cuckoo search the cuckoo uses a method called Levy flight mechanisms to amend its nest choices.

$$y_i^{(t+1)} = y_i^{(t)} + \alpha \oplus \text{Levy}(\lambda) \tag{17.2}$$

$\alpha > 0$ for the size of step. In the chaotic cuckoo search the algorithm depends on a random based method which uses chaotic variables. In the chaotic cuckoo search it takes the property of non-repetition and it is ergodic in nature [7]. It also helps to accelerate the convergence. It also tends to normalize all the chaotic maps which have the variation [0,2]. So, basically it helps to obtain the best cuckoos in the process of the algorithm (in Figure 17.6).

17.1.9 Parallel Cuckoo Search

The parallel cuckoo search is based on parallel computing. OpenMP is one of the ways to implement the parallel cuckoo search. The advantage of the parallel cuckoo search is that if you can carry out any step of the cuckoo search by parallel computing then the cuckoo search is known as parallel computing. For example, the main population can be divided into sub-populations to run in the same space [8].

17.1.10 Application of Cuckoo Search

Cuckoo search is used in almost every field for better optimization, i.e., in science, maths, technology, medicine, [10] environment etc. (some of the most popular areas are shown in Figure 17.7).

17.2 Glow Worm Algorithm

17.2.1 Introduction to Glow Worm

Glow worms are actually both the adult beetles (in Figure 17.8) and the larvae (maggots). Both the adults and the larvae follow a process known as biolumines-cence in which they can produce luminosity in the organs in their abdomens. They are also able to radiate their glow as a pluse or as a continual glow [11].

17.2.2 Glow Worm Swarm Optimization Algorithm (GSO)

This is also known as the GSO algorithm. In its first phase the glow worms are randomly deployed in the space of the solution. In the next step we count every solution as an objective which carries luciferin. The effect of luciferin is correlated with the strength, i.e., fitness of each handler. We measure the fitness according to the luciferin level. Every agent only goes to the neighbour if the neighbour has higher luciferin than its own [12]. The decision radius is affected by the density of the glow worms' neighbour and the local-decision domain can be determined:

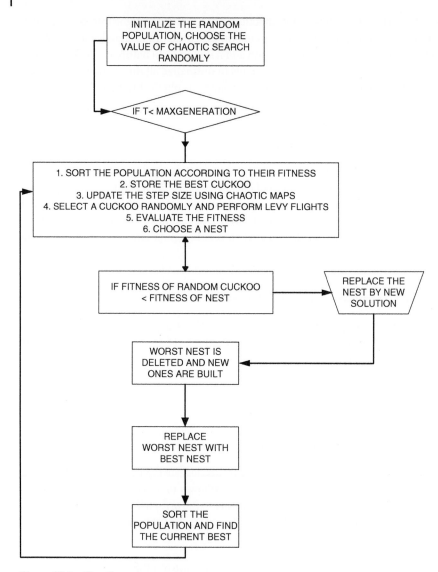

Figure 17.6 Chaotic cuckoo search.

in the case when the algorithms find that the neighbourhood density is low, then it shows that the local-decision domain becomes enlarged in structure to explore more near-neighbours; otherwise, it will be reduced to allow the swarm to disband into smaller groups (Figure 17.9). This procedure is repeated until the termination condition is reached. It follows five rules.

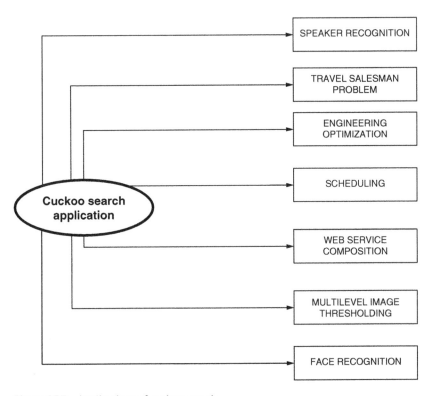

Figure 17.7 Applications of cuckoo search.

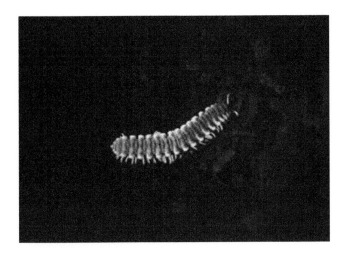

Figure 17.8 A glow worm.

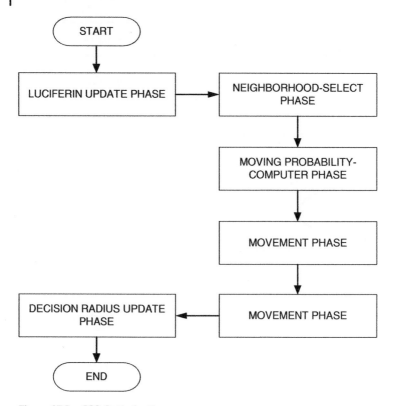

Figure 17.9 GSO Optimization.

- Step 1 *Luciferin update phases*
  In the GSO algorithm luciferin is represented as fitness of glow worms. So the update of fitness depends on the previous luciferin value

$$l_i(t+1) = (1 - \rho)\, l_i(t) + \gamma \, \text{Fitness}\left(x_i(t+1)\right) \tag{17.3}$$

  Here, (t) : amount of luciferin for glow worm i in time t,
  $(t+1) \in$: location of candidate i at time $t+1$,
  Fitness $(x_i(t+1))$ presents the value for the strength at glow worm i's location where the time is $t+1$.

- Step 2 *Neighborhood-select phase*
  In this phase the neighbors of glow worm i in time t are those who glow brightest so, we can write it as

$$N_i(t) = \left\{ j\, :\, d_{ij}(t) < r_d^i(t)\,;\, l_i(t) < l_j(t) \right\} \tag{17.4}$$

  $d_{ij}(t)$: Euclidean distance in between the i and j candidates in time t
  $N_i(t)$: neighbors

$r_d^i(t)$: decision radius of the glow worm i at time t
- Step 3 *Moving probability-computer phase*
  In this phase the glow worm follows a rule of probability.

$$P_{ij}(t) = \frac{l_j(t) - l_i(t)}{\sum_{k \in N_i(t)} l_k(t) - l_i(t)} \tag{17.5}$$

$P_{ij}(t)$: probability of glow worm i which is moving towards glow worm j
- Step 4 *Movement phase*
  In this phase the glow worm j, which belongs to the neighbor with the probability

$$x_i(t+1) = x_i(t) + s \frac{\left(x_j(t) - x_i(t)\right)}{\left\| x_j(t) - x_i(t) \right\|} \tag{17.6}$$

- Step 5 *Decision radius update rule*
  In this phase the glow worm updates the decision radius

$$r_i^d(t+1) = \min \left\{ r_s, \max \left\{ 0, r_d^i(t) + \beta \left(n_t - | N_i(t) | \right) \right\} \right\} \tag{17.7}$$

r_s : sensory radius
n_t : parameter which is used to control the neighbor number.

17.3 Wasp Swarm Optimization

17.3.1 Introduction to Wasp Swarm and Wasp Swarm Algorithm (WSO)

When a group of insects move in large numbers we call it a swarm (Figure 17.10). A wasp swarm is a dangerous situation because the insects can sting and they have the potential to quickly become aggressive. Their main function is to sting continuously until they find themselves safe [13].

The algorithm is described below.

- Step 1
  Initialize by giving the number of option
- Step 2
  Give your objective function
- Step 3
  Provide the strength function (according to the application)
- Step 4
  For every possible solution
  Calculate the strength
  Calculate the probability of each solution with each strength

Figure 17.10 Wasp swarm.

- Step 5
  Pick an individual
  Calculate the cost
- Step 6
  Output optimal solution with minimum cost (individual picked for cost)

17.3.2 Fish Swarm Optimization (FSO)

The FSO algorithm concept is where the fish in a swarm are searching for food (Figure 17.11). From the point of view of optimization this type of behaviour is the phase of learning which leads the swarm of fish change direction and begins an exploration for a new food source. Swarms of fish have the property of finding a search space. They also have the property of behavior to move randomly in a swarm and to chase food. When a region stagnates then they change to a new region for looking for food [14].

Fish swarm optimization algorithm follows certain conditions

1) Each fish is represented as a solution candidate for the problem.
2) It needs to optimize the density of the food which is the objective function.
3) The final condition is it to find food within the design space.

17.3.3 Fruit Fly Optimization (FLO)

This is proposed by Pan in 2011 [31]. The small flies that you may see in your kitchen are probably fruit flies. Fruit flies actually have no brain instead they

Figure 17.11 Swarm of fish.

have hundreds of neurons instead. They usually have red eyes. They have certain behaviours which lead them to give a unique identification [15].

1) The fruit flies can smell the food using the osphresis organ and are especially attracted to ripe food.
2) They have 760 unit eyes in their compound eye which allows them to exploit a wide range of food sources.

- Step 1 *Initialization:*
  In the search space, the flies are randomly distributed.
- Step 2 *Path construction phase:*
  The fruit flies can judge the distance of the food according to the concentration of smell

$$D_i = \sqrt{x_i^2 + y_i^2} \tag{17.8}$$

$$S_i = \frac{1}{D_i} \tag{17.9}$$

D_i Is the distance between the food and the fruit flies (x,y)
S_i Is the concentration of the smell judgement value.

- Step 3 *Fitness calculation :*

$$\text{Smell}_i = \text{function}\left(S_i\right) \tag{17.10}$$

Smell is the concentration of each individual. The best smell will be the max (smell).

- Step 4 *Movement Phase:*
  Fruit flies fly towards the location of the source of food upon identifying the best smell.

17.3.4 Cockroach Swarm Optimization

This algorithm was introduced by Zhaohui and Haiyan for global optimization problems. This algorithm mimics the behaviour of the cockroach. It particularly mimics chase swarming behavior, dispersion, and their ruthless social behaviour [16].

- Step 1 *Initialization:*
  The population is distributed randomly in the region.
- Step-2 *Chase-swarming behaviour:*
  The cockroaches can take fixed size steps. They are distributed randomly. They take steps according to their personal best and the global best. And each individual will chase the local optimum individual.
- Step-3 *Dispersion behaviour:*
  The random behaviour of the cockroach is to stay in a random vector of dimension which can be set in a range.
- Step-4 *Ruthless character:*
  It replaces the current best with another individual, and the individual will be selected randomly.
- Step-5
  If it meets the termination condition then it stops, otherwise it again repeats the chase-swarming behavior.

17.3.5 Bumblebee Algorithm

Bumblebees follow a set of rules to evolve in this artificial world. The rules that they consider for evolving gives us a solution for some problems. In the bumblebee algorithm the fitness function as the lifespan of the bumblebee is considered while in the nature-inspired algorithm they consider either some cost function or some fitness function. The algorithm is described below [20].

- Step 1
  Insert the random solutions
- Step 2
  Generate all of the solutions for your given problem
- Step 3
  Set the value of the maximum generation

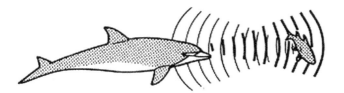

Figure 17.12 Dolphin and the prey.

- Step 4
  Like the bumblebees nest give dimensions
  1) Take *n* no of bumblebees and make a colony
  2) Randomly locate the food cells
  3) Every individual bumblebee is a generated random solution
  4) Give a range to each bumblebee as their life span (give the span considering the weight of the solution)
- Step 5
  If the current population is greater than the maximum generation
  1) Find all the solutions
  2) If you get the global optimum
  3) Exit the algorithm
- Step 6
  If global optimum is not found then
  1) Life counters of bumblebees need to decrease
  2) Again you have to create new bumblebees.

17.3.6 Dolphin Echolocation

Dolphins are known for their natural friendly behaviour. They find food using echolocation (Figure 17.12). In the figure you can see that the dolphins produce a sound and the sound travells a certain range, if inside that range it collides with an object then the sound will be returned to the dolphin, i.e., the sound echoes towards the dolphin. The dolphin then tracks the sound of the echo and moves towards the food source [19].

The dolphin directs itself towards the prey by tracking the convergence factor. The convergence factor depends on the standard_deviation of the prey, upper_limit and lower _limit (i.e. upper size and lower size of the prey).

$$\text{Convergence\_factor} = 1 - \frac{\text{standard\_deviation}}{(\text{upper\_limit-lower\_limit})/2} \tag{17.11}$$

$$\text{predefined\_probability (loop)} = \text{predefined\_probability}_1 + \left(1 - \text{predefined\_probability}_1\right) \frac{\text{loop}^{\text{power}} - 1}{(\text{loop\_number})^{\text{power}} - 1} \tag{17.12}$$

Figure 17.13 Frog leap.

The dolphin echolocation algorithms steps are defined below.

- Step 1
  Select a confined location in which the dolphin can move only
- Step 2
  Calculate the predefined probability
- Step 3
  Find the fitness for every location
- Step 4
  Create the best fitness matrix, leading curve, smooth best fitness curve

17.3.7 Shuffled Frog-leaping Algorithm

This algorithm shows the characteristics of frogs that live together (a frog leap is shown in Figure 17.13). The frogs find their food by locating the biggest stones. The frogs can communicate with each other so that they can indicate the location of food and can change the distance of the leap [16].

The shuffled frog-leaping algorithm is described below.

- Step 1
  Initialize the number of the location and the number of frogs inside each location
- Step 2
  According to the behavior of the frogs, rank them
- Step 3
  Rank the location according to the rank of the frogs in that location
- Step 4
  Evolution of frogs inside every location
- Step 5
  Shuffle the frogs in all locations
- Step 6
  Sort all locations
- Step 7
  Find the convergence and assign it to every frog.

Figure 17.14 Seed in fertile soil with proper care.

17.3.8 Paddy Field Algorithm

This algorithm is inspired by paddy fields (in Figure 17.14, Figure 17.15). The seeds that have access to fertile soil, water and sunlight tend to grow taller and give more seeds compared with the plants that grow in infertile soil. So in order to get the optimum solution the conditions must satisfied so that the optimum solution is achieved [17].

Figure 17.15 Seed in infertile soil.

The algorithm consists of six steps.

- Step 1 *sowing* :
  Initially the seeds are distributed randomly in the space.
- Step 2 *selection:*
  When the plants are produced by the seed then the best plants are carried out by using the threshold method.
- Step 3 *seedling:*
  Each plant produces new seedlings which tend to be healthier. The health of a plant is counted as the number of seedlings grown in that plant.
- Step 4 *pollination:*
  From all the seedlings that have been produced only few of them are actually considered for pollination. The seedlings that are considered for pollination are dependent on the plants that are present in its neighborhood. The higher the number of the neighborhood of a plant then the greater chance it has of getting its seedlings pollinated.
- Step-6 *dispersion:*
  Then again the seeds are dispersed and the sowing process is started.

17.4 Real World Applications Area

The IoT has arrived at its peak time, we now have smart cities, smart buildings, smart homes, smart kitchens, smart appliances, security, smart homes, smart health, smart factories, smart machines, smart supply chains, smart transportation, smart manufacturing, autonomous vehicles, smart consumer devices, etc. [24]. The infrastructure and management will be based on the smart systems which have sensors and crowd sourcing as information sources. The future success of the IoT depends on the optimization and new approaches for application of interfaces, allocation of behavior, evolution of technology, and system security [25]. The success of the IoT will only be realized through major development of systems engineering processes and tools that will be able to overcome the challenges of the IoT network.

Bioinspired algorithm taxonomy such as evolutionary, swarm based, ecology, network, and immune based are applied to overcome challenges such as resource constraint, scalability, heterogeneity, mobility and security. We found that three domains where swarm intelligence have better solutions. They are, vehicle routing problem for connected cars with ant colony optimization, data routing from a widespread sensor network with ACO and cloud computing techniques for data optimization [23].

There are also some areas where nature inspired algorithms are widely used, for example: benchmarking optimization, cluster computing, image processing, economic dispatch problems, engineering design and applications, power and energy optimization, swarm intelligence in routing in wireless sensor networks, swarm intelligence in vehicle routing in IoT, swarm intelligence in prediction of remaining useful life for batteries in IoT [30], swarm intelligence for energy efficient sensor movement in wireless sensor networks, swarm intelligence for node search for wireless sensor networks [28], swarm intelligence for radio frequency identification (RFID) network planning problem in IoT, swarm intelligence for service optimization problem in IoT, swarm intelligence for clustering in wireless sensor networks [27], swarm intelligence for sensor deployment in wireless sensor networks [29], swarm intelligence for optimal controller design for battery energy storage, swarm intelligence for collaborative mobile sensing in wireless sensor networks, swarm intelligence for energy management in IoT [26], swarm intelligence for node and sink nodes localization of sensor networks, Swarm intelligence for trajectory optimization for an autonomous mobile robot in IoT, swarm intelligence for search route in routing processing of IoT [22].

Summary

This chapter presents a detailed explanation of the cuckoo search algorithm, different types of cuckoo search, variants in cuckoo search, different application fields in cuckoo search, glow worm optimization, wasp swarm algorithm, fish swarm optimization, fruit fly optimization, bumblebee optimization, cockroach swarm optimization, dolphin echolocation, shuffled frog leaping algorithm, and paddy field algorithm. These can be used in many fields of engineering.

Bibliography

1 Mareli, M. and Twala, B., 2018. An adaptive Cuckoo search algorithm for optimisation. *Applied computing and informatics*, 14(2), pp.107–115.
2 Abdelaziz, A.Y. and Ali, E.S., 2015. Cuckoo search algorithm based load frequency controller design for nonlinear interconnected power system. *International Journal of Electrical Power & Energy Systems*, 73, pp.632–643.
3 Mohanty, P.K. and Parhi, D.R., 2016. Optimal path planning for a mobile robot using cuckoo search algorithm. *Journal of Experimental & Theoretical Artificial Intelligence*, 28(1-2), pp.35–52.
4 Gandomi, A.H., Yang, X.S. and Alavi, A.H., 2013. Cuckoo search algorithm: a metaheuristic approach to solve structural optimization problems. *Engineering with computers*, 29(1), pp.17–35.

5 Rodrigues, D., Pereira, L.A., Almeida, T.N.S., Papa, J.P., Souza, A.N., Ramos, C.C. and Yang, X.S., 2013, May. BCS: A binary cuckoo search algorithm for feature selection. In *2013 IEEE International Symposium on Circuits and Systems (ISCAS2013)* (pp. 465–468).

6 Ouaarab, A., Ahiod, B. and Yang, X.S., 2014. Discrete cuckoo search algorithm for the travelling salesman problem. *Neural Computing and Applications*, 24(7-8), pp.1659–1669.

7 Wang, G.G., Deb, S., Gandomi, A.H., Zhang, Z. and Alavi, A.H., 2016. Chaotic cuckoo search. *Soft Computing*, 20(9), pp.3349–3362.

8 Xu, X., Ji, Z., Yuan, F. and Liu, X., 2014, January. A novel parallel approach of cuckoo search using mapreduce. In *2014 International Conference on Computer, Communications and Information Technology (CCIT 2014)*. Atlantis Press.

9 Yang, X.S. ed., 2013. *Cuckoo search and firefly algorithm: theory and applications* (Vol. 516). Springer.

10 Yang, X.S. and Deb, S., 2014. Cuckoo search: recent advances and applications. *Neural Computing and Applications*, 24(1), pp.169–174.

11 Krishnanand, K.N. and Ghose, D., 2009. Glowworm swarm optimisation: a new method for optimising multi-modal functions. *International Journal of Computational Intelligence Studies*, 1(1), pp.93–119.

12 Reddy, S.S. and Rathnam, C.S., 2016. Optimal power flow using glowworm swarm optimization. *International Journal of Electrical Power & Energy Systems*, 80, pp.128–139.

13 Fan, H. and Zhong, Y., 2012. A rough set approach to feature selection based on wasp swarm optimization. *Journal of Computational Information Systems*, 8(3), pp.1037–1045.

14 Jiang, M. and Zhu, K., 2011, June. Multiobjective optimization by artificial fish swarm algorithm. In *2011 IEEE International Conference on Computer Science and Automation Engineering* (Vol. 3, pp. 506–511).

15 Li, H.Z., Guo, S., Li, C.J. and Sun, J.Q., 2013. A hybrid annual power load forecasting model based on generalized regression neural network with fruit fly optimization algorithm. *Knowledge-Based Systems*, 37, pp.378–387.

16 ZhaoHui, C. and HaiYan, T., 2010. Cockroach swarm optimization. In *2010 2nd International Conference on Computer Engineering and Technology*.

17 Hassanzadeh, I., Madani, K. and Badamchizadeh, M.A., 2010, August. Mobile robot path planning based on shuffled frog leaping optimization algorithm. In *2010 IEEE International Conference on Automation Science and Engineering* (pp. 680–685).

18 Premaratne, U., Samarabandu, J. and Sidhu, T., 2009, December. A new biologically inspired optimization algorithm. In *2009 international conference on industrial and information systems (ICIIS)* (pp. 279–284).

19 Kaveh, A. and Farhoudi, N., 2013. A new optimization method: Dolphin echolocation. *Advances in Engineering Software*, 59, pp.53–70.

20 Marinakis, Y. and Marinaki, M., 2014. A bumble bees mating optimization algorithm for the open vehicle routing problem. *Swarm and Evolutionary Computation*, 15, pp.80–94.

21 X.-S. Yang; S. Deb (December 2009). Cuckoo search via Lévy flights. World Congress on Nature & Biologically Inspired Computing (NaBIC 2009). IEEE Publications. pp. 210–214.

22 Zedadra, O., Guerrieri, A., Jouandeau, N., Spezzano, G., Seridi, H. and Fortino, G., 2018. Swarm intelligence-based algorithms within IoT-based systems: A review. *Journal of Parallel and Distributed Computing*, 122, pp.173–187.

23 Said, O., 2017. Analysis, design and simulation of Internet of Things routing algorithm based on ant colony optimization. International Journal of Communication Systems, 30(8), p.e3174.

24 Chamoso, P., De la Prieta, F., De Paz, F. and Corchado, J.M., 2015. Swarm agent-based architecture suitable for internet of things and smartcities. In *Distributed Computing and Artificial Intelligence, 12th International Conference* (pp. 21–29). Springer, Cham.

25 Zedadra, O., Savaglio, C., Jouandeau, N., Guerrieri, A., Seridi, H. and Fortino, G., 2017, December. Towards a reference architecture for swarm intelligence-based internet of things. In *International Conference on Internet and Distributed Computing Systems* (pp. 75–86). Springer, Cham.

26 Lin, Y.H. and Hu, Y.C., 2018. Residential consumer-centric demand-side management based on energy disaggregation-piloting constrained swarm intelligence: Towards edge computing. *Sensors*, 18(5), p.1365.

27 Zahedi, Z.M., Akbari, R., Shokouhifar, M., Safaei, F. and Jalali, A., 2016. Swarm intelligence based fuzzy routing protocol for clustered wireless sensor networks. *Expert Systems with Applications*, 55, pp.313–328.

28 Zungeru, A.M., Ang, L.M. and Seng, K.P., 2012. Classical and swarm intelligence based routing protocols for wireless sensor networks: A survey and comparison. *Journal of Network and Computer Applications*, 35(5), pp.1508–1536.

29 Saleem, M., Di Caro, G.A. and Farooq, M., 2011. Swarm intelligence based routing protocol for wireless sensor networks: Survey and future directions. *Information Sciences*, 181(20), pp.4597–4624.

30 Thangaraj, M., Ponmalar, P.P., Sujatha, G. and Anuradha, S., 2016, July. Agent based Semantic Internet of Things (IoT) in smart health care. In *Proceedings of the The 11th International Knowledge Management in Organizations Conference on The changing face of Knowledge Management Impacting Society* (p. 41). ACM.

18

Optimization Techniques for Intelligent IoT Applications in Transport Processes

Muzafer Saračević[1,], Zoran Lončarević[2], and Adnan Hasanovic[3]*

[1] University of Novi Pazar, 36300 Novi Pazar, Dimitrija Tucovica bb, Serbia
[2] ITS - Studies for Information Technologies, 11000 Belgrade, Savski nasip 7, Serbia
[3] University of Novi Pazar, 36300 Novi Pazar, Dimitrija Tucovica bb, Serbia

Abstract

In this chapter, we outline the optimization techniques for intelligent IoT applications in transport processes. The travelling salesman problem (TSP) has an important role in operational research and in this case it was implemented in the design of the IoT (Internet of Things) application. The chapter describes some specific methods of solving, analysis and implementation of a possible solution with an emphasis on a technique based on genetic algorithms. In this chapter we connect the TSP optimization problem in transport and traffic with IoT-enabled applications for a smart city. In the experimental part of the chapter we present specific development and implementation of the application for TSP with testing and experimental results.

Keywords *Transport problem; optimization; IoT applications; Travelling salesman problem; navigation system*

18.1 Introduction

IoT-enabled smart city use cases span multiple areas: from contributing to a healthier environment and improving traffic to enhancing public safety and optimizing street lighting. The utilization of IoT traffic regulation is used to reduce costs and increase passenger satisfaction while reducing the number of traffic accidents. Smart cities are utilizing new digital technologies and are related to traffic management through smart traffic lights, parking space design,

Fog, Edge, and Pervasive Computing in Intelligent IoT Driven Applications, First Edition.
Edited by Deepak Gupta and Aditya Khamparia.

gathering information on their waste storage, rational use of refreshments, to build energy-efficient communications in the future with sufficient local government. All of these services involve reducing pollution, moving to clean transportation and heating. Future solutions will be based on main memory and environmentally sound vehicles and their connection to infrastructure facilities such as gas stations, parking lots, garages and others. Wider employment of advanced information technology, except within vehicle communications, will find its position within vehicle infrastructure, capability, and communication.

The main business value that enables IoT technology is not in the data, but in a system that facilitates real-time operations, enabling greater efficiency, reliability and resource utilization in the optimal time. The most important and functional applications of fog computing are used to gather information to support other applications and services. This technology could also improve the progress in eliminating daily safety hazards. For instance, smart traffic lights can recognize remote ambulance or police rotary lights and change the traffic lights on the way to the hospital or the crime scene, by turning on the green lights, or proactive signal checks so the emergency vehicle doesn't have to slow down. Another interesting example would be a smart transportation distribution with load balancing applications, in order to prevent larger losses in the distribution itself. In addition, managing smart factories accomplished using fog computing might monitor the needs of progress and maintenance. On the other hand, edge computing can resolve the challenges of delay and enable companies to better leverage the capabilities of cloud computing architecture. Data centers are edging end-users towards the required extensive content, and the data-sensitive applications are getting them closer.

This chapter presents how to implement and design an application for TSP (travelling salesman problem) and we relate the above technique in optimizing transportation and traffic with IoT and smart city applications.

This chapter is organized as follows. The second part lists some similar research in the field of transport problems and optimization, as well as the possible applications of TSP in traffic and transport. In addition, certain research into edge/cloud computing and IoT analytics have been provided. The third part presents optimization techniques for TSP resolution where the emphasis is on metaheuristics related to genetic algorithms (GA). We used the aforementioned heuristics to implement our TSP application, which is described in the fourth section of this chapter. This section describes the implementation details and provides specific testing of the proposed solution. The fifth section gives the experimental results. The final section of this chapter provides concluding considerations and suggestions for further research.

18.2 Related Works

The transport problem is generally one of the problems in the field of operations research. By solving this task, we may encounter distribution, collection or other activities on the transport network. In their paper (Salonikias *et al.* 2016, pp. 15–18) the authors present access control issues in utilizing fog computing for transport infrastructure. The authors study the operational characteristics of proposed intelligent transportation systems paradigm utilizing fog computing and identifying corresponding access control issues. In their paper (Munoz *et al.* 2018, pp. 1421–1423) the authors present integration of IoT, transport SDN, and edge/cloud computing for dynamic distribution of IoT analytics and efficient use of network resources. They present and experimentally validate the first IoT-aware multilayer transport software defined networking and edge/cloud orchestration architecture that deploys an IoT-traffic control and congestion avoidance mechanism for dynamic distribution of IoT processing to the edge of the network based on the actual network resource state.

In their paper (Huang *et al.* 2012, pp. 223–225) the authors present a project based on the framework of the IoT for intelligence navigation. They expect to achieve the greatest objective of intelligence navigation system. Considering the safety of the users, their server managers can monitor the location of the users anytime. Also, in their paper (Kim *et al.* 2018, pp. 433–436) the authors describe an IoT platform-based smart parking navigation system with the shortest route and anti-collision. Specifically, the authors propose a smart parking platform with IoT and mobile applications. Because of the experiment, the authors have confirmed that these smart parking lot IoT platform systems are working. In their paper (Godavarthi *et al.* 2017, pp. 1–5) the design and the implementation of vehicle navigation system in urban environments using internet of things are described.

There are many methods for solving TSP with applications. In their paper (Laurik 2015, pp. 89-101) the authors show mathematical modeling of the travelling salesman problem that is concerned with the theory of graphs and combinatorics. This paper deals with three aspects: an android application for the Google maps navigation system; solving the travelling salesman problem and optimization through genetic algorithm. The genetic algorithm is used to determine the optimum route on Google maps and thus solves the travelling salesman problem. In their paper Kim *et al.* (2015, pp. 1654-1661) survey the City VRP (vehicle routing problem) literature categorized by stakeholders and summarize the constraints, models, and solution methods for VRP in urban cities. Through this review, the authors identify the state-of-the-art of city VRP, highlight the core challenging issues, and suggest some potential research areas in this field that has remained underexplored. Also, in their paper (Miller *et al.* 2010, pp. 994-997) the authors present the intelligent transportation systems travelling salesman problem, which is a heuristic algorithm loosely based on the traditional TSP with the edge weights that can change constantly. This problem has direct application to the transportation sector where vehicles leave from a source and need to visit a certain set of locations before returning back to the source. In their paper (Valeriu 2006, pp. 202–205) the authors present a new model of TSP examined – a synthesis of classical TSP and classical transportation problems.

18.3   TSP Optimization Techniques

The classic optimization process of TSP starts from one initial solution, while the existing solution is replaced by better ones from the immediate environment and the closest local optimum is always located. The probability of accepting a worse solution decreases as the algorithm progresses. TSP technique knows exactly which cities should be visited and their distance, the only problem being that it is required to visit each city only once and return to the departure city (Johnson and McGeoch 1995, pp. 54). The solution to the problem is in the fact that a travelling salesman determines the order of the cities while travelling the best possible route, which would mean the shortest and fastest route (Applegate *et al.* 1998, pp. 645–655; Little *et al.* 1963, pp. 972–987). Moving from city to city takes place at the same speed, or a shorter path is faster. There are a number of algorithms for solution, and with the advent of computers, the success of optimizing and solving this problem has increased significantly. Problems of designing traffic routes are by their nature combinatorial (Pielić 2008, pp. 28). From the large number of possibilities, it is necessary to choose the best possible solution. The optimal solution is the route to which the minimum transport costs correspond, the travel time or the length of the route itself depending on which of these parameters has the most significant role in a particular problem, or the price of the route is minimal (Held and Karp 1970, pp. 1138–1659).

In many studies, the method of optimizing TSP is described by a colony of ants (ant colony optimization – ACO technique, see Barán and Gomez 2005, pp. 1–6; Ngassa and Kierkegaard 2007, pp. 18). Other works include heuristics such as simulated hardening and taboo search. Taboo search is a local search method for combinatorial optimization that shows very good results in solving a similar problem, which is the vehicle routing problem. The goal of troubleshooting is to find the shortest route that starts in a particular node, passes through all other nodes and ends in the start node. The sizes we optimize do not have to be just the distance (it can be travel costs, travel time or some other dimension). Simulated hardening uses a stochastic search approach, and this method allows the search to continue in the direction of adjacent solutions, although the target function in that direction produces poorer results. In contrast to the classic optimization process, the "simulated hardening method" has a global optimum.

Determining the most favorable road used by a group of vehicles while serving a user set represents a general problem of vehicle routing. The first approach to solving this problem is the search for an exact solution to the problem. The practical application of this approach is very limited because optimal solutions can be found only for a smaller number of users. The number of possible routes for the general case of vehicle routing is growing rapidly, so it is not possible to expect that this approach in general generates real-time solutions that are required in practice:

1) A heuristic approach in solving such a problem is the use of experience, intuition and self-esteem. Unlike exact methods, heuristic methods do not represent knowledge of the structure or relationships within the model of the problem we are solving. Heuristic methods represent rules for selecting, filtering and rejecting solutions, and serve to reduce the number of possible paths in the problem-solving process. Heuristic algorithms are often based on the construction of routes where the construction and improvement of routes is done iteratively.

2) Meta-heuristics in practice is a set of algorithms that are used to solve several different optimization problems where the algorithm itself changes very little depending on the problem being solved (Saračević *et al.* 2013a, pp. 39–45; Saračević *et al.* 2013b, pp. 8). A meta-heuristic approach to solving TSP is often based on a local search driven by processes that take over from nature (such as genetic algorithms).

Heuristics, which was used for resolving TSP, is a technique based on genetic algorithms from the category of modern meta-heuristics (Pepic *et al.* 2017, pp. 3; Borovska 2006, pp. 4; Nilsson 2003, pp. 9; Pilat and White, 2002, pp. 282–285; Lin and Kernighan 1973, pp. 498–505). This problem is based primarily on graph theory. Heuristic algorithms for solving TSPs are constructive algorithms that try to easily and quickly find a solution that is good enough in a real short time (Christopher and White 2004, pp. 29). When solving the problem, we tend to find the shortest route that starts in a particular node, passes through all other nodes and ends in the starting node. The sizes we optimize do not have to be just the distance. These may be travel expenses, travel times or some other dimension. In the case of a merchant passenger, there may be more stringent requirements, such as for example to pass through each node exactly one time. The vehicle routing problem is one of the transport problems that fall into the category of linear programming problems. The task of the VRP is to organize transport for a given number of suppliers and orders of some goods so that its price is optimal, that is, its task is to determine the most favorable route. Using some of the VRP resolution methods, the cost of transport is considerably measured. In the event that there is only one vehicle and if there are no additional restrictions, then VRP becomes TSP, where it takes a vehicle to visit all the points of the graph with minimum cost (e.g., time or capacity).

18.4 Implementation and Testing of Proposed Solution

We used genetic algorithms to improve the initial solution. Particularly in modeling and application (Saračević *et al.* 2012, pp. 73–75; Saračević *et al.* 2010, pp. 3–5), we introduced the following operator combination, which relates to genetic algorithms:

- elimination (*natural selection*),
- mutation (*2opt method*),
- crossing (*greedy sub-tour crossover or GSX*).

The following processes are presented:

- **Initialization** is a once-in-a-lifetime process at the start of an algorithm. This section's task is to make a starting population that will evolve into better populations through further processes of natural selection. One of the approaches is to create a random population. For TSP problems, the random population is a set of random permutations of $\{1,2,\ldots, n\}$, where n is the number of cities.
- **Selection** is a process that takes place between every two consecutive generations. In this section the individuals that are sufficiently fit to move into the next generation are exempt, or likewise, those that are not strong enough, and therefore will not belong to the next generation or offspring. Selection in this case is a function that determines the eligibility of an individual taking into account the length of the distance between two cities, that is, the length of the road that merchant travellers would have to cross if they were to visit cities in comparison with the given permutation of that individual.
- **Crossing** is equivalent to biological crossbreeding. We select two individuals and form another two. In this process, we create as many individuals as we need for the generation to again contain M individuals, i.e., as much as we have eliminated in the previous step.
- **Mutation** is the last step in one cycle. In this step, the percentage of pi individuals is randomly selected, and they somehow change, again simulating a mutation in biological evolution. The significance of this step is again to increase the diversity of the population, because mutation produces new individuals with little or no correlation to other individuals in the same population.

Figure 18.1 presents a use case diagram for TSP based on genetic algorithms.

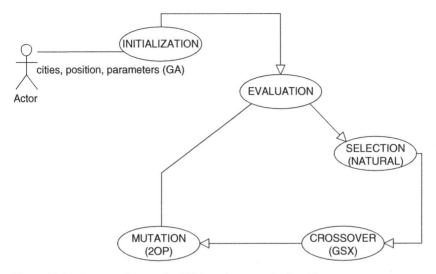

Figure 18.1 Use case diagram for TSP based on genetic algorithms.

The information needed for a good strategy in resolving the TSP are:

- starting point that represents the warehouse,
- quantity of goods to be collected or delivered,
- time period in which the user needs to be served,
- time needed for the delivery or collection,
- time that depends on the type of vehicle and the applied technology,
- a subset of available vehicles that can be used with a single user due to the possible access to the input/output.

The design of this application for the travelling salesman problem is presented in Figure 18.2, which contains the relevant sections (panels). The application has the ability to adjust the necessary parameters and visually display the optimal solution with the diagram for the budget for improving the solution. The largest part of the application contains a panel where the nodes are displayed.

It is important to understand the relationship between the individual classes that make up the TSP application. The implemented solution consists of the following three classes:

- **class GCity** – this class stores the coordinates of the cities. It also contains methods used to calculate the distance between cities. The GCity class stores in the variables x_poz and y_poz the position of the cities and points, which in this case represent the cities in x, y coordinates. This class also contains methods that return the values mentioned. The collection of cities in this class is used to create a path that is implemented in the *GeneticAlgorithm* class.

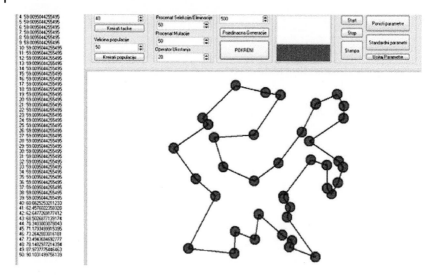

Figure 18.2 Application for TSP.

- **class GeneticAlgorithm** – the class implements most of the functionality of the genetic algorithm (GA). This class contains the three methods: selection, crossing, mutation. It basically represents the implementation of TSP using the GA technique.

- **class GUI_Tsp** – the class implements the user interface and performs a general initialization, like plotting a graphical solution. In particular, this class represents the executable class of the implemented solution and serves to pass parameters to the previous classes, and also serves to display the solution via graphical elements. The basic method of this class is the init method in which the components and actions of the graphical environment are defined. Another very important method in this class is startThread, used to run threads in the background. This method initializes the distribution of cities by x, y coordinates for a predefined number of cities. At each startup, this method allows for a different arrangement of cities in the panel. This method returns the positions of the cities and then it passes them to the *Genetic Algorithm* class and the executive class.

- **Run Method** represents the main loop of the background thread that allows updating the map, or folder, to be displayed to the user. The program then starts a loop that will continue to execute until the same solution is found for 100 generations in a row. Finally the statistics are calculated and displayed. When the main loop is left, the final solution is displayed.

At the bottom of the application there is a status bar displaying the current number of generations executed and a diagram of a graphical representation of relations between the initial lengths of the route and obtained optimum length of the route since the last generation. In the upper part of the application, there is a section of the required parameters input where there the following options are given:

- input of number of points,
- number of population,
- the percentage of selection, mutation, crossing,
- number of generations during the execution,
- individual performance review,
- resolution and options for the total number of generations.

Example. For specific testing, we input the following parameters:

- *number of points* = 40,
- *number of population* = 50,
- *number of generations* = 50,
- *parameters for GA operators: selection* = 50, *crossing*=50, *mutation*=20.

Figure 18.3 shows how the application works, i.e., while being tested in a few generations, we can see how we get the optimal solution.

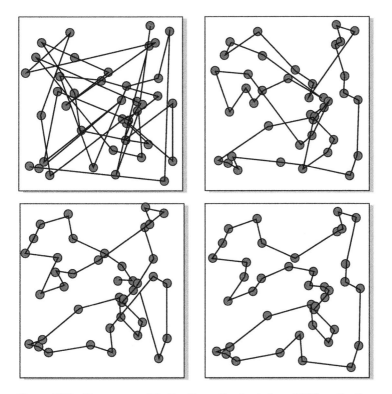

Figure 18.3   The process of finding the optimal solution of TSP application.

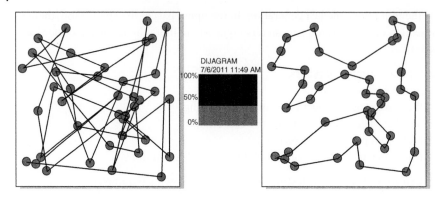

Figure 18.4 Display of the optimized solution and percentage of route improvements.

We specified the comparison between the initial and optimal solution and we can see that the length of time decreased to 73.04%: from 234.69 to 53.05 (see Figure 18.4).

18.5 Experimental Results

The test results indicate that up to a weight level of five cities with parameter setting such as the above, the solution is already found in the second generation which is somewhat expected. The solution is always 100% correct, which means that the path is always optimal. The program finds problems with 10 cities almost instantaneously for the first 20 generations. Testing procedures were conducted in 10, 20, 30, 50 and 100 cities.

The application was tested in the same way for each of the 10–100 city levels, which means ten random program tests with always the same parameter setting. Parameter settings refer to population settings of 1000, 5000, and 10 000 populations, as well as applying three mutation values for each population selection of 10%, 50%, and 100%.

From Table 18.1, we can deduce a few things relevant to genetic algorithms. With the increase in the number of cities, in the example we tested, the total distance between cities increases, and there is a drastic increase in the execution time of the genetic algorithm. The algorithm almost always finds the optimal solution up to the level of 30 cities. It is important to note that the execution time of the algorithm is inversely proportional to the number of populations.

We will not start the specific analysis from a five-city setup because the algorithm always provides the ideal solution in just a few ms. In the example tested, the average execution of the algorithm is 52.1 ms, that is, for the aforementioned lower resolution, the execution time is 29.3 ms. The following are examples and test results of an application for two test examples.

Table 18.1 Average results of TSP_GA algorithm evaluation.

Number of cities	Mean path length	Mean minimum path length	Number of generations of solution finding	Time to find solution (ms)
5	1270	781	1–2	29.3–65.1
10	1540.06	1223.5	3–13	38–4707
20	2346.2	2018.1	38–101	257–27 840
30	3018.4	2679.2	65–234	409–58 019
50	4129.35	3648.85	120–416	967–47 551
100	6952.5	6433.3	416–9019	5529–581 936

Table 18.2 Test results for 50 cities.

Mutation	Generations	A	B
10.00%	1000	4141	5206.2
50.00%		4241	5102.6
100.00%		4083	5083.5
10.00%	5000	4068	4622.3
50.00%		3965	4443.2
100.00%		4025	4543
10.00%	10000	3993	4580.5
50.00%		4244	4630.1
100.00%		4182	4567.5
	AVG	4104.667	4753.211

18.5.1 Example Test with 50 Cities

The example test presented is to display test results when the number of the cities is 50 (Table 18.2). The ratio between the mean minimum (A) and the mean path (B) in the stated example is 86.36%, which means that the percentage deviation from the minimum mean path is 13.64%.

In Graph 18.1, we can see the influence of the percentage of mutations on the length of the paths by 10, 50 and 100% respectively. The graph also displays the ratio of the average shortest path to the average path length values presented for the three population levels.

There is a noticeable decline, i.e., a smaller number of generations is needed to find a solution when the number of generations is 1000 and 10 000 for the number

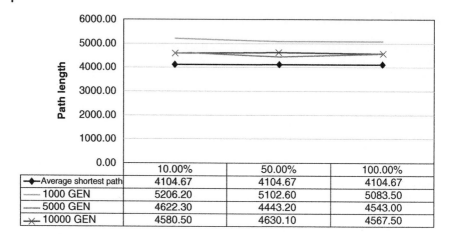

	10.00%	50.00%	100.00%
◆ Average shortest path	4104.67	4104.67	4104.67
—— 1000 GEN	5206.20	5102.60	5083.50
—— 5000 GEN	4622.30	4443.20	4543.00
✕ 10000 GEN	4580.50	4630.10	4567.50

Graph 18.1 Effect of mutation on determining the path length.

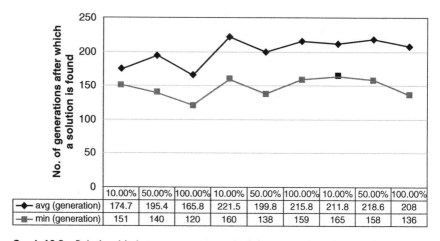

	10.00%	50.00%	100.00%	10.00%	50.00%	100.00%	10.00%	50.00%	100.00%
◆ avg (generation)	174.7	195.4	165.8	221.5	199.8	215.8	211.8	218.6	208
■ min (generation)	151	140	120	160	138	159	165	158	136

Graph 18.2 Relationship between average and minimum number of generations.

of populations of 1000 and 10 000. The same downward trend in the number of generations required for a higher percentage of mutations was not observed in a population of 5000. The ratio of the average and the minimum number of generations is shown in Graph 18.2.

18.5.2 Example Test with 100 Cities

Testing at 100 cities in most cases gives quite good results, at the cost of large computing time spending. However, this is conditional, given that for some other algorithms this time could be expressed in millennia.

Table 18.3 Experimental results for population n=100.

Mutation	Generations	Method	Length of the path	Solution after (n) generation	Time
10.00%	1000	MIN	9065	524	6690
50.00%		AVG	9895.75	1487.25	19 657.75
100.00%		MAX	11287	6849	83 821
10.00%	5000	MIN	6527	760	65 854
50.00%		AVG	7007.571	3125.143	161 876.86
100.00%		MAX	7425	9019	384 019
10.00%	10000	MIN	6492	760	118 301
50.00%		AVG	6779.8	3148.8	581 936.2
100.00%		MAX	7180	5848	1 472 790

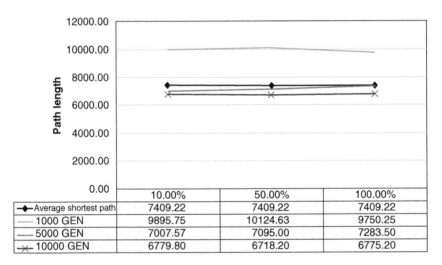

	10.00%	50.00%	100.00%
Average shortest path	7409.22	7409.22	7409.22
1000 GEN	9895.75	10124.63	9750.25
5000 GEN	7007.57	7095.00	7283.50
10000 GEN	6779.80	6718.20	6775.20

Graph 18.3 Effect of mutation on path length determination.

Table 18.3 presents the results obtained by the aforementioned testing and relates the relationships between the minimums with the average paths and their mean values.

The deviation of the average path from the average minimum path in this case is 6.65%, which is for the time being the best result in terms of deviation from the average minimum path. Graph 18.3 presents the effect of the mutation on determining the length of the paths, that is, the effect of the mutations on the individual values of the populations.

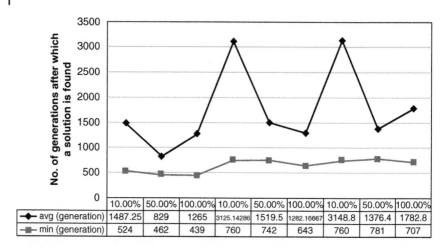

	10.00%	50.00%	100.00%	10.00%	50.00%	100.00%	10.00%	50.00%	100.00%
avg (generation)	1487.25	829	1265	3125.14286	1519.5	1282.16667	3148.8	1376.4	1782.8
min (generation)	524	462	439	760	742	643	760	781	707

Graph 18.4 Ratio of average and minimum number of generations.

In addition, it also shows that the average shortest path for generations of 5000 and 10 000 populations is a worse solution than the solution obtained by applying these parameters.

Obviously the effect of mutations on path length is similar to the tests mentioned previously. Graph 18.4 presents the ratio of the average and the minimum number of generations in relation to their minimum values of the generations after which the solution was found. We can also see the average number of generations and the minimum number of generations after which a solution was found.

18.6 Conclusion and Further Works

There is almost no business, not even an individual consumer, who does not use some kind of cloud system, where the new trend is the increasing use of fog computing. This new technology uses the concept of the Internet of Things (IoT) and combines data coming from IoT devices. In addition, edge computing can solve the challenges of delays and enable companies to better leverage the capabilities of cloud computing architecture. To support the needs, computing power and memory are fed into edge computing to reduce data transfer time and increase availability.

In this chapter, we have presented the manner of implementation and design of an application for two problems in transport optimization (TSP). We have linked the mentioned problems in optimizing transportation and traffic with IoT and smart city. We used genetic algorithms to improve the initial solution for TSP. Genetic algorithms gain another advantage because in practice it is not necessary to find an optimal, a good enough path and one in the vicinity of the optimum. In addition to all this, it is possible to combine genetic algorithms with the methods of searching for a local optimum in order to convert the individual as quickly as possible to its optimal potential which is potentially global. The experimental results are shown through tables with numerical data and through graphs. In the tables are listed specified experimental results of testing the application to the problem of routing the vehicle to test example, in which is presented the said ratio of the speed of all the tour points for the random path and the optimal path obtained by the method described. In addition to the numerical data table, the graphs present the shortening of time for the optimal path, where the efficiency of the applied method can be clearly seen from the aspect of the speed of sighting of all points.

In further research, the TSP solution model using GA heuristics could find application in the construction of some applications for smart cities and smart maps according to the research model of some authors. There are some travelling, public transportation and discovery applications on the smart maps. In paper (Maggioni *et al.* 2014, pp. 528–533), the authors provide a standard methodology to extend routing instances from the literature incorporating real data provided by sensors networks. In order to test the methodology, authors consider a routing problem specifically designed for city logistics and smart city applications, the multi-path TSP. In (Ozcan *et al.* 2018, pp. 1–4), the authors show that model can be applied to make smart cities to be created by heuristic algorithms on the real maps to perform some tasks through TSP. The Hill Climbing heuristic search algorithm which is generally used for mathematical optimization problems in artificial intelligence field has been preferred in this study. Our further research in this area could be based on these principles.

Bibliography

Applegate, D., Bixby, R., Chvatal, V. and Cook, W. (1998). On the solution of travelling salesman problems. *Documenta mathematica* 3: 645–656.

Barán, B. and Gomez, O. (2005). Omicron ACO: A New Ant Colony Optimization Algorithm. *CLEI Electronic journal* 8 (1): 1–8.

Borovska, P. (2006). Solving the Travelling Salesman Problem in Parallel by Genetic Algorithm on Multicomputer Cluster. *Proceedings of the International Conference on Computer Systems*. Veliko Tarnovo, Bulgaria (15–16 June 2006).

Christopher, M. and White, G. (2004). *A Hybrid Evolutionary Algorithm for Traveling Salesman Problem*. USA: Stillwater, Oklahoma State University.

Godavarthi, B., Nalajala, P. and Ganapuram, V. (2017). Design and Implementation of Vehicle Navigation System in Urban Environments using Internet of Things. *IOP Series: Materials Science and Engineering* 225: ID=012262.

Held, M. and Karp, R. (1970). The Traveling Salesman Problem and Minimum Spanning Trees. *Operations Research* 18 (6): 1138–1162.

Huang, C.H., Ciou, Y.J., Shen, P.Y. et al. (2012). The application of IoT for intelligence navigation. *Proceedings of the International Conference: Intelligent Signal Processing and Communications Systems (ISPACS)*, New Taipei City, Taiwan (4-7 November 2012).

Johnson, D.S. and McGeoch, L.A. (1995). *The traveling salesman problem: a case study in local optimization*. USA: Amherst College, Department of Mathematics and Computer Science.

Kim, D., Park, S., Lee, S. and Roh, B. (2018). IoT Platform Based Smart Parking Navigation System with Shortest Route and Anti-Collision. *Proceedings of 18th International Symposium on Communications and Information Technologies (ISCIT)*, Bangkok, Thailand (26-29 September 2018).

Kim, G., Ong, Y., Heng, C. K. et al. (2015). City Vehicle Routing Problem (City VRP): A Review. *IEEE Transactions on Intelligent Transportation Systems* 16 (4): 1654–1666.

Laurik, H. (2015). An Android Application for Google Map Navigation System, Solving the Travelling Salesman Problem, Optimization throught Genetic Algorithm. *Proceedings of the International Conference FIKUSZ*, Budapest, Hungary (13 November 2015).

Lin, S. and Kernighan, B.W. (1973). An effective heuristic algorithm for the traveling-salesman problem. *Operations research* 21(2): 498–516.

Little, J.D.C., Murty, K.G., Sweeney, D.W. and Karel, C. (1963). An algorithm for the traveling salesman problem. *Operations research* 11(6): 972–989.

Maggioni, F., Perboli, G. and Tadei, R. (2014). The Multi-path Traveling Salesman Problem with Stochastic Travel Costs: Building Realistic Instances for City Logistics Applications. *Transportation Research Procedia* 3: 528–536.

Miller, J., Kim, S. and Menard, T. (2010). Intelligent Transportation Systems Traveling Salesman Problem (ITS-TSP) - a specialized tsp with dynamic edge weights and intermediate cities. *Proceedings of the International Conference: 13th International IEEE Conference on Intelligent Transportation Systems*, Funchal, Madeira Island, Portugal (19 – 22 September 2010).

Munoz, R., Vilalta, R., Yoshikane, N. et al. (2018). Integration of IoT, Transport SDN, and Edge/Cloud Computing for Dynamic Distribution of IoT Analytics and Efficient Use of Network Resources. *Journal of lightwave technology* 36 (7): 1420–1428.

Ngassa, J. and Kierkegaard, J. (2007). ACO and TSP. Bachelor thesis. Computer Sciences Roskilde University, Denmark.

Nilsson, C. (2003). Heuristics for the traveling salesman problem. Techical Report. Linköping University, Sweden.

Ozcan, S.C. and Kaya, H. (2018). An Analysis of Travelling Salesman Problem Utilizing Hill Climbing Algorithm for a Smart City Touristic Search on OpenStreetMap. *Proceedings of the International Conference: 2nd International Symposium on Multidisciplinary Studies and Innovative Technologies (ISMSIT)*, Ankara, Turkey (19–21 October 2018).

Pepic, S., Lončarević Z. Miodragović G. and Aleksandrov S. (2017). Solution implementation genetic algorithms of traveling salesman problem in java programming language, *Proceedings of the International Conference ITOP2017: Information technology, education and entrepreneurship*, Serbia (08–09 April 2017.).

Pielić, M. (2008). Traveling salesman problem with help of genetic algorithms (in Croatian). Bachelor thesis. Faculty of Electrical Engineering and Computing, University of Zagreb, Croatia.

Pilat, ML. and White, T. (2002). Using genetic algorithms to optimize ACS-TSP. *Lecture Notes in Computer Science* 2463: 282–287.

Salonikias, S., Mavridis, I. and Gritzalis, D. (2016). Access Control Issues in Utilizing Fog Computing for Transport Infrastructure. Critical information infrastructures security. *Lecture Notes in Computer Science* 9578: 15–26.

Saračević, M., Elfić, E., Plojović, Š. and Mašović, S. (2013a) . Implementation of transportation problem by using the method of meta-heuristics approach, *Economic Challenges* 3: 39–48.

Saračević, M., Mašović, S. and Medjedović, E. (2010). Application of object-oriented analysis and design in navigation systems and transport networks. *Proceedings of the International Conference: Research and Development in Mechanical Industry*, Serbia (16–19 September 2010).

Saračević, M., Mašović, S. and Plojović, Š. (2012). UML modeling for traveling salesman problem based on genetic algorithms. *Southeast European Journal of Soft Computing* 1 (2): 72–79.

Saračević, M., Plojovic, Š., Ujkanovic, E. and Bušatlic S. (2013b) . Implementation of transportation problem by using the GA method. *Proceedings of the International Conference on Economic and Social Studies*: ICESoS, Bosnia and Herzegovina (10–11 May 2013).

Valeriu, U. (2006). Traveling Salesman Problem with Transportation. *The Computer Science Journal of Moldova* 14: 202–206.

19

Role of Intelligent IOT Applications in Fog paradigm: Issues, Challenges and Future Opportunities

Priyanka Rajan Kumar* and Sonia Goel

Punjabi University Patiala.

Abstract

In a world full of digital innovation technologies including 5G wireless, Internet of things, embedded artificial intelligence is developing. There are billions of applications and devices that need to be managed. These are working to measure, monitor, process, analyze and react seamlessly on a huge amount of data. There is a need for a new communication architecture that is to deal with latency, bandwidth, accessibility, rising cost, sensitive data security

In today's scenario all the devices are connected to communicate huge and heterogeneous amounts of data. With each new day, the way the world interacts is changing. With the increasing use of sensor and IoT devices, the generation of data is also increasing. With this increased data, the computing and storage is also becoming important. The fathomless placement of smart, interconnected devices is expected to reach 50 billion units by 2020. This exponential rise is fueled by proliferation of networks of mobile devices, smart sensors, wireless sensors and actuators. There is a need for new concepts and technologies to manage this growing squadron of IoT devices.

The Internet of everything solutions are connecting every object. This has generated a large amount of data. This amount of data cannot be processed by a centralized cloud environment. There are applications where data needs real-time response and low latency. The data being sent to the cloud for processing and then coming back to the application generating data can seriously impact on the performance. This delay can cause delays in decision making and this is not bearable in real-time

*Corresponding Author: srpriyankass@gmail.com

Fog, Edge, and Pervasive Computing in Intelligent IoT Driven Applications, First Edition.
Edited by Deepak Gupta and Aditya Khamparia.

applications. To handle such scenarios, fog computing emerged as a solution. Fog computing extends the cloud near to the edge of network to decrease latency as well as bandwidth requirements. It acts as an intermediate layer between the cloud and devices generating data

Keywords *Fog Computing; Internet of Things (IoT); Security; Confidentiality; Integrity; Authentication; Availability; Scalability*

19.1 Fog Computing

Fog computing: Fog computing, or fogging, is a term coined by CISCO [1]. The idea of fog computing is to extend the cloud nearer to the IoT devices. The primary aim was to solve the problem of data processing. It acts as an intermediate layer between IoT and cloud. According to statistics, by 2020 around 40% of the whole world data will come from sensors. 90% of the world's data has been generated in the last two years. Every day 2.5 quintillion of data is generated. This will lead to total expenditure of $1.7 trillion by 2020. It has also been observed that by 2020 the total number of connected devices will be 250 million. There will be around 30 billion devices. These devices will be sending a huge amount of data for processing to the cloud [2]. There is a need for architecture to handle this large data with reduced latency and bandwidth. This has given rise to the concept of fog computing. The fog is a programming model for Internet applications that are geographically distributed and latency sensitive.

19.1.1 Need of Fog computing

Fog computing plays the role of bridging the gap between the sensor nodes and the cloud layer. Cloud computing has been an advance that has rapidly developed itself by its effective design of cloud services of executing extensive computing tasks, and storage of huge and versatile data. But cloud computing has some constraints that have aroused a need for computing at the edge of the network, known as edge computing [3].

1) **Security:** Organizations have sensitive data that is confidential [4]. The data on the cloud is open and out of the control of the user. Fog computing analyzes and processes data at the local end rather than publishing it far away.
2) **Connectivity restraint:** In cloud computing, data is on the cloud rather than on local systems. There is a requirement for high speed Internet connectivity to send and receive data on the cloud. Fog computing processes the data on the edge of the network and gives real-time results.

CLOUD

FOG

EDGE

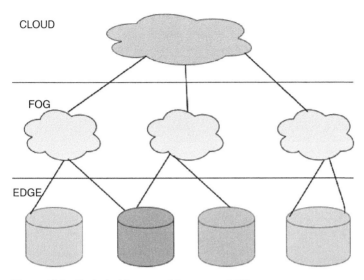

Figure 19.1 Basic Architecture of Fog network [8].

3) **Bandwidth:** Cloud computing consumes more bandwidth as a large amount of data is to be sent to the data centers for storage [5]. Fog computing reduces the bandwidth consumption as the data is processed at the edge devices and a smaller volume of data is sent to the cloud for processing and storage.

4) **Latency:** Cloud computing has larger latency due to longer distance from device generating data and the cloud. Fog computing gives real-time responses and reduces the latency due the smaller distance between device and the fog node processing data [6].

5) **Scalability:** Cloud computing is less scalable due to centralized sources. Fog computing enhances scalability due to distributed processing of data on fog nodes.

Fog computing is not an alternative to cloud computing, it is an extension of it.

19.1.2 Architecture of Fog Computing

The cloud is up there in the sky, secluded and purposely abstract. Fog is near to the ground, where data is generated [8]. A basic architecture of fog network is shown as a simple three level hierarchy in Figure 19.1.

Fog is not a substitute for the cloud, it is complementary to the cloud. Both fog and cloud are needed in the IoT infrastructure. Cloud services are extended to IoT devices through fog. Fog is a layer between cloud and IoT devices. In fog architecture many fog nodes can be present. Sensor data are processed in the fog before being sent to the cloud. Fog provides computing at the edge of IoT devices. It reduces latency, saves bandwidth and saves the storage in the cloud.

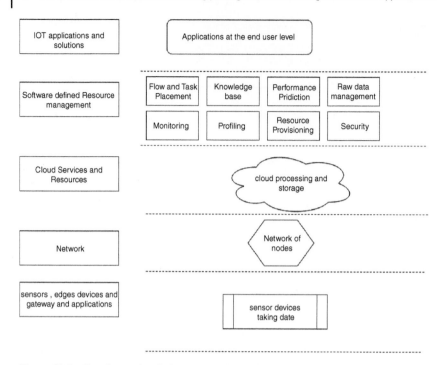

Figure 19.2 Fog Computing Reference Architecture.

19.1.3 Fog Computing Reference Architecture

In the reference architecture for fog computing, at the first layer are the sensor devices, IoT devices and gateways. This layer also has various applications on the sensor devices for accomplishing their objectives. The sensor layer coordinates with the network layer for communication between the layers and with the cloud. The cloud service and resources layer is above the network layer. It processes the various task sent by devices and also manages the storage of data. The software layer is above the cloud layer. This layer manages the whole communication architecture and delivers quality of service to fog computing paradigm. This layer furnishes intermediary services for the best use of resources at the fog as well as cloud. The services are flow and task placement, knowledge base, performance prediction, raw resource and management, monitoring, profiling, resource provisioning and security. The topmost layer is IoT applications and solutions which provide support services to end users. The reference architecture is depicted in Figure 19.2

In the software defined resource management layer, a number of services are implemented for the optimization of cloud and fog resources. The sub-blocks are performing their respective functions:

Table 19.1 Processing on Fog [9].

	Fog nodes close to devices	Fog aggregate nodes	Cloud
Analysis duration	Fraction of seconds	Seconds to minutes	Hours to weeks
Geographical coverage	Very local	Wider	Global
IoT data storage duration	Transient	Hours, days	Months to years

Flow and task placement: Its function is to provide fog, cloud and network resources for the accomplishment of various task as well as maintaining their proper execution.

Knowledge base: This is to store information regarding various applications and their respective demand for resources and provides services of decision making.

Performance prediction: This uses information of knowledge-base to realize the performance of cloud resources. It is used for resource provisioning to determine the availability resources when the number of tasks is large in number.

Resource provisioning: This is an energetic allotment of resources to various tasks depending on time. These are fog and network resources that are required for a particular application.

Security: This provides authentication, authorization and cryptographic requirements of services and applications.

19.1.4 Processing on Fog

Fog computing is computing at the edge of network. This concept came with an increased latency and bandwidth consumption in the IoT-cloud infrastructure where the data was being sent to the cloud for further processing and storage from IoT devices. In fog computing, processing and storage resources are provided near the IoT devices that are on the edge of the network. This results in reduced bandwidth consumption as well as latency. This makes real-time decisions on sensitive data easier and quicker [9]. It gives rise to increased efficiency and quality of service. Table 19.1 depicts the processing on fog nodes in terms of analysis duration, geographical coverage and type of data on IoT devices.

19.2 Concept of Intelligent IoT Applications in Smart Computing Era

1) **Real Time Health Analysis:** Patients with chronic illness can be monitored in real-time. Stroke patients' data can be immediately analyzed and during

emergency can alert the respective doctors. Historical data analysis can predict future dangers to patients [5, 10].

2) **Smart grid:** Energy load balancing applications may run on network edge devices, such as smart meters and micro-grids [11]. The devices automatically swap between various energy sources depending on energy demand, availability and price.

3) **Real Time Rail Monitoring:** Fog nodes can be deployed to railway tracks for real-time monitoring of track conditions. For high speed trains, sending data into the cloud for analysis is inefficient. Fog nodes provide fast data analysis, improve safety and reliability [12].

4) **Pipeline Optimization:** Gas and oil are transported through pipelines. Real-time monitoring of pressure, flow and compressor is necessary. Terabytes of data is created. Sending this data to the cloud for analysis and storage is not efficient. In this latency can be reduced with fog computing [13].

5) **Real-Time Windmill And Turbine Analysis:** Wind direction can be analyzed in real-time with fog computing.

19.3 Components of Edge and Fog Driven Algorithm

The fog computing approach is a lightweight thin layer where computation processing is done closer to the origin of the data. Fog computing provides distributed processing of real-time data which provides quicker response. The fog layer sits between the cloud and the end devices generating data. The fog nodes are resource constrained. There are algorithms which are used to support the resource constrained environment of fog computing. The algorithms are classified into four types:

1) **Discovery:** This is the process of identifying the fog resources for the purpose of assigning tasks either from the cloud or from users. Fog computing research assumes that fog resources are discovered but it is not that simple [14]. It has to discover the best resource for task deployment in a distributed computing environment.

2) **Benchmarking:** An approach for describing the performance of computing system based on performance metrics. The metrics are set using standard performance evaluation tools. These metrics set some benchmarks for the dynamic computing system. The cloud has a time-consuming benchmarking processes and has high costs. A more lightweight benchmarking with containers is proposed to reduce costs as well as processing overheads.

3) **Load-Balancing:** The tasks are distributed within the resources on the basis of priority, fairness and other criteria so that throughput can be increased with the

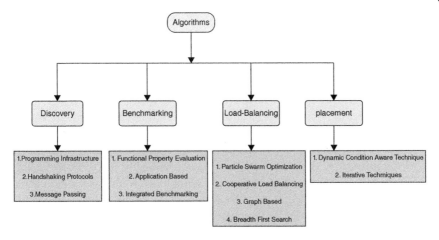

Figure 19.3 Fog Driven Algorithms.

available resources. The fog data centers are placed in the fog network which are responsible for distributing tasks to the fog nodes. The load-balancing algorithms play an important role in the distribution of tasks among fog nodes.

4) **Placement:** The process of placing incoming tasks of computation to a suitable fog resource. The placement algorithms are designed to address this and to look for the availability of resources in the fog layer and for changes in the environment. The aim is to place tasks to the appropriate resource to achieve the optimal network solution.

Figure 19.3 represents various fog driven algorithms which are used to manage resources in the fog layer for quicker real-time response and optimal network performance.

19.4 Working of Edge and Fog Driven Algorithms

The algorithms facilitating fog computing resource employment are classified into several types. The four types of algorithms and their respective working techniques are as follows:

1. **Discovery:** This finds the fog resources that are available so that tasks from cloud and user devices can be deployed on them. It discovers the resources in the network that can be employed for distributed processing tasks. There are three techniques that are used to identify the availability of resources
   (I) Programming Infrastructure: The programming infrastructure has provided a platform to fog resources to work in cloud–fog system.

Foglets, a programming infrastructure, is proposed for geo-distributed computational sequences represented by fog nodes and the cloud [15]. It provides API for data abstraction and manages application components of fog nodes. An additional protocol for the selection of one fog node from the available is also employed.

(II) Handshaking Protocols: This presents a lightweight discovery protocol for a set of homogeneous fog resources [16]. It has a master node which has all the information regarding the availability of nodes and the processing overheads. It is a manager node that communicates availability and assigns task to the fog nodes as per availability. It has the benefit of being lightweight, and consumes less time for overall execution of the process. The protocol works on a single collection of nodes. It has to be enhanced for the larger fog environment. The protocol works in an open environment. The security of the master node should also be taken into consideration.

(III) Message Passing: Message passing is done for all the nodes in the network for the availability of resources by processing nodes that are connected to the Internet [17]. In the sensor network end devices may not be connected to the Internet. The user can communicate with any node for the processing of tasks through message passing and make use of simulation-based validation.

2. **Benchmarking:** This is an approach for setting standards for resources. The CPU, memory, storage etc. considered as metrics for benchmarks. The challenge with benchmarking is that in the dynamic computing system the existing techniques are slow and consume more time. There is a requirement for a lightweight technique for benchmarking as the fog nodes are resource constrained [18, 19]. Fog benchmarking can be classified into three types:

Most fog benchmarking research evaluates power, CPU and memory of the fog processors [20].

(I) Functional Property Evaluation: The functions of the fog nodes are evaluated. The fog nodes are assigned tasks depending upon the functions performed by them.

(II) Application Based: In application-based benchmarking, standards specific to applications that will use fog nodes are set.

(III) Integrated benchmarking: In this an integrated standard for cloud and fog resources is set. This will provide the performance standards for deploying applications across the cloud and fog.

Limitations of benchmarking:

1) The application of a specific benchmark that can identify varying task loads are not available. The scientific benchmark cannot be applied to fog [21]. It can make use of voice driven benchmarks [22] and IoT applications [23].

2) It is very time consuming in a resource constrained fog environment.

3) It is not sufficient to benchmark only the fog resources but an integrated approach for both fog and cloud resources is required. This will provide a combination of both fog and cloud resources for maximizing the overall performance of the system.

3. **Load-Balancing:** The load balancing algorithm is significant for distributing task to the fog nodes. The fog data centers are placed in the fog network. These data centers are responsible for distributing task to the fog nodes. The load balancing algorithms play a significant role for the distribution of tasks among the fog nodes. The four techniques used at the fog for applying load balancing algorithms are:

   (I) Particle Swarm Optimization: The particle swarm optimization technique is used for optimal load balancing and providing a reduced latency solution in the fog environment. This will increase the performance of the fog network as real-time decision making is done at the resource constrained layer. A particle swarm optimization–constrained optimization (PSO–CO) is applied to decrease latency and increase quality of service [24].

   (II) Cooperative Load Balancing: In this technique a buffer is assigned to every data center. The requests from the clients are stored in the buffer, when the number of requests exceeds a threshold value it transfers the request to the adjacent data center to balance the load. The data centers are assumed to be connected with each other by a high-speed link. A CooLoad [25] Cooperative Load balancing has been proposed for optimal resource allocation.

   (III) Graph Model: In this technique a graph of the fog network is prepared where the fog node acts as a vertex and the data dependency of the task is represented as the edge. This graph uses dynamic repartitioning using the previous request status of the data center as input and provides resources to minimize delay in processing the request.

   (IV) Breadth First Search: In the breadth first search method every data center is evaluated with the current data load and the maximum capacity of the data center. The authentication algorithm is used to authenticate the data centers. An efficient load balancing algorithm with the authentication method for data centers is proposed [26].

4. **Placement:** This is the process of placing an incoming task of computation to a suitable fog resource. The placement algorithms are designed to fulfil the need of resources in the fog layer and to accomplish the processing task in the distributed resource constrained environment. The techniques are:

   (I) Dynamic Condition Aware Technique: This method solves the problems encountered in the static load balancing algorithms that do not consider the dynamic conditions of the network such as priorities of the users, location of database and the user and system load. This method considers the dynamic conditions of the network for resource allocation. The parameters values are predicted and the overall cost is estimated and the optimal solution is provided.

(II) Iterative Technique: The iterative method is further classified into two types:

(i) Iterative over resources: This method provides a hierarchal fog computing model in which both the fog and cloud layers are considered. It starts from the fog layer for the allocation of resources and moves to the cloud. This method iterates from fog to cloud for the resources.

(ii) Iterative over problem space: This method provides a number of iterations over the problem space. The iterations are done to achieve the optimum performance solution in the fog network. The fog nodes are evaluated on the basis of the three parameters CPU, memory and storage. The placement is done periodically to achieve high performance. If the resources are not available then the request is forwarded to the nearest fog network. The method is evaluated by extended iFogSim [27].

19.5 Future Opportunistic Fog/Edge Computational Models

Various techniques can be used to achieve intelligence and enhance the performance of the fog network. Fog computing is an architectural solution for reducing the load on the cloud network by bringing the processing of the data to the edge of the network. To address the requirements of energy efficiency and latency for time-sensitive IoT applications, fog computing can apply some intelligent features in their processing [28]. The two approaches can be applied to achieve intelligence in the fog network.

(I) **Device-driven Intelligence:** The device-driven approach of intelligence can be implemented using smarter devices with smarter functionalities such as sensing, computing, storage and communications. The devices with such capabilities include IoT gateways, local servers and portable data aggregating nodes [29, 30]. These devices allow information extraction with fine granularity to support decision-making and resource management.

(II) **Human-driven Intelligence:** Humans play an important role in data sources in the network. The behavior pattern can be exploited to train the network to be smarter. The system can be envisioned to learn from the behavior of human beings while serving people [31, 32]. This will lead to improved efficiency in terms of resource scheduling and conservation of battery power of the devices.

19.5.1 Future Opportunistic Techniques

The techniques that provide future opportunities to fog computing are:

(I) **Data Analytics:** The data can be analyzed in a meaningful manner with the help of machine learning algorithms and neural networks and deep learning techniques. This can be done on small-scale data that is collected locally.

(II) **Optimization:** Decision-making is the process of taking the best solution from the available solutions that minimize cost or maximize the utility-based objective. There are various optimization algorithms which, when applied, will lead to optimized decisions.

(III) **Multi-agent learning:** Reinforcement or learning is one of the important parts of artificial intelligence. This will enable computing devices to take action based on interactions and feedback from the environment to maximize rewards.

These intelligent techniques present new opportunities when applied to fog computing.

19.6 Challenges of Fog Computing for Intelligent IoT Applications

The fog computing architecture has emerged as a concept for low-latency and resource-rich processing to complement edge and cloud computing. Fog computing provides a solution to processing the enormous data being generated by the sensor devices. As it provides a hands on solution to the cloud in terms of data, computation and storage functionality. But there are still some challenges being faced by fog. The challenges of fog computing are shown in Figure 19.3.

1) **Security:** The data to be processed is sensitive and needs to be protected [33]. The fog is acting as middle layer in the IoT–fog–cloud communication infrastructure. It is processing the real-time data at the edge of the network, thus protecting the data of utmost importance to the organization on the network gateways rather than sending it away from the network for processing. Although it gives improved system performance there are still some security goals which need to be achieved. The security goals can be classified as primary and secondary. The primary goals are known as standard security goals such as confidentiality, integrity, authentication and availability

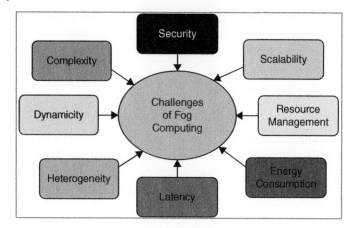

Figure 19.4 Challenges of Fog Computing.

(CIAA). The secondary goals are data freshness, self-organization, time synchronization and secure localization.

2) **Complexity:** The smart applications from which the processing is done have versatile data to be processed. These applications need processing at different levels of hierarchy. The complexities can be the nature of the applications, the data to be processed and the time of response that are to be managed at the fog layer [34].

3) **Scalability:** The fog is a distributed computing application that provides scalability to the centralized cloud solution [35]. The fog node or the processing gateways manage the data to be processed depending upon the sensitivity requirements of data, that is: time sensitive data, less time sensitive data and data which is not time sensitive.

4) **Dynamicity:** A dynamic system is one that constantly changes and progresses. The fog computing paradigm handles and processes the constantly changing requirements of data [36]. The challenge here is to handle the varying needs of processing and storage of the data in order to achieve better operational efficiency and higher performance for the real-time as well as non-real-time data.

5) **Resource management:** The IoT devices or the sensors which are generating data are resource constrained devices in terms of processing power and computational efficiency [37]. The fog computing works for the optimization of resources to achieve better decision making and improved operational efficiency.

6) **Heterogeneity:** The IoT devices generate a huge and heterogeneous amount of data to be processed and stored. The fog nodes should be able to make decisions regarding the level of sensitivity and priority of the data under consideration. This heterogeneous data generated by varying applications is complex to

process and store as these applications have varying processing requirements [38].

7) **Energy consumption:** Fog computing extends cloud computing by processing data at the edge of the network. This processing will reduce the network traffic. But as the processing is being done at the fog nodes, the nodes should be able to manage the processing of data in terms of critical and sensitive data from the normal data so as to optimize the energy consumption of the network [39]. The IoT devices are resource-constrained, while some tasks such as image and speech processing are resource-intensive. This is done by off-loading the task between various fog nodes in order to reduce the computational complexity and energy consumption.

8) **Latency:** Latency is the delay that is caused when sending the data for processing and taking decisions depending on the processing results. Fog has reduced the latency of the data by processing at the edge of the network. This latency can be further reduced to a minimum by segregating the processing of fog nodes according to the sensitivity of the application generating data so that mission-sensitive data can be processed in real-time to increase efficiency.

19.7 Applications of Cloud Based Computing for Smart Devices

The applications of cloud computing for smart devices are apparent due to the convenience provided to people and society at large. Cloud computing has facilitated society in the following ways:

1) **Home automation:** Smart devices such as mobile phones, PDA, and smart watches can be used for home automation. The cloud helps to monitor the environmental conditions, the data being sensed by the smart devices is sent to the cloud and the home appliances are automated to work according the processed data signals from the cloud.

2) **Factory automation:** The factory environment consists of various processes to accomplish the production of goods and services. The sensor devices in the industry premises send the sensed data to the cloud and the signals are processed in the cloud. Various processes such as job monitoring, and job scheduling are automated via smart devices in the factory.

3) **Border surveillance:** Cloud computing helps in surveillance of border situations: the signals generated by the cloud are used to control border conditions when intrusion of unauthorized parties are detected

4) **Healthcare application:** The patients are monitored through smart and wearable devices and the signals are sent to the cloud for further processing and

storage. These signals are sent to the medical facilities to identify any requirement for the patients for expert advice from doctors. This provides improved medical aid to people in society at large by making them connected to medical services everywhere all the time instead of being tied to a wired monitoring system at their bedside .

5) **Traffic monitoring:** Smart devices installed in vehicles and on the road can sense road traffic and give information to drivers so they can act accordingly.

Bibliography

1 Tang, B., Chen Z., Hefferman, G., Wei, T., He H. and Yang, Q. (2015) A hierarchical distributed fog computing architecture for big data analysis in smart cities. *Proceedings of The ASE BigData & Socialinforrmatics 2015.*

2 Sagiroglu S.and Sinanc D. (2013) Big data: A review In: Collaboration Technologies and Systems (CTS), *International Conference on IEEE.*

3 Malik Hayat, B., Cheema Nawaz, S., Iqbal I., Mahmood, Y., Ali, M., and Mudasser, (2018) A. From Cloud Computing to Fog Computing(C2F): The Key technology provides services in health care big data. *MATEC Web of Conferences MEAMT 2018.*

4 Vaquero, L.M. and Merino Rodero, L. Finding your way in the Fog (2014) *ACM SIGCOMM Computing Communication Revolution 44 (5):27–32.*

5 Cao, Y., Chen, S., Hou, P., and Brown, D. (2015) FAST: A Fog computing assisted distributed analytics system to monitor fall for stroke mitigation. *In Proceedings IEEE International Conference Network Architecture Storage (NAS)* (August 2015) : 2–11.

6 Deng, R., Lu, R., Lai, C.and Luan , T.H. (2015) Towards Power Consumption delay tradeoff by workload allocation in cloud-fog computing. *In Proceedings IEEE International Conference Communication (ICC)* (June 2015) : 3909–3914.

7 Bonomi, F., Milito, R., Zhu, j. and Addepalli, S. (2012) Fog Computing and its role in the Internet of Things. *In Proceedings First Workshop Mobile Cloud Computing (MCC).* (February 2012) Helsinki, Finland. : 13–16.

8 Yi, S., Qin, Z., and Li, Q. (2015) Security and Privacy issues of Fog Computing : A survey *In Proceeding International conference, Wireless Algorithms System Applications (WASA).* : 685–695.

9 Patil, P.V. (2015) Fog Computing *International Journal of Computer Applications* , National conference on Advancements in Alternate Energy Resources for Rural Applications (AERA-2015).

10 Stantchev, Vladmir,(2015) Smart Items, Fog and Cloud Computing Servitization in Healthcare. *Sensors and Transducers.*

11 Paul, A., Pinjari,H., Hong, W.H., Cheol, H. S., and Seungmin Rho, S. (2018) Fog Computing-Based IoT for Health Monitoring System. *Hindawi Journal of Sensors.*

12 Bonomi, F., Milito, R., Natarajan, P., and Zhu, j. (2014) Fog Computing : A platform for internet of things and analytics in Big Data and Internet of Things. A Roadmap for Smart Environments. Springer :169–186.

13 CISCO (2015) Fog Computing and the Internet of Things: Extend the Cloud to Where the Things Are. (2016) http://www.cisco.com/c/dam/en_us/solutions/trends/iot/doc/computing-solutions.pdf

14 Varghese, B., Wang, N., Bharbuiya, S., Kilpatrick, P and Nikolopous, D.S (2016) Challenges and Opportunitiesin Edge computing, *In IEEE International Conference on Smart Cloud. : 20–26.*

15 Saurez,E., Hong, K., Lillethum, D., Ramchandran, U., and otten-walder (2016) Incremental Deployement and Migration of Geo-distributed Situation Awareness Applications in the Fog *in Proceedings of 10th ACM International Conference on Distributed and Event-based Systems : 258–269.*

16 Varghese, B., Wang, N., Bharbuiya, S., Kilpatrick, P and Nikolopous, D.S Edge-as-a-Service: Towards Distributed Cloud Architectures, *in International Conference on Parallel Computing,* Ser. Advances in Parallel Computing. IOS Press (2017): 784–793.

17 Kolcumarkm, R., Boyle, D., and McCann, A., (2015) Optimal Processing node discovery algorithm for distributed computing in IOT , *in 5th International Conference on the Internet of Things*:72–79.

18 Varghese, B., Subba, L.T., Thai, L., and Barker, A., DocLite: ADocker-Based Lightweight Cloud Benchmarking Tool, *in the 16th IEEE/ACM International SympoComsium on Cluster, Cloud and Grid Computing* (2016): 213–222.

19 Kozhirbayev, Z. and Sinnott, R.O., (2017) A Performance Comparison of Container-based Technologies for the Cloud, *Future Generation Computer Systems,* 68:175–182.

20 Morabito, R., Virtualization of Internet of Things Edge Devices with Container Technologies: A Perfromance Evaluation, *IEEE Access* (2017) 5:8835–8850.

21 Cherrueau, R., Pertin, D., Simonet, A., Lebre A., and Simonin, M., Towards a Holistic Framework for Conducting Scientific Evaluations of OpenStack, in *Proceedings of the 17th IEEE/ACM International Symposium on Cluster, Cloud and Grid Computing (*2017):544-548.

22 Sridhar, S. and Tontino, M.E (2017) Evaluating Voice Interaction Pipelines at the Edge, in *IEEE International Conference on Edge Computing* :248–251.

23 Krylovskiy, A. (2015) Internet of things Gateways Meet Linux Containers: Performance Evaluation and Discussion, in *IEEE 2nd World Forum on Internet of Things (WF-IOT)*: 222–227.

24 Parsopoulos, K.E, Vrahatis, M.N. (2002) Particle Swarm Optimization Method for Constrained Optimization Problems, *Intelligent Technologies -Theory and Application: New Trends in Intelligent Technologies,* 76(1):214–220.

25 Beraldi, R., Mtibaa, A. and Alnweiri (2017) Cooperative Load Balancing Scheme for Edge Computing Resources, in *the 2nd International Conference of Fog and Mobile Edge Computing, IEEE* : 94–100.

26 Puthal, D., Obaidat, S.M., Nanda, P., Prasad , M., Mohanty, S.P. and Zomaya, A.Y. (2018) Secure and Sustainable Load Balancing of Edge Data Centes in Fog Computing, *IEEE Communication Magazine,* 56(5):60–65.

27 Gupta, H., Dastjerdi, V., Ghosh, S.K., and Buyya, R.(2017) iFogSim: A Toolkit for Modeling and Simulation of Resource Management Techniques in Internet of Things, Edge and Fog Computing Environments,*Software: Practice and Experience* 47(9):1275–1296.

28 La, Q.D., Ngo, V. M., Dinh, Q.T., Quak, S. K.T., Shin, H. (2019) Enabling Intelligence in Fog Computing to achieve energy and latency reduction. *Digital Communication and Networks* 5 (2019) 3–9.

29 Huang, H., Cai, Y., Yu, H. (2016) Distributed -Nueron -Network based machine learning on Smart gateway network towards real-time indoor data analytics. *In Proc:eeding Conference Design, Automation & Test in Europe. EDA Consortium* (2016) : 720–725.

30 Rahmani A.M., Thanigaivelan, N.K., Gia, T.N, Granados, J., Nagesh B., Lijberg, P., Tenhumen (2015) Smart E-health Gateway Bringing Intelligence to Internet of Things Based Ubiquitious Healthcare System *In Proceedings IEEE CCNC* (2015) : 826–834.

31 Chen, S., Liu, T., Gao, F., Ji, J., Xu, Z., Qian, B., Guan, H. X., Butler (2017) A Human-centric Smart Home Energy Management System, *IEEE Communication.* Magzine 55(2) : 27–33.

32 Feng, S., Setoodeh, P., Haykin, S., (2017) Smart home; cognitive interactive people centric Internet of Things IEEE Communication Magzine 55(2) : 34–39.

33 Saad, Khan., Parkinson, Simon and Qin, Yongrui, (2017) Fog computing security: a review of current applications and security solutions *Journal of Cloud Computing: Advances, Systems and Applications.*

34 Schumacher M, Femandez-Bugloioni E, Hybertson D, Buschmann F, Sommerland P (2013) Security Patterns: Integrating security and systems engineering. Wiley

35 Krishnana YN, Bhagwat CN, Utpat AP Fog computing-network based cloud computing. *In: Electronics and Communication Systems (ICECS), 2nd International Conference On. IEEE,* 2015.

36 Sarkar, S., Chatterjee, S., and S. Misra,εAssessment of the suitability of fog computing in the context of Internet of Things. *IEEE Trans. Cloud Computing.*

37 Truong, N.B., Lee, G.M., and Ghamri-Doudane, Y. (2015) Software defined networking-based vehicular Adhoc network with Fog Computing In Proceedings IFIP/IEEE (May 2015) :1202–1207.

38 Mahmud R., Koch, F.K., Rajkumar Buyya, (2018) Cloud-Fog Interoperability in IoT-enabled Healthcare Solutions ICDCN'18, Varanasi, India, January 4–7, 2018.

39 Mahmood, A., Muhammad, A., Bilala and Hussain, Shujaat and Kang Byeong Ho and Cheong, Taechoong and Lee, Sungyoung, Health Fog: a novel framework for health and wellness applications. *The Journal of Supercomputing* 72(10) : 3677–3695.

20

Security and Privacy Issues in Fog/Edge/Pervasive Computing

Shweta Kaushik[1,] and Charu Gandhi[2]*

[1] Department of Computer Science Engineering, ABES Institute of Technology, New Delhi, India
[2] Department of Computer Science, JIIT, Noida, India

Abstract

This chapter is concerned with security and privacy handling issues occurring in pervasive and edge boundary systems using voice recognition, and sound using intelligent IoT mining techniques. Fog computing is a promising registering worldview that extends distributed computing to the edge of systems. Like distributed computing yet with unmistakable qualities, fog computing also faces new security challenges other than those acquired from distributed computing. This chapter studies existing writing on fog figuring applications to identify the basic security holes. Comparable innovations like edge figuring, cloudlets, and micro-server farms have additionally been incorporated to give an all-encompassing survey process. Most of the fog applications are driven by the need for usefulness and end-client necessities, while the security perspectives are frequently overlooked or considered in retrospect. This chapter additionally identifies the effect of those security issues and potential arrangements, giving future security-applicable bearings to those responsible for structuring, creating, and maintaining fog frameworks.

Keywords *authentication; data storage; access control; security*

Corresponding author: Shweta.kaushik@abes.ac.in

Fog, Edge, and Pervasive Computing in Intelligent IoT Driven Applications, First Edition.
Edited by Deepak Gupta and Aditya Khamparia.

20.1 Introduction to Data Security and Privacy in Fog Computing

Security and privacy of data is needed at each layer while dealing with fog computing to ensure data correctness and authenticity. It depends on the following parameters as shown in Figure 20.1:

Trust: Trust assumes a significant role in cultivating relations dependent on past cooperation between fog nodes and edge gadgets. A fog node is considered to be the most basic segment for what it's worth responsible for guaranteeing security and namelessness for end clients [1]. In addition, the fog node should be considered as a trusted point for an appointment, as it should be guaranteed that the fog node executes the worldwide masking process on their discharged information and performs the health exercises without damaging the received information. This requires that all hubs that are a part of the fog system should have a specific degree of trust in each other. A notoriety-based trust model [2] has been effective in web-based business, distributed (P2P), client surveys and online interpersonal organizations. Damiani *et al.* [3] proposed a powerful notoriety framework for asset determination in P2P systems utilizing an appropriate surveying calculation to evaluate the unwavering quality of an asset before downloading. In structuring a haze processing notoriety-based notoriety framework, we may need to handle issues, for example, (1) how to accomplish tirelessly, one of a kind, and unmistakable character, (2) how to treat deliberate and accidental misconduct, (3) how to direct discipline and reclamation of notoriety. There are additionally many trusted models dependent on uncommon equipment, for example, secure element, disclosed in execution environment or trusted platform module, which can give trust utility in fog registering applications.

- **Authentication:** In a distributed computing arrangement, server farms are normally possessed by cloud administration suppliers. In this case, fog specialist co-ops can consist of various gatherings because of different arrangement decisions: (1) Internet specialist organizations or remote transporters, who have control of home doors or cell base stations, may assemble haze with their current foundations; (2) cloud specialist organizations, who need to extend their cloud administrations to the edge of the system, may likewise construct mist frameworks; (3) end clients, who possess a nearby private cloud and need to diminish the expense of proprietorship, might want to transform the neighbourhood private cloud into fog and rent the required information or resources on the neighbourhood private cloud.

  Unstable confirmation conventions between fog stages and end-client gadgets have been distinguished as the fundamental security worry of fog registering [4]. The creator's case that the IoT gadgets, particularly in intelligent networks, are inclined to information altering and take off the attacks and can be forestalled with the assistance of a public key infrastructure (PKI), Diffie–Hellman key trade, intrusion identification systems and checking for adjusted information estimation. In addition, the creators show the high significance and effect of MITM assault on fog processing by propelling a stealth assault on video calls among 3G and the WLAN clients inside a fog arrangement.

Many researcher's results show that if any attack did not cause any obvious change in memory and CPU utilization of the fog hub, subsequently, it is very hard to recognize and remove those affected fog nodes. The creators suggest that the danger of such assaults can be forestalled by ensuring correspondence channels between the fog stage and the client through actualizing verification plans.

In light of the present condition of validation in the fog stage, fog stages are utilizing thorough verification and secure correspondence conventions according to their detail and prerequisites. In a fog stage, both security and execution factors are considered to be related, and devices, for example, the encryption procedures known as completely homomorphic [5] and Fan-Vercauteren fairly homomorphic [6] can be utilized to confirm the information validity. These plans comprise a cross breed of symmetric and open key encryption calculations, just as different variations of quality-based encryption. As homomorphic encryption licenses typically activate without decoding the information, the decrease in key circulation will maintain the protection of information. Other research work also gives a comparative system to ensure that understanding frameworks paying little attention to fog figuring, is called the efficient and privacy-preserving aggregation [7]. The framework performs information accumulation dependent on the homomorphic Paillier cryptosystem as the homomorphic capacity of encryption makes it feasible for neighbourhood organization portals to play out a procedure on figure content without unscrambling.

- **Data storage:** In the fog processing environment, client information is collected and the client's command over the information is given over to the fog nodes, which presents the same security dangers as in distributed computing. To start with, it is difficult to guarantee information trustworthiness, since the re-appropriated information could be lost or erroneously adjusted. Second, the transferred information could be mishandled by unapproved parties for different interests. To address these dangers, auditable information stockpiling administration has been proposed with regards to distributed computing to secure the information. Strategies, for example, homomorphic encryption and accessible encryption are consolidated to give honesty, confidentiality and variability for the distributed storage framework to permit a customer to check the information saved on untrusted servers. Need *et al.* [8] have proposed security saving open reviewing for information saved on the cloud, which depends on an outside evaluator (TPA), utilizing a homomorphic authenticator and irregular veil system to ensure protection against TPA. To guarantee the quality of stored information, earlier capacity frameworks use deletion codes or system coding to manage information violation location and information fixes, while *Cao et al.* [9] have proposed a plan utilizing LT code, which gives less capacity cost, a lot quicker information recovery, and similar correspondence cost. Yang *et al.* [10] have given a decent review of existing work towards information stockpiling, inspecting administrations in distributed computing. In haze registering, there are new difficulties in structuring a safe stockpiling framework to accomplish low-inertness, bolster dynamic activity and manage the exchange between fog and cloud.

- **Network security**: Because of the prevalence of distance in haze organizing, remote system security is a major worry to haze organizing. Instances of assaults are sticking assaults, sniffer assaults, etc. Typically, in organizations, we need to trust the configurations physically created by a system director and detach the executive's traffic from standard information traffic [11]. However, mist hubs are taken to the edge of the Internet, which uniformly carry overwhelming weight to the system, predicting the expense of keeping enormous scale cloud servers containing information that are circulated throughout the system edge without any maintenance for access criteria. The work of SDN can facilitate the execution, what is more, the group, and increment arrange versatility and lessen costs, in numerous parts of mist registering.

  In what way can SDN help the mist arrange security? (1) Network monitoring and intrusion detection system (IDS): CloudWatch [12] can use OpenFlow [13] to route traffic for security observing applications or IDS. (2) Traffic isolation and prioritization: traffic seclusion and prioritization can be utilized to keep an assault from blocking the system or commanding shared assets, for example, CPU or circle I/O. SDN can without much of a stretch use VLAN ID/tag to disengage traffic in the VLAN gathering and isolate malevolent traffic. (3) Network resource access control: Klaedtke *et al.* [14] proposed an entrance control plot on an SDN controller dependent on OpenFlow. (4) Network sharing: a fog-upgraded switch in-home system can be opened to visitors if the system sharing to visitors is deliberately planned with security concerns. Research work in [15] has proposed the technique OpenWiFi, in which the visitor WiFi confirmation is moved to the cloud to build up visitor personality; allowing varying visitors, and bookkeeping to be done for authorized end users to appoint commitment of visitors, at each arrival.

- **Access control**: Access control has been a solid device to guarantee the security of a framework and protect the client. Conventional access control normally tends to be in a similar trust area. Due to the idea of distributed computing, entrance control is normally cryptographically actualized for re-appropriated information. The symmetric key-based arrangement is not versatile in key administration. A few open key based arrangements are proposed attempting to accomplish fixed control over the information. To address different security issues [16] and to control security conventions client based, job based and trait-based access control conventions have been proposed. Each has its own focal points and limitations. The client-based access control strategy is simple to execute as just a list of approved clients is overseen but it is not reasonable for haze and cloud conditions which have an enormous number of clients. The job-based technique approves the client dependent on their jobs in a framework. In attribute-based access control strategy, the client's id, and attributes it has approach decide the authority of the client for that information. The characteristic set is particular to a client, it could be a blend of id, current area, date of joining, etc. The entrance approach is a tree which has inner hubs as limit doors (AND, OR) and the leaf hubs are characteristics. The utilization of resources would be permitted to get to the information only if the set of traits coordinates the early strategy. The attribute-based technique assists with saving information on an untrusted cloud server as only the client with the coordinating property set would have the option to unscramble the information. Yu *et al.* [17] have proposed an engrained information acquisition to control combined built on attribute based encryption (ABE). Work [18] proposes an arrangement-based asset get to control haze processing, to help secure coordinated effort and interoperability between heterogeneous assets. In fog processing, how to configure and how to get control over data spreading for customer in fog cloud, simultaneously meet the planning objectives and asset necessities will be done through testing

– **Intrusion Detection:** Intrusion detection methods are broadly sent in a cloud framework to moderate assaults, for example, insider assault, flooding assault, port filtering, assaults on VM also, hypervisor [19], or in a clear network framework to investigate cost estimations and recognize irregular estimations that could have been damaged by attackers. In mist processing, IDS can be sent on a mist hub framework side to distinguish interference by checking and breaking down the log, getting control and client login data. They can likewise be sent at the fog side to distinguish malignant assaults, for example, Denial of Service (DoS), port filtering, and so on. In mist processing, there are new opportunities to explore how mist registering can help with interruption identification on both the customer side and the concentrated cloud side.

• **Virtualization:** With improvements in existing process power on implanted gadgets, there is an opportunity to push additional applications and insights to the system edge. The capacity to run applications at the edge allows another level of virtualization to happen where an assortment of virtual programming based gadgets would all be able to work inside a single piece of equipment. In the equivalent way that virtualization has motivated the way in which cloud scale servers run, fog programming stages meet a few capacities into a private device that will comprise of different characters/functionalities executing equally. Nearby grid fog computing platforms empower different virtualized gadgets to exist together on numerous supported equipment targets. Virtualization of capacities and multi-target backing can diminish operational unpredictability, increase the life of accessible resources, and rely upon the fundamental job of cloud pay-as-you-go, the suppliers give you the capacity to obtain the offered virtual machines and in certain circumstances, will construct them the same as a real PC and you can buy them for a set time, taking advantage of this capacity with no constraints about how they run. What you are buying is the presence of these managers. Suppliers in return guarantee to give you these services with no hindrance at a significant level [20].

- **Privacy:** The security of private data, for example, information, area or use, are important considerations when end clients are utilizing administrations such as distributed computing, remote systems, IoT. There are additional difficulties for saving such protection in haze figuring since mist hubs are in the region of end-clients and can gather more sensitive data than remote clouds lying in the center of the system. Security protection methods have been proposed in numerous situations.

Information Privacy In the fog organization, security protection calculations can be running in the middle of the fog and cloud while those calculations are generally asset restricted toward the end gadgets. The fog hub at the edge, for the most part gathers sensitive information created by sensors and end gadgets. In structures, for example, homomorphic encryption can be used to permit total security safeguarding at the nearby entryways without decoding. Differential security [21] can be used to guarantee the confidentiality of protection of a unrestricted single section in the informational collection in the event of assessable studies.

Use Privacy Another protection issue is the utilization design with which a fog customer uses the fog administrations. For instance, in the intelligent framework, checking the smart meter will reveal a great deal of data about a family unit, for example, when there is no-one at home, what time the TV is turned on, which totally breaches the client's security. Despite the fact that security safeguarding instruments have been proposed in smart metering [22], they can't be straightforwardly applied in mist computing, because of the absence of a confessed in outsider or no partner gadget like a battery. The haze hub can, without much difficulty, gather insights into end-client utilization. One potential arrangement is that the fog customer makes fake undertakings and offloads them to different fog hubs, concealing genuine tasks among

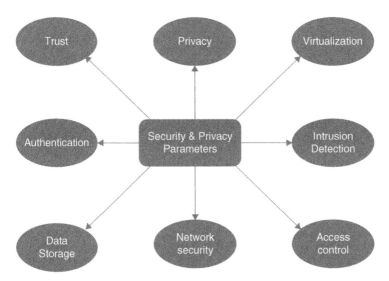

Figure 20.1 Security & Privacy parameters in Fog Computing.

spurious ones. However, this arrangement will increase the fog customer's payment and waste assets and energy. Another arrangement would be planning a smart method for packaging the application to ensure the offloaded asset utilizations don't reveal private data.

Area Privacy In fog processing, area protection principally refers to the area security of the mist customers. A haze customer, for the most part, offloads its assignments to the closest mist hub, the mist hub, to whom the undertakings are offloaded, can assume that the haze customer is close by and more distant from different hubs. Moreover, if a mist customer uses various haze administrations in different areas, it might reveal the way to the haze hubs, assuming the mist hubs communicate. For whatever length of time such a fog customer is connected to an individual or a significant item, the area security of the individual or the article is in danger. On the off chance that a fog customer in every case carefully picks its closest mist server, the fog hub can realize that the mist customer using its registering assets is close by. The best way to protect area security is through personality confusion so that despite the fact that the haze hub knows a mist customer is close by it can't recognize which customer it is.

20.2 Data Protection/ Security

Data security in fog computing environments needs to keep information secure and avoid unapproved access. Protection of information can be attained by encoding, encryption, interpretation of information, cryptographic methods and so on which changes the classified information into limitless structure. It typically incorporates security of information protection from any harmful action performed by a malicious assailant. In a fog environment, the proprietor's data is stored in a remote area and may include some sensitive or classified data. Clients can access the information simply using appropriate ID and approval processes as proposed by the proprietor.

Security Issues: Apart from protection and trust support between various communicating parties, there are other security issues, for example, approval, access control, respectability, non-revocation, organization security, classification and so on requires high reasonability. To give approval and access control for clients, the information manager can choose one of three tools: job based access control, client based access control or attribute based access control to enable the clients to access the required information. To demonstrate that information honesty is maintained with no vulnerabilities the manager can encode the information with a computerized signature. Only the approved clients receive the confirmation key to check the integrity of the received information from the specialist organization to guarantee that recovered information is pure.

In distributed computing, various security issues need serious consideration, as numerous innovations which incorporate OS, database, organizing, virtualization, simultaneousness control, load adjusting, memory the board etc. are covered. For all of these systems security requirements are contrasted and relevant for legitimate treatment of fog conditions. Utilization of virtualization in fog also has a few security concerns, for example, mapping among physical and virtual machines should be completed safely. What is more, arranging the interface of the client with the fog environment should also be secure for information transmission to and from the cloud specialist. And finally, for recognition of any malware movement information mining methods must be applied – a methodology normally embraced for interruption identification framework.

Cloud Security Alliance [23] have identified twelve basic security issues, including different analysts. These issues legitimately influence the disseminated, shared and on-request nature of distributed computing. Being a virtualized domain like cloud, the fog stage can likewise be influenced by similar dangers. This investigation considers the following twelve security classes to plan a precise audit:

1) **Advance Persistent Threats (APT)** are digital assaults where the point is to reduce an organization's framework with the desire to steal information and protected innovation.

2) **Access Control Issues (ACI)** can bring about poor administration and allow an unapproved client to obtain information and authorization to introduce programming and design changes.

3) **Record Hijacking (RH)** is the where an assault intends to seize the client data for malignant reasons. Phishing is a potential method for account seizing.

4) **Denial of Service (DoS)** are where real clients are kept from utilizing a framework (information and applications) by overpowering a framework's limited resources.

5) **Information Breaches (IB)** are when sensitive, safeguarded or secret information is released to or taken by an aggressor.

6) **Information Loss (IL)** is where information is inadvertently (or perniciously) erased from the framework. This does not necessarily result from a digital assault and can occur through a catastrophic event.

7) **Unsteady APIs (UA)** Many cloud/fog suppliers uncover application programming interfaces (APIs) for client use. The security of these APIs is significant to the security of any actualized applications.

8) **Framework and Application Vulnerabilities (FAV)** are exploitable bugs emerging from programming promotion arrangement mistakes that an aggressor can use to invade and bargain a framework.

9) **Deadly Insider (DI)** is a client who has approved access to the system and framework, yet has purposefully chosen to act malignantly.

10) **Inadequate Due Diligence (IDD)** regularly emerges when an association rushed the selection, plan, and usage of any framework.

11) **Misuse and Nefarious Use (MNU)** frequently emerges when assets are made freely available and vindictive clients use those assets to embrace malignant action.

12) **Common Technology Issues (CTI)** occur due to sharing frameworks, stages or applications. For instance, hidden equipment parts might not have been intended to offer solid confinement properties.

The following table lists a wide-range of fog applications, giving specific consideration to their potential security suggestions. As fog figuring is still in its early stages, comparative advances have additionally been discussed to make the overview increasingly all-encompassing and valuable. The fog framework is evaluated by dividing the openly accessible script that have been gathered into the lower subsections. Throughout this section, the twelve classes showed in Figure 20.2 are considered and a synopsis is given in Table 20.1.

20.3 Great Security Practices In Fog Processing Condition

While utilizing fog computing to store and process any classified information, different security practices should be taken into account for verifying the information from any assault or hazard discussed below and shown in Figure 20.3.

Table 20.1 Security issues with possible impact and Solution.

S. No.	Possible Attack	Threat	Impact	Solution
1	**Web security issues**	• SQL injection • Session/account hijacking • Malicious redirections • Cross-site scripting • Insecure object references • Cross-site request forgery	As all data, services and virtual machines are executing in a virtualized environment, its conciliation will have opposing effect on all fog services, data and users	• Multi-factor authentication • Process isolation • Intrusion detection system • User-based access control model • Role-based access control model • User data isolation • Attribute based encryption
2	**Wireless security issues**	• Active impersonation • Message replay attacks and distortion issues • Data loss or breach • Illegal resource consumption • Sniffing attacks	Usage of different vulnerable wireless access points can cooperation communication privacy, consistency, accuracy, availability and reliability	• Authentication • Wireless security protocols • Encrypted communication • Secure routing • Key management service • Private network
3	**Internal/external communication issues**	• Man-in-the-middle attack • Single-point of failure • Inefficient rules/policies for access control • Insecure APIs and services • Session/account hijacking • Application vulnerabilities	Attacker can acquire any user secret or sensitive information by snooping and can also get access to unauthorized fog resources for their own purposes	• Encrypted communication • Transport layer security • Mutual/multi-factor authentication • Certificate trapping • Partial encryption • Isolating compromised nodes
4	**Malware protection**	• Virus • Worms • Trojans • Spyware • Ransomware • Performance reduction	Infected nodes will lower the performance of the entire fog platform and allow backdoors to the system and corrupt or damage data permanently	• Anti-malware programs • System restore points • Rigorous data backups • Intrusion detection system

Table 20.1 (Continued)

S. No.	Possible Attack	Threat	Impact	Solution
5	**Data security related issues**	• Data replication and sharing • Illegal data access by unauthorized users • Data altering and erasing attacks • Malicious insiders • Multi-tenancy issues • Denial of service attacks • Data ownership issues	High probability of illegal resource, services and database access, where attacker can negotiate both user and fog system's data	• Policy enforcement • Network monitoring • Encryption and secure key management • Security inside design architecture • Data classification • Or masking
6	**Virtualization issues**	• Hypervisor attacks • Privilege escalation attacks • VM-based attacks • Inefficient resource policies • Weak or no Logical Segregation • Side channel attacks • Service abuse	As all services and virtual machines are executing in a virtual environment, its compromise will have adverse effect on all fog services, data and users	• Multi-factor Authentication • Process isolation • Intrusion detection system • Attribute/identity-based encryption • User or role-based access control model

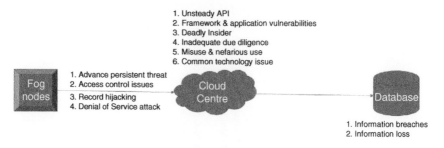

Figure 20.2   Security issues in Fog Computing.

Figure 20.3   Security practice in Fog Computing.

1) **Protection against Internal and External dangers:** security hacking organizations offer help to improve the security system of a client by successfully examining logs and cautions from devices working non-stop. Watching group correspond information from various security devices to give security examiners the data they require to decline wrong positives and respond to authentic risks against the undertaking.

2) **Early identification:** Early location or exposure of administrations decisions and reports new security vulnerabilities immediately after they show up. For the most part, the security risks and dangers are associated with outside sources and a caution or report is provided to the information owner in regards to this. These security vulnerabilities reports connect point by point to give a view of the vulnerabilities and the stage impacted, alongside partial information.

3) **Intelligent log centralization:** Intelligent log centralization and an assessment course of action involving checking, is essentially dependent on the relationship and organization of log passages. Such assessment has the effect of setting a standard of operational execution and gives a list of security dangers. An alarm can be raised in the event of an incident to move the defined standard parameters past a predefined edge limit.

4) **Discovery of vulnerabilities and the executives:** Recognition of vulnerabilities and organizing engaged automated confirmation and organization of the security level of the information framework. The organization performs intermittent follow-up of undertaking performed by security specialists administering information structure security and gives reports that can be used to execute a course of action for interminable difference in the system's security level.

5) **Intervention, forensics and beneficial area administrations** Immediate or quick translation when a hazard is recognized is essential to direct the effect of this hazard. This requires a group of security engineers with good knowledge of various advancements and developments, alongside the ability to support use and foundation on a 24/7 premise. Whenever a recognized hazard is found, it is important to look at the issue to discover what it is, how much effort it will take to settle the issue and what effects are presumably going to be seen.

20.4 Developing Patterns in Security and Privacy

These days, fog computing is used in different spaces and all areas have their own requirements with regards to security, protection and trust. They utilize their own components, semantics and interface. While applying any new innovation, some basic issues with respect to security and protection require quick consideration for adoption as shown in Figure 20.4.

1) **Authentication and Identity Management:** Fog administrations enable clients to share information with different clients over the web. Personal information can help validate clients depending on their attributes and accreditations. People using the board will likewise help in safeguarding the sensitive and private data of the owner from an attacker using authorization methods. While clients are associating with any front-end administration, that administration must guarantee that their information is hidden from the other administrations with which it communicates. The use of validation procedures guarantees that only the approved client for an application can get the required information.

2) **Access Control and Accounting:** Heterogeneity and average assortment of information or administrations, and essential in dispersed processing

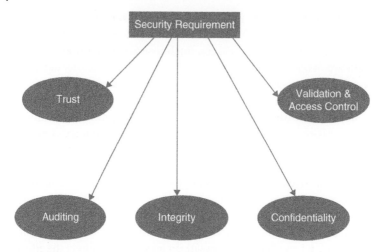

Figure 20.4 Security Requirement in Fog Computing.

conditions, demanding fine-grained access control systems. Explicitly, control should be adequately adoptable to discover dynamic setting, characteristics or qualification-based access essentials and to maintain the rules of least benefit. Such access control administrations may need to arrange security protection requirements imparted through complex rules.

ABE is an advanced cryptographic procedure to give end-clients a versatile, adaptable and fine-grained method of control. The idea of ABE was first proposed by Sahai, furthermore, Waters used another strategy for foggy character-based encryption. ABE has two variations, they are ABE (KPABE) and CP-ABE strategies.

Notwithstanding, the maximum substantial downside of the usage of ABE in fog figuring is the computational rate inside the encryption, also, the interpreting level is straightforward relative to the unpredictability of the strategy. Fog hubs, the threshold of the cloud and toward end-clients, are possibly the first-class desire for a re-appropriating intermediary, which may be applied to make immense counts to decrease the computational overhead required in IoT devices with limited assets. The principal association of the existing plans is to disseminate the computations of the CP-ABE encryption and unscrambling level, so the limited IoT devices can rent a part of the usage responsibilities to the hubs of the system. Louise *et al.* has considered a cloud-based total layout for medical WSNs, in which the sensor hubs redistribute cryptographic sports to a reliable passage that scrambles CP-ABE-primarily based total facts earlier than sending it to the cloud.

To perform cryptographic content update benefits in fog figuring, the cloud service provider must have the option to confirm the client's test prior to tolerating the altered cryptographic content. ABS is a developing mark calculation to guarantee cryptic client validation

In view of ABS, Ruj *et al.* have proposed another decentralized way to control the framework for the safe checking and composing of information in the cloud, which bolsters the confirmation of unknown clients. Right now, cloud confirms validity without knowing the client's personal details previously saving information. His *et al.* proposed an expressive plan of ABS in IoT, which utilizes a credit tree to guarantee that only a client with the proper credentials which meet the entrance approach can approve the message.

3. **Trust administration and arrangement joining** There are numerous specialist co-ops existing together in the fog environment, who team up to give different administrations from a single spot. Each cooperative may have their own security and protection approaches. Therefore, we should address heterogeneity among their approaches. A security supplier should be cautious while overseeing access control approaches, to ensure that coordination of strategy doesn't break any security or protection arrangement. Accordingly, a trust machine should be created to proficiently seize a traditional association of parameters required for agreement with the executives. Likewise, whilst aiming for coordination of strategies, special protection problems with recognition of semantic heterogeneity management interoperability and technique evaluation wishes to cope cautiously. "

The trust establishment with the executives is firmly recognized with the person to allow entry to manipulate the board structure. Despite the truth that there may be no dependable means of agreement [24], scientists understand the importance of agreeing with the executives. Numerous plans have been proposed to supervise the agreement close to the IoT in a way to manipulate looking up IoT devices. Reference [25] presented a fuzzy notoriety based total agreement with the executives solution for IoT distant sensor systems, wherein they take into consideration the parcel sending/conveyance percentage and power usage due to the Quality of Services (QoS) measurements to evaluate the agreement with connection. Recently, the IoT has received a lot of attention due to the excessive extensibility through coordinating casual groups concept to IoT. In this manner, many agreements with the executive's preparations have evolved depending on the IoT worldview. For example, [26] provided agreement with the executive's solutions for IoT worldview. A comparative agreement was given in [27] with the executives conference for IoT without a focused revealed power. All such preparations make use of the

QoS measurements in addition to the proprietorship amongst devices and owners and social connections among customers to evaluate the agreement to guard against assaults.

In any case, those agreeing with the executive's preparations are likewise primarily based totally beyond the verifiable person presumption that customers and expert groups need to all agree with their character providers so they may distinguish all differences in a comparable protection area. Inside a comparable protection area, customers and expert co-ops agree with and rely upon a comparable person provider, conceding that their personal information may not be undermined or abused through the character provider or outsiders. As a rule, person providers are established upon vulnerabilities which discover people's information vaults to be taken through notion aggressors.

Security Requirement	Description with arrangement
Validation and access control	To give the proprietor's information to approved client as per client's entrance criteria. Specialist organization must separate the approved and unapproved clients and enable only approved clients to access the necessary information as per their job, ability, personality and so on.
Confidentiality and privacy	To ensure classification and security, the information should be transmitted and stored in a scrambled structure as opposed to exceptional. Only the approved clients can get the information after unscrambling, specialist co-op must be ignorant of this decoding procedure.
Integrity	Data owner must sign its information and only the approved client has the confirmation key to check that the obtained information is clean with no modification done by a malicious assailant during transmission.
Auditing	To keep up the interoperability highlight in fog based framework, any vindictive movement will be checked by reviewer and inspecting reports are sent to alert the owner. The owner will take action as required.
Trust	To guarantee that the information is stored in the right place and recovered accurately without presenting any weakness. This tends to be accomplished by an agreement marked between the individual parties.

20.5 Conclusion

The reason for this examination was to audit and break down genuine fog process-ing applications to distinguish their conceivable security blemishes. It was found that most fog applications don't think about security as a major aspect of the frame-work, but instead centre around usefulness, which brings about many fog stages being helpless. Writing likewise subtleties that fog processing has a wide poten-tial and scope of utilizations that all interest an elevated level of security to ensure the CIA of the client information. Haze stages are a generally new worldview, and this investigation can support peruses and engineers to anticipate safety efforts and their difficulties while conceiving the plan of new fog frameworks.

Bibliography

1 Elmisery, A. M., Rho, S., & Botvich, D. (2016). A fog based middleware for automated compliance with OECD privacy principles in internet of healthcare things. *IEEE Access*, 4, 8418–8441.

2 Jøsang, A., Ismail, R., & Boyd, C. (2007). A survey of trust and reputation systems for online service provision. *Decision support systems*, 43(2), 618–644.

3 Damiani, E., di Vimercati, D. C., Paraboschi, S., Samarati, P., & Violante, F. (2002, November). A reputation-based approach for choosing reliable resources in peer-to-peer networks. In *Proceedings of the 9th ACM conference on Com-puter and communications security* (pp. 207–216).

4 Stojmenovic, I., Wen, S., Huang, X., & Luan, H. (2016). An overview of fog computing and its security issues. *Concurrency and Computation: Practice and Experience*, 28(10), 2991–3005.

5 Gentry, C. (2009, May). Fully homomorphic encryption using ideal lattices. In *Proceedings of the forty-first annual ACM symposium on Theory of comput-ing* (pp. 169–178).

6 Bos, J. W., Castryck, W., Iliashenko, I., & Vercauteren, F. (2017, May). Privacy-friendly forecasting for the smart grid using homomorphic encryp-tion and the group method of data handling. In *International Conference on Cryptology in Africa* (pp. 184–201). Springer, Cham.

7 Lu, R., Liang, X., Li, X., Lin, X., & Shen, X. (2012). EPPA: An efficient and privacy-preserving aggregation scheme for secure smart grid communications. *IEEE Transactions on Parallel and Distributed Systems*, 23(9), 1621–1631.

8 Wang, C., Wang, Q., Ren, K., & Lou, W. (2010, March). Privacy-preserving pub-lic auditing for data storage security in cloud computing. In *2010 proceedings ieee infocom* (pp. 1–9). IEEE.

9 Cao, N., Yu, S., Yang, Z., Lou, W., & Hou, Y. T. (2012, March). LT codes-based secure and reliable cloud storage service. In *2012 Proceedings IEEE INFO-COM* (pp. 693–701). IEEE.

10 Yang, K., & Jia, X. (2012). Data storage auditing service in cloud computing: challenges, methods and opportunities. *World Wide Web*, 15(4), 409–428.

11 Tsugawa, M., Matsunaga, A., & Fortes, J. A. (2014). Cloud computing security: What changes with software-defined networking?. In *Secure Cloud Computing* (pp. 77–93). Springer, New York, NY.

12 Shin, S., Xu, L., Hong, S., & Gu, G. (2016, August). Enhancing network security through software defined networking (SDN). In *2016 25th international conference on computer communication and networks (ICCCN)* (pp. 1–9). IEEE.

13 McKeown, N., Anderson, T., Balakrishnan, H., Parulkar, G., Peterson, L., Rexford, J., ... & Turner, J. (2008). OpenFlow: enabling innovation in campus networks. *ACM SIGCOMM Computer Communication Review*, 38(2), 69–74.

14 Klaedtke, F., Karame, G. O., Bifulco, R., & Cui, H. (2014, August). Access control for SDN controllers. In *Proceedings of the third workshop on Hot topics in software defined networking* (pp. 219–220).

15 Thompson, N. A., Yin, Z., Luo, H., Zerfos, P., & Pal Singh, J. (2007, September). Authentication on the edge: distributed authentication for a global open wi-fi network. In *Proceedings of the 13th annual ACM international conference on Mobile computing and networking* (pp. 334–337).

16 Sookhak, M., Yu, F. R., Khan, M. K., Xiang, Y., & Buyya, R. (2017). Attribute-based data access control in mobile cloud computing: Taxonomy and open issues. *Future Generation Computer Systems*, 72, 273–287.

17 Sookhak, M., Yu, F. R., Khan, M. K., Xiang, Y., & Buyya, R. (2017). Attribute-based data access control in mobile cloud computing: Taxonomy and open issues. *Future Generation Computer Systems*, 72, 273–287.

18 Dsouza, C., Ahn, G. J., & Taguinod, M. (2014, August). Policy-driven security management for fog computing: Preliminary framework and a case study. In *Proceedings of the 2014 IEEE 15th International Conference on Information Reuse and Integration (IEEE IRI 2014)* (pp. 16–23). IEEE.

19 Modi, C., Patel, D., Borisaniya, B., Patel, H., Patel, A., & Rajarajan, M. (2013). A survey of intrusion detection techniques in cloud. *Journal of network and computer applications*, 36(1), 42–57.

20 Durairaj, M., & Kannan, P. (2014). A study on virtualization techniques and challenges in cloud computing. *International Journal of Scientific &Technology Research*, 1(1).

21 Van Tilborg, H. C., & Jajodia, S. (Eds.). (2014). *Encyclopedia of cryptography and security*. Springer Science & Business Media.

22 Rial, A., & Danezis, G. (2011, October). Privacy-preserving smart metering. In *Proceedings of the 10th annual ACM workshop on Privacy in the electronic society* (pp. 49–60).

23 Oberoi, P., & Mittal, S. (2017). SURVEY OF VARIOUS SECURITY ATTACKS IN CLOUDS BASED ENVIRONMENTS. *International Journal of Advanced Research in Computer Science*, 8(9).

24 Sicari, S., Rizzardi, A., Grieco, L. A., & Coen-Porisini, A. (2015). Security, privacy and trust in Internet of Things: The road ahead. *Computer networks*, 76, 146–164.

25 Chen, D., Chang, G., Sun, D., Li, J., Jia, J., & Wang, X. (2011). TRM-IoT: A trust management model based on fuzzy reputation for internet of things. *Computer Science and Information Systems*, 8(4), 1207–1228.

26 Chen, R., Bao, F., & Guo, J. (2015). Trust-based service management for social internet of things systems. *IEEE transactions on dependable and secure computing*, 13(6), 684–696.

27 Bao, F., & Chen, I. R. (2012, September). Dynamic trust management for internet of things applications. In *Proceedings of the 2012 international workshop on Self-aware internet of things* (pp. 1–6).

21

Fog and Edge Driven Security & Privacy Issues in IoT Devices

Deepak Kumar Sharma[1], Aarti Goel[1], and Pragun Mangla[2]

[1]Department of Information Technology, Netaji Subhas University of Technology
[2]Department of Electronics and Communication, Netaji Subhas Institute of Technology, New Delhi, India

Abstract

With exponential growth of Internet of Things (IoT) devices, the amount of data produced has also increased explosively. Traditional cloud computing has come to bottleneck due to its limitations in bandwidth, resources, latency and time sensitive data which is generated within milliseconds. Therefore, upcoming technologies such as fog and edge computing techniques are required to handle the data efficiently. The fog computing paradigm improves storage and computing and network facilities of cloud computing that reduce the delay or service latency to end users. It has the capacity to process a high number of nodes and supports geographical distribution. Edge computing also provides features such as content perception, parallel processing and real time computing. Hence increasing the network performance and reducing latency. However, all the new techniques have their own security and privacy issues and challenges that need to be studied. Similarly, edge computing also has some additional features due to which it requires special attention. This chapter is categorized into two sections: the first section is focused on fog computing and the second section deals with edge computing. We will first introduce the basics of fog and edge computing, its architecture, working, advantages and use cases, and then majorly focuses on their security and privacy issues separately. In the end offered solutions and research opportunities in both fields are discussed.

Keywords *Fog computing; Edge computing; Security; Privacy; Malware Attacks; Authentication; Access control*

Fog, Edge, and Pervasive Computing in Intelligent IoT Driven Applications, First Edition.
Edited by Deepak Gupta and Aditya Khamparia.
© 2021 John Wiley & Sons, Inc. Published 2021 by John Wiley & Sons, Inc.

21.1   Introduction to Fog Computing

Due to the explosive rise in the field of the Internet of Things (IoT), the quantity of data produced has increased significantly and the cloud will not be able to store and process huge quantity of data. This gave birth to a three-layered architecture to handle increased communication and data processing called fog computing also known as fog networking or fogging. It is similar to cloud computing but the basic difference between the two is that fog computing indicates that it is present near the Earth whereas cloud computing suggests it is in the sky. In other words, fog computing is close to the devices which produce or generate the data. Devices can only upload data sporadically to the cloud and therefore sensitive data loses its value. For example – one may face a medical emergency situation and the patient may die due to the latency delay. If the data produced by the IoT devices are sent for further processing, it will give rise to congestion and hence consume all the bandwidth.

In order to solve this issue, a fog layer is added between the cloud and the end user to carry out part of the storage and data processing, thus reducing congestion. In other words, fog computing is a communication and programming paradigm that reduces the distance between the cloud and the IoT devices by acting as an interface between them.

21.1.1   Architecture of Fog

The most important task of fog computing is to make latency predictable and low in cases of time-sensitive IoT applications. The whole architecture of fog environment comprises a total of six layers, namely – physical and virtualisation, monitoring, pre-processing, temporary storage, security and transport layer as shown in Figure 21.1 [1].

Now let us discuss each layer in detail. The physical and virtualisation layer is a first layer at the bottom level which consists of various nodes that can be categorised into virtual, physical nodes [1]. Maintenance and management of the nodes are done in accordance with their types and service requests. Geographically distributed sensors of various kinds are given the task of analysing the environment and provide the gathered data to further layers via gateways [7]. The task of the second layer, i.e., the monitoring layer, is utilization of the resources and the availability of sensor, fog nodes and network elements are being observed. Due to the fact that fog environment utilizes many tools with varying power consumption levels, it becomes important to monitor energy management measures in a timely and effective way [8, 9].

The task of management of data is performed by the pre-processing layer. The major role of this layer is to analyse the data and then its filtering and trimming is

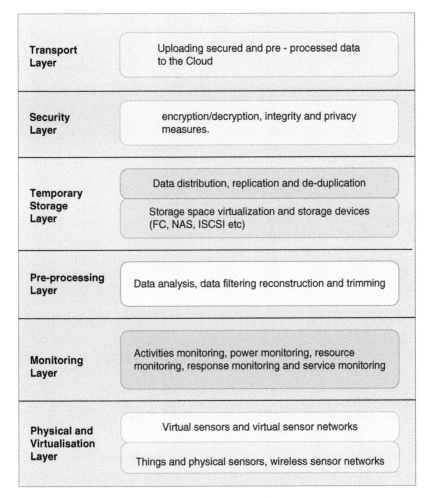

Figure 21.1 Layered architecture of fog computing.

done in order to extract the required information. The pre-processed data is then temporarily stored in the next layer, namely the temporary storage layer. After the data is sent to the cloud, the temporary data can be removed since it is no longer needed [9, 10].

In the second layer, namely the security layer, coding and decoding of data takes place and the integrity measures are also applied (if necessary) in order to secure it from tampering. Finally, in the transport layer, uploading of pre-processed data into the cloud takes place in order to extract and develop essential services [9, 10]. For efficiency purposes, only a part of the gathered data is being sent into the cloud.

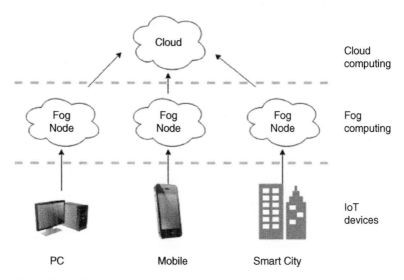

Figure 21.2 Fog computing is closer to end devices.

We can also say that the gateway device establishing a bridge between the IoT and cloud observes the data prior to dispatching it to the cloud. This gateway is referred to as a smart gateway [11]. Information generated from IoT devices as well as sensor networks goes through smart gateways into the cloud. The cloud stores the data and creates services for the users [6].

21.1.2 Benefits of Fog Computing

As illustrated in Figure 21.2, the fog environment layer is located between the cloud and end users, expanding the cloud computing model to the edge of the network. The fog computing brings networking services, storage and processing closer to the end user devices. The fog environment comprises fog nodes that can be placed at any location in a connection. Devices having storage, network connectivity and computing can act as a fog node, for example – routers, switches, industrial controllers, embedded services etc. [1].

The major advantages of the fog environment can be defined and summarized as follows:

1) **Improved business flexibility**: When combined with the optimum tools, fog applications are easily developed and placed. They can even schedule the device to act in accordance with the needs of the customer [5].
2) **Geographical distribution**: Fog provides boundless applications and services, since fog nodes are distributed and can be deployed anywhere.

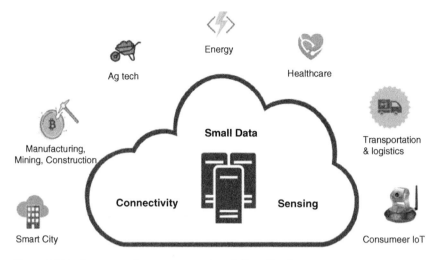

Figure 21.3 Fog computing supports various IoT applications.

3) **Lower operating expense**: Since the fog layer process selects information locally rather than pushing it into cloud, it reduces traffic as well as saves network bandwidth [5].
4) **Interoperability:** The ability to exchange information between a variety of domains and service providers has become easier due to fog computing.
5) **Heterogeneity:** Since fog nodes can easily work on various platforms, it has made deployment of fog nodes produced by various manufacturers possible and thus support heterogeneity.
6) **Awareness of location and reduced latency**: Due to the ability of fog to get its node deployed in different locations, it supports location awareness. Due to the fact that the gap between fog and end devices is less, it supports reduced latency when handling the time sensitive data of the user devices.
7) **Scalability:** Fog supports extended storage resources and distributed computing which can work with large-scale sensor networks monitoring the enclosing environment.

21.1.3 Applications of Fog with IoT

As illustrated in Figure 21.3, a number of IoT based applications are supported by fog computing and plays a critical role in their working. Some of the major applications are:

1) **Smart Traffic Lights**
   The fog environment helps traffic signals to start movement on roads based on sensing the flashing lights. Sensors can sense the presence of cyclists and

pedestrians and calculates the speed and distance of surrounding transport. A sensor light turns on/off depending on the movement [5]. Smart traffic lights can behave as fog nodes that are in synchronization creating a web of fog nodes to produce and escalate an alert or warning message to surrounding transport. With facilities like Wi-Fi, 3G, 4G, smart traffic lights can improve the interactions between fog and vehicles.

2) **Healthcare and Activity Tracking**

The fog environment is heavily used in healthcare since the data generated in the healthcare sector can be time-sensitive. It provides event responses and real-time processing to the time sensitive data which needs immediate attention [15]. Also, the intercommunication of devices for various purposes like remote storage, processing and previous medical record retrieval needs a secure and trustworthy network [16].

3) **Smart Home**

A large number of devices that have a direct connection at home are sensor based, but at the same time they work on different platforms since they are taken from different manufacturers which hinders their ability to work in accordance with each other. Also, few particular functions require a huge quantity of storage and computation. Fog computing integrates all the different platforms and gives more power to smart home applications agilely [4, 14].

4) **Wireless Sensor and Actuator Networks**

In this system actuators act as fog nodes providing a variety of ways to control user end devices [1]. These WSNs consumes very little energy, require reduced bandwidth and reduced processing power [7].

5) **Augmented Reality**

Augmented reality (AR) can be stated as systems that provide the real world with information [17]. Since computing devices are on the rise and have become much faster, smaller and ubiquitous, AR is becoming more and more critical. Applications of AR are quite latency sensitive because even a small delay in response can be very expensive to the user and can even create problems in the future [13]. When supported by fog computing through their fog and cloud servers, AR is able to perform real time applications [1].

21.1.4   Major Challenges for Fog with IoT

It is a fact that fog computing offers many benefits in a variety of IoT applications but still there are various challenges being faced as shown in Figure 21.4.

1) **Scalability** Data generated by billions of IoT devices requires a huge amount of resource, mainly storage and processing power. Hence it becomes important for fog servers to support and provide all the required resources. As a result a real challenge lies in the ability to give a progressive response to these ever increasing IoT devices and applications.

2) **Dynamicity** This refers to the ability to dynamically alter the workflow composition of IoT devices. This changes the elemental properties and efficiency of IoT devices. Furthermore, these devices suffer greatly from ageing of their hardware and software. The repercussions can be seen in their workflow behaviour [1].

3) **Latency** The creation of fog computing was due to the need for low latency for time-sensitive applications which the cloud was not able to provide, but still there exist some factors that result in high latency on platforms of fog and edge computation that leads to user discomfort [20].

4) **Security** Fog nodes are protected by the same physical and cyber-security configuration used in cloud computing [5]. But the environment of fog itself is quite vulnerable to security attacks. Therefore, we need different security mechanisms for fog due to its attributes such as large-scale geo-distribution and heterogeneity [21]. Security of fog nodes can be improved by cryptography as proposed by many research studies [4, 19].

5) **Complexity** Choosing the best components for IoT devices based on software, hardware and personal requirements has become quite difficult since these are being designed by different producers and manufacturers which leads to increases in the difficulty of operation [19].

6) **Heterogeneity** Due to the fact that sensors and IoT devices are designed by different manufacturers, heterogeneity in terms of computing power, storage and communication becomes a major challenge [4].

7) **Resource Management** Resource capacity of the fog-end devices cannot be compared with that of traditional service even if they are equipped with additional computing power and storage, hence judicious use and management of fog resources is needed now for more efficient operation of fog platforms [12, 13].

8) **Energy Consumption** When compared to the typical cloud model, fog can be seen as requiring more energy due to the huge number of fog nodes that are used to create the fog environment [22].

21.1.5 Security and Privacy Issues in Fog Computing

Since the security measures that are being applied on the cloud cannot be applied directly to the fog because of the difference in size and the nature of data it is dealing with. The following security and privacy issues are being faced by fog computing as shown in Figure 21.5. Some of them are listed below.

1) **Trust**

The primary relationship between fog nodes and IoT devices is being set by authentication but it alone is not sufficient since these IoT devices are vulnerable to malicious attacks as well as malfunctioning. This deficiency is tackled by trust since it can help the fog nodes to determine whether the IoT

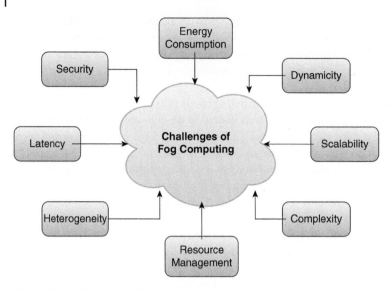

Figure 21.4 Challenges of fog computing.

devices making the request are genuine or not. At the same time, the IoT devices that send their time sensitive data to these fog nodes can also check the security level of the fog nodes [30]. But the question that arises is how to gauge the level of trust provided by fog computing and what are the major characteristics and attributes that supports the idea of trust in fog computing services?

The traditional trust models that are currently being used by cloud computing service can be used by fog computing also because of the mobility issues and absence of centralized management. It is a fact that fog computing provides attributes in order to gauge the level of trust but still the question arises, "who is going to watch over these attributes?"

There are a number of options available in the case of the trust-management model, of them the rating based trust model is the most used by e-commerce websites since the reputation of the service providers is based upon the ratings given by a wide variety of consumers with different opinions. Because of the dynamic property of the fog nodes, this is not well suited in this case but can be used in order to select a fog service as it can be helpful in judging the reliability of the fog service provider. One of the strongest factors being considered while building a trust-model in cloud computing is service level agreement (SLA) [21]. In case of the fog layer, a known third party or an expert is needed to watch over SLA verification [21].

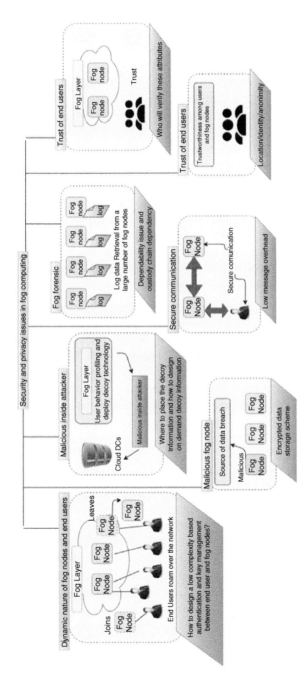

Figure 21.5 Open research challenges in fog security and privacy issues.

2) **Authentication**

IoT devices using fog services are required to be authenticated by the fog network, it is the primary step in building a connection between both as it provides access to services provided by the fog network and also helps in blocking unauthorised and corrupted nodes from entering the network. Since the IoT devices that participate in networking are limited by factors such as processing, power and storage, authentication processes that use public key infrastructure (PKI) and certificates are not well suited [21].

3) **Secure Communications in Fog Computing**

The interaction between IoT devices and fog nodes takes place only when IoT devices face requests regarding storage and processing requirements from the fog nodes, any other kind of communication does not come under the environment of fog computing. In order to improve or manage the network, fog nodes can also communicate between themselves and can even work in a distributed manner in order to perform a particular task. The two kinds of communication that we need to secure in order to have safe fog computing environment are:

1) communication involving fog nodes and IoT devices

2) communication among fog nodes within the same network

Interaction between an IoT device and a fog node starts when a fog node requests storage and processing. At the same time the IoT device is not even aware of the existence of the network, hence the use of symmetric cryptographic techniques to provide security to the communication is not possible. In fact, the use of asymmetric cryptography is also not possible due to a unique set of challenges that it forces upon the IoT environment. The other type of communication involving only fog nodes requires end to end security as there is a possibility of the existence of corrupted nodes involved in multi-hop path.

4) **End User's Privacy**

The major attribute that fog computing works upon is the high degree distribution of fog nodes that reduces the net pressure of the complete data centre. In the case of fog computing the task of security preservation is a major challenge as the fog node communicating with the end user can possibly collect the user's personal information such as usage of entities, location, identity, etc. As we are aware of the fact that the fog nodes are highly distributed in a large geographical area, it becomes difficult to control them centrally. If security of even a single edge node is compromised, it can become an entry point for an intruder. Once inside the network, intruders can steal any sensitive information about the end user that the network can offer. Intruders can launch a number of attacks within the network and can even analyse the user's habits from the pattern of fog environment usage and its services. Even the highly secured systems that are very well designed can be exploited through their side channels.

5) **Malicious Attacks**

Fog computing can be vulnerable to vicious malicious attacks if not supported by strong security measures. One of the most common attacks that come under this category is a denial-of-service (DoS) attack that exploits mutual authentication between the fog network and the end-user device. It comes into action when a device requests an undefined storage service, during which the corrupted node can make repeated storage requests and hinder or stall the legitimate requests made by other devices. Defence or security strategies that are being used by other networks such as cloud computing are not applicable in the case of fog computing because of the openness and the huge size of the network. Since authentication of all the devices by the fog nodes is not possible, they rely on outside parties, for example a certification authority that uses some form of credentials to authenticate a device. But even the certification authority is not very effective in providing security in this case. The reason is that the credentials of the user will be used by a processing node to judge whether the request has been made by a legitimate node or not.

21.2 Introduction to Edge Computing

Edge computing (EC) is considered as an extension to cloud computing, but it is, in fact, different in several basic ways. Let us understand edge computing with the help of real-life examples such as fingerprint recognition, facial recognition, CCTV camera etc. Users simply put fingerprints on edge and machine uses artificial intelligence and machine learning to recognise it. Modern day security has become so advanced that cameras have an intelligent system built in to understand the sensitivity of data before sending it to the server. Cameras record everything when something is happening, when there is movement, someone arriving, but if nothing happens or there is no motion then data will not be sent anywhere, as a result of which bandwidth is reduced. Secondly, we do not need to send every bit of data to a cloud server, however the bit of information we really need can be sent to the cloud.

Edge computing allows users to analyse generated data from smart devices on their device instead of transferring it to a central storage unit or server far away. It also allows some of the workload to be offloaded from the cloud or user's device in order to speed up the applications which require low latency response. Data being generated from smart devices is stored in the device instead of sending it to the cloud which takes some time thus reducing the latency. Additionally, it provides fast response, fast processing and reduced network traffic.

In edge computing, the edge interface comes in between the IoT devices and the cloud server. This edge could be directly connected with an IoT device or configured as close as possible to the device. The objective is to offload some of the computing logic and data to the edge network so the data will not travel all the way back to the central server every time, for the device to know that the function needs to be executed. Therefore, edge computing networks can greatly reduce latency and enhance performance. This approach offers speed and flexibility for handling data and an exciting range of possibilities for organizations.

21.2.1 Architecture and Working

EC is a new approach to network architecture where first, cloud computing continues to play an important role, and secondly the new possibilities offered by IoT devices which are capable of processing the data are forcing companies to rethink their approach to IT infrastructure.

For edge computing we can use any desktop, laptop, low end, and high-end servers based on the requirements. Even todays we can find edge specific servers in the market place manufactured by some hardware companies. IoT and 5G together enable manufacturers to embed sensors and microchips into products and services. Smart home systems can now be controlled by a mobile phone, smart meters can monitor energy use and smart home appliances can monitor consumption and reorder groceries. Edge computing devices are a game changer in edge computing.

Rapid advantages in IT have enabled manufacturers to embed sensors and microchips into products and services. Currently the data that needs storage, analysis and decision making is sent to the data centres or servers in the cloud [25]. In edge computing architecture, the lowest level data sources consist of several edge or physical devices enabled with sensing and communication capacity. With edge computing, data is not sent to the cloud but acted on at the source, creating real time insights. This happens at the edge of the device or at the nearest location, where data is generated. As a result, data processing improves significantly by reducing the data travelling between devices and centralized cloud data centres.

21.2.2 Applications and use Cases

Figure 21.6 represents applications and use cases of edge computing.

1) Vision Recognition with Deep Learning:
   Vision recognition application can scan images and tell us what these images are about with its nearest probabilities. They use deep learning and neural algorithms; this requires complex and heavy logic computation. Doing it through a cloud server will consume a huge amount of Internet bandwidth and server resources. On the other hand, doing it through edge computing heavy logic

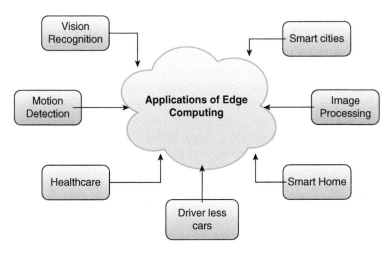

Figure 21.6 Applications and use cases of edge computing.

computation and data processing will happen in local networks which will be quicker and more responsive.

2) Motion Detection:

In a traditional cloud computing scenario, every camera will stream their raw footage to the cloud; the cloud server will process the raw footage and identify any motion in the footage. When servers find motion in the footage, they will send a report to the security control unit. This process continuously happens on the cloud and requires huge data processing and consumption of Internet bandwidth. This problem can be solved with edge computing where we can offload the tasks of footage processing and motion identification activities to edge devices. This will significantly reduce the overburdening of data streaming on the cloud.

3) Wearable healthcare devices

Real time information is so important that it cannot be compromised with things such as tracking patient health condition, monitoring of hospital equipment, active drugs tracking, improved fitness and wellness for users of wearable devices. These wearable healthcare devices rely on cloud servers for patients' critical data processing and edge devices can help in real time data processing and mitigate any risk of devices not being able to communicate with cloud servers.

4) Driverless cars

IoT devices play a major role in this, where each IoT devices sensor does a specific task such as location tracking, speed control, machinery condition checks, fuel condition monitoring, tracking traffic signals, other vehicle movements and many more. All of the sensors communicate with the cloud server where data is computed, structured and instructions are given to the vehicle. What if data processing is delayed due to Internet or latency problems? Let us see how edge computing can help. In edge computing, all the sensor data is sent to edge devices for processing which will happen in real time. Edge can further send this processed data to the cloud and can help in controlling the vehicle more efficiently. Real time autonomous vehicle control and monitoring, vehicle schedule and route tracking and monitoring, logistics tracking, equipment efficiency improvement from data collection on the roads, location-based advertising in public transport through ads on vehicles, warehouse package tracking, vehicles condition tracking are some of the use cases.

5) Smart Home IoT devices communication

The number of home applied IoT devices is growing rapidly day by day. Just assume if each device communicates separately to the cloud network then there will be huge consumption of Internet bandwidth and most importantly they will be dependent on the cloud performance. By enabling edge computing we can easily control all the IoT devices locally. For this there can be many approaches. One of the approaches is that we can group devices and each group can be controlled by edge computing. Even groups can inter-communicate through mesh networking topology.

Edge computing can help in work site safety conditions, proactive equipment maintenance, operations monitoring and workers activity tracking. It can also help in doing real time tracking of work site safety conditions, tracking equipment conditions, critical sensor-based drilling device conditions, monitoring and art.

6) Image processing in digital imaging devices

Image processing happens on the cloud server. In this scenario edge computing can take care of the required image processing locally and later upload to the cloud server.

7) Smart cities

Traffic and lane congestion information can be shared in real time with the traffic control units and can be uploaded to the server. Through edge computing we can monitor real time air quality, water supply and we can even identify the root cause of water leakage problems. Similarly, in logistics we can easily track the movement and the exact location of containers. We can even control sensor based refrigerated containers, temperature and other conditions. In agriculture edge computing can also help in many ways by remotely handling farm equipment, water supply, analysing soil condition and many more.

Figure 21.7 Characteristics of edge Computing.

21.2.3 Characteristics of Edge Computing

Figure 21.7 below interprets the characteristics of edge computing.

1) **Dense geographical distribution**
   Edge computing provides computation, analysis closer to the user. This can be achieved by off-loading the processing logic to the local edge environment. Machines connected to the IoT and the cloud will have to make accurate, split second decisions without help from humans. To do that successfully machines will use edge computing.

2) **Mobility support**
   In the retail sector we can do real-time inventory tracking and optimization using RFID and billing of material handling. Edge computing turns sensors into mini data centres, which allow self-driving cars to safely re-route if an obstacle appears, could allow robots to someday perform rescue missions after a disaster. Edge computing helps in transportation by enabling real time

monitoring of sensor data for airline maintenance, safety equipment monitoring, satellite performing functions such as telecom and weather patterns, ship navigation, tracking of shipping containers and its conditions and many others.

3) **Location awareness**

This can be achieved by undertaking critical operations in the local environment without fearing network disconnection or response timeout. Business data and processes are the most critical things and edge computing can play a vital role in the overall service performance. Edge computing can do real-time equipment monitoring, production line tracking, warehouse monitoring, worker productivity monitoring, safety equipment monitoring and improvement tracking and production control tracking.

4) **Proximity**

Edge computing can significantly reduce Internet bandwidth usage and associated cost since data processing happens at the edge network, the server resources are free most of the time and can be utilized in other cloud specific operations. Hence it can reduce server resource utilization and its associated cost.

5) **Low latency**

Low levels of latency are vital for a great digital customer experience. Since data is not travelling every time to a cloud server, edge networks can greatly reduce latency and enhance data processing performance. Edge computing can help in air quality monitoring, traffic and lane congestion monitoring, smart parking, smart water meters, water leakage, water quality monitoring and structure monitoring.

6) **Context awareness**

Edge computing places networked computing resources as close as possible to where data is created. It allows devices that would have relied on the cloud to process some of their own data. This is the core objective of edge computing since data computation happens locally that can achieve real time data processing.

7) **Heterogeneity**

Edge computing provides heterogeneity in the network by connecting IoT devices manufactured by different organisations together.

21.2.4 Challenges of Edge Computing

1) Security and privacy

5G and IoT lead to inefficiency in addressing problems like security, high latency, resource allocation and bandwidth limitation in cloud computing. The latest technologies such as edge computing and fog computing have come into the picture for data management of IoT devices and smart applications [26]. Some of the attacks on edge computing are eavesdropping, DoS attacks and data tampering attacks [23]. Data at the edge can be troublesome especially when it is being handled by different devices that might not be as secure as centralized cloud-based systems. Some of the security and privacy issues are:

a) Authentication: authentication is the process of verifying the identity of a user.

b) Authorization: authorization is something which comes after authentication. It is a process of deciding what user can access once authenticated.

c) Identity synchronization.

d) Identity proxy.

2) Colocation cloud data centres

The cloud provider is required to set up or collaborate with local data centres which will bring many challenges in terms of data virtualization and replication.

3) More local hardware maintenance

Increasing the amount of local hardware at edge devices means more investment and more maintenance cost.

4) Network connectivity and electrical power management

Uninterrupted network connectivity and electrical power management will be required because different devices require different processing power and network connectivity.

21.2.5   How to Protect Devices "On the Edge"?

Moving to a hyper-connected world service and data by the cloud, millions of things are connected every day. Edge computing is needed for IoT and cloud computing but has loopholes for cyber security. The solution to this is a highly integrated security gateway to enable computation, security and connectivity at the edge.

21.2.6   Comparison with Fog Computing

Till now we have studied and understood fog and edge computing architecture, working, applications, use cases, challenges and security issues. Now let us understand edge computing with respect to fog computing: edge is the concept whereas fog is a standard that defines how edge computing should work.

Both these technologies can help the organization to reduce reliance on cloud-based platforms to analyse data. The main difference between fog and edge computing is that the way data processing takes place. In fog computing, local area network (LAN) acts as a gateway, whereas in edge computing, computing is done at devices [26]. You may also refer to Figure 21.1 for more clarity on fog and edge computing architecture.

Both the fog and edge are concerned with similar things and look somewhat similar. In other words, fog and edge computing complement each other.

Bibliography

1 Fog Computing and the Internet of Things: A Review Hany F. Atlam 1,2,* ID , Robert J.Walters 1 and Gary B. Wills 1

2 Ai, Y.; Peng, M.; Zhang, K. Edge cloud computing technologies for internet of things: A primer. Digit. Commun. Netw. 2017, in press. [CrossRef]

3 Bonomi, F.; Milito, R.; Zhu, J.; Addepalli, S. Fog computing and its role in the internet of things. In Proceedings of the First Edition of the MCC Workshop on Mobile Cloud Computing-MCC '12, Helsinki, Finland, 17 August 2012; pp. 13–15.

4 Yi, S.; Hao, Z.; Qin, Z.; Li, Q. Fog computing: Platform and applications. In Proceedings of the 3rdWorkshop on Hot Topics in Web Systems and Technologies, HotWeb 2015, Washington, DC, USA, 24–25 October 2016; pp. 73–78.

5 Fog Computing and the Internet of Things: Extend the Cloud to Where the Things Are. White Paper. 2016.

6 Aazam, M.; Huh, E.N. Fog computing and smart gateway based communication for cloud of things. In Proceedings of the 2014 International Conference on Future Internet of Things Cloud, FiCloud 2014, Barcelona, Spain, 27–29 August 2014; pp. 464–470.

7 Liu, Y.; Fieldsend, J.E.; Min, G. A Framework of Fog Computing: Architecture, Challenges and Optimization. IEEE Access 2017, 4, 1–10.

8 Mukherjee, M.; Shu, L.; Wang, D. Survey of Fog Computing: Fundamental, Network Applications, and Research Challenges. IEEE Commun. Surv. Tutor. 2018, PP.

9 Aazam, M.; Huh, E.N. Fog computing micro datacenter based dynamic resource estimation and pricing model for IoT. Proc. Int. Conf. Adv. Inf. Netw. Appl. AINA 2015, 2015, 687–694.

10 Muntjir, M.; Rahul, M.; Alhumyani, H.A. An Analysis of Internet of Things (IoT): Novel Architectures, Modern Applications, Security Aspects and Future Scope with Latest Case Studies. Int. J. Eng. Res. Technol. 2017, 6, 422–447.

11 Aazam, M.; Hung, P.P.; Huh, E. Smart Gateway Based Communication for Cloud of Things. In Proceedings of the 2014 IEEE Ninth International Conference on Intelligent Sensors, Sensor Networks and Information Processing, Singapore, 21–24 April 2014; pp. 1–6.

12 Mouradian, C.; Naboulsi, D.; Yangui, S.; Glitho, R.H.; Morrow, M.J.; Polakos, P.A. A Comprehensive Survey on Fog Computing: State-of-the-art and Research Challenges. IEEE Commun. Surv. Tutor. 2017, 20, 416–464.

13 Dastjerdi, A.V.; Gupta, H.; Calheiros, R.N.; Ghosh, S.K.; Buyya, R. Fog Computing: Principles, architectures, and applications. In Internet of Things: Principles and Paradigms; Morgan Kaufmann Publishers Inc.: San Francisco, CA, USA, 2016; pp. 61–75.

14 Atlam, H.F.; Attiya, G.; El-Fishawy, N. Integration of Color and Texture Features in CBIR System. Int. J. Comput. Appl. 2017, 164, 23–28.

15 Nikoloudakis, Y.; Markakis, E.; Mastorakis, G.; Pallis, E.; Skianis, C. An NF V-powered emergency system for smart enhanced living environments. In Proceedings of the 2017 IEEE Conference on Network Function Virtualization and Software Defined Networks (NFV-SDN), Berlin, Germany, 6–8 November 2017; pp. 258–263.

16 Dastjerdi, A.V.; Buyya, R. Fog Computing: Helping the Internet of Things Realize Its Potential. IEEE Comput. Soc. 2016, 112–116.

17 Kim, S.J.J. A user study trends in augmented reality and virtual reality research: A qualitative study with the past three years of the ISMAR and IEEE VR conference papers. In Proceedings of the 2012 International Symposium on Ubiquitous Virtual Reality, ISUVR 2012, Daejeon, Korea, 22–25 August 2012; pp. 1–5.

18 Zao, J.K.; Gan, T.T.; You, C.K.;Méndez, S.J.R.; Chung, C.E.;Wang, Y.T.; Mullen, T. Augmented brain computer interaction based on fog computing and linked data. In Proceedings of the 2014 International Conference on Intelligent Environments, IE 2014, Shanghai, China, 2–4 July 2014; pp. 374–377.

19 Luan, T.H.; Gao, L.; Li, Z.; Xiang, Y.; Wei, G.; Sun, L. Fog Computing: Focusing on Mobile Users at the Edge. arXiv 2015, arXiv:1502.01815.

20 Choi, N.; Kim, D.; Lee, S.; Yi, Y. Fog Operating System for User-Oriented IoT Services: Challenges and Research Directions. IEEE Commun. Mag. 2017, 55, 2–9.

21 Mukherjee, M.; Matam, R.; Shu, L.; Maglaras, L.; Ferrag, M.A.; Choudhury, N. Security and Privacy in Fog Computing: Challenges. IEEE Access 2017, 5, 19293–19304.

22 Ni, J.; Zhang, K.; Lin, X.; Shen, X. Securing Fog Computing for Internet of Things Applications: Challenges and Solutions. IEEE Commun. Surv. Tutor. 2017, 20, 601–628.

23 Praveen Kumar, Nabeel Zaidi, and Tanupriya Choudhur "Fog Computing: Common Security Issues and Proposed Countermeasures" 5th International Conference on System Modeling & Advancement in Research Trends, 2016, pp. 311–315

24 Wazir Zada Khan, Ejaz Ahmed, Saqib Hakak, Ibrar Yaqoob, Arif Ahmed, "Edge computing: A survey".

25 Ashkan Yousefpour, Caleb Fung, Tam Nguyen, Krishna Kadiyala, Fatemeh Jalali, Amirreza Niakanlahiji, Jian Kong, Jason P. Jue, " All one needs to know about fog computing and related edge computing paradigms: A complete survey".

26 Shalin Parikh, Dharmin Dave, Reema Patel, Nishant Doshi, "Security and Privacy Issues in Cloud, Fog and Edge Computing"

27 Daojing He, Sammy Chan, and Mohsen Guizani "Security in the Internet of Things Supported by Mobile Edge Computing" IEEE Communications Magazine, 2018, pp. 56–61

28 Saad Khan, Simon Parkinson and Yongrui Qin "Fog computing security: a review of current applications and security solutions" Journal of Cloud Computing: Advances, Systems and Applications, 2017

29 R. Rapuzzi, M. Repetto, "Building situational awareness for network threats in fog/edge computing: Emerging paradigms beyond the security perimeter model"

30 R. K. L. Ko et al., "TrustCloud: A framework for accountability and trust in cloud computing," in Proc. IEEE World Congr. Services, Jul. 2011 pp. 584588.

31 Y. W. Law, M. Palaniswami, G. Kounga, and A. Lo, "WAKE: Key management scheme for wide-area measurement systems in smart grid," IEEE Commun. Mag., vol. 51, no. 1, pp. 3441, Jan. 2013.

Index

Fog, Edge, and Pervasive Computing in Intelligent IoT Driven Applications, First Edition.
Edited by Deepak Gupta and Aditya Khamparia.
© 2021 John Wiley & Sons, Inc. Published 2021 by John Wiley & Sons, Inc.

Printed and bound by CPI Group (UK) Ltd, Croydon, CR0 4YY